AMATEUR RADIO GENERAL CLASS LICENSE STUDY GUIDE
3RD EDITION

BY JAMES KYLE, K5JKX,
KEN SESSIONS, K6MVH & JOSEPH J. CARR, K4IPV

TAB BOOKS Inc.
BLUE RIDGE SUMMIT, PA. 17214

THIRD EDITION

FIRST PRINTING

Copyright © 1981 by TAB BOOKS Inc.

Printed in the United States of America

Reproduction or publication of the content in any manner, without express permission of the publisher, is prohibited. No liability is assumed with respect to the use of the information herein.

Library of Congress Cataloging in Publication Data

Kyle, James, writer on electronics.
    Amateur radio general class license study guide.

    Includes index.
    1. Radio—Amateurs' manuals. I. Sessions, Ken W. II. Carr, Joseph J. III. Title.
TK9956.K93 1981   621.3841   81-9148
ISBN 0-8306-0044-2   AACR2
ISBN 0-8306-1351-X (pbk.)

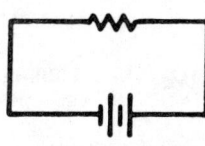

Contents

About This Book   6

Preface   7

1 A Starting Point   9
Current and Voltage—Electricity and Magnetism—Electromagnetic Fields—Electromagnetic Energy—Electromagnetic Propagation—Resistance—Storage and Charge—Capacitance—Electric Fields—Ohm's Law—Alternating Current—Wavelength—Inductance—Measuring ac and dc—Circuits

2 Basics of ac Theory   35
Current and Voltage Relationship—Capacitive Reactance—Inductive Reactance—Calculating Reactance—Impedance—Combining Reactances—Skin Effect—Resonance—Circuit Q

3 Power, Decibels and Harmonic Frequencies   63
Power—What Is a Decibel?—Harmonics

4 Matching, Filtering and Amplifying   85
Impedance Matching—Filters—Amplification

5 Tubes and Transistors   107
Vacuum Tube Operation—Diodes—Triodes—Tetrode—Pentode—Factors Limiting a Vacuum Tube's Usefulness—Vacuum Tube Applications—Transistors—Elementary Semiconductor Theory—Transistor Amplifiers—Transistors As Switches—Tube Transistor Correspondence

6 Power Supplies   161
Converting ac to dc—Power Supply Performance Rating—Improving Performance—Regulators—Current Limiting

7 Amplifiers — 185
Amplifier Definition—Amplifier Classifications—Amplifier Malfunctions

8 Transmitters and Their Operation — 207
Transmitter Operation—Oscillator Operation—Amplifier Operation—Power Measurement

9 Modulation and Modulators — 233
What is Modulation?—Modulation and Bandwidth—AM and FM Modulation—Modulation Measurement

10 The Listening Post: Radio Receivers — 255
Receiver Types—The Superhet

11 Single Sideband — 265
SSB Transmission—SSB Reception—SSB Transmitter Circuits—SSB Receiver Circuits

12 Antennas: The Common Denominator — 273
Antenna Operation—Frequency and Wavelength—Directivity—Antenna Efficiency—Polarization—Antenna Types—Antenna Arrays—Impedance—SWR—Lightning Protection

13 Radio Propagation Phenomena — 299
Ham Band Differences—Wave Progagation—Atmospheric Layers—Typical Ham Band Propagation—FCC Rules—Eliminating TVI—Improving the Ham Band Operation

14 Rules and Regulations Q and A — 319
Questions and Answers 1-16—Questions and Answers 17-32

15 Technical Q and A — 327
Questions and Answers 1-21—Test Taker's Cameo—Questions and Answers 22-104—Questions and Answers 105-150—Questions and Answers 151-232

16 Learning the Code — 381
Memorizing the Code—Learning Aids—Study Habits—The FCC Test

17 Applying for Your License — 385
Instructions for Form—Novice—General/Advanced/Extra Class—Where to Take Your Exam—Examination Points

Appendix A Element 3 Syllabus — 401

Appendix B International Morse Code — 407

Appendix C TVI/BCI/Hi-Fi-I — 409

Appendix D	Additional Reading	439
Appendix E	Part 97 FCC Rules and Regulations	440
Appendix F	International Prefixes	493
Appendix G	International "Q" Signals	507
Appendix H	Citizens Band and Amateur Radio	511
Appendix I	Symbols, Codes and Alphabets	515
Appendix J	Log Sheets	521
Index		527

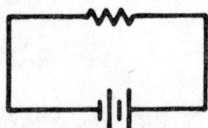

About This Book

The material in this book was originally published as a monthly series in the amateur journal *73* magazine. This Third Edition contains all the necessary original information plus the additional textual material necessary to make the study guide applicable to today's requirements, as defined by FCC Rules & Regulations, Part 97; FCC study questions for the General/Technician written examinations; and the all-new Element 3 amateur radio study syllabus.

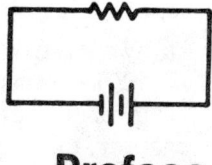

Preface

There was a time when a ham license wasn't too difficult to get. One could order a "question and answer" book, do a whole lot of memorizing, then rush down to the nearest FCC office to take the amateur license exam before the facts began to swirl out of mind. But that process doesn't work any more. The FCC questions are tougher than ever, and the answers for the most part just aren't little capsules that can be committed to memory. Nowadays, the FCC requires you to have some real knowledge of the electron world as it applies to amateur radio.

That's where this book comes in. The staff of *73* magazine looked over the myriads of questions the FCC publishes, and asked, "How can a person easily learn enough to answer the questions correctly?" In other words, how can a nontechnically-oriented individual become a ham radio operator without going for a college degree in electronics? The answer seems painfully simple, but it was the result of many months of research by *73* magazine staffers: Present the information in gradual, easy doses, starting with the least complicated elements of the exam and working up to the more complex theoretical questions.

To do this, the FCC-published list of study questions was separated into related groups. Then each of the groups became the subject of a chapter in this book. The elements of theory learned in the first chapter are necessary to understand the elements of the second chapter, and so forth. By the end of the book, it is reasoned, the reader will be equipped with sufficient knowledge to tackle the FCC General- class license exam, regardless of the wording used in the actual exam questions.

The concept was put into operation at *73* as a trial; the chapters were run in the magazine as monthly installments. When the series was completed, the letters started to pour in. Amateurs everywhere praised the course. Many had tried and failed several times in the attempt to get into ham radio—but with the help of this material they made it. The study course was the right approach.

You'll make it, too. Start right at the beginning, with Chapter 1. After you have completely digested the first chapter, take a second look at the questions listed in the text at the start of the chapter. If you can't answer them right off the top of your head, do a little rereading and try again. When you have the material in one chapter sufficiently ingested, go on to the next. By the time you reach the last page of the book, you'll be ready to take—and pass—the FCC's General class amateur radio exam. 73!

James Kyle   K5JKX
Ken Sessions   K6MVH
Joseph J. Carr   K4IPV

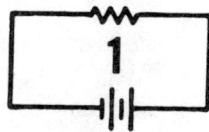

A Starting Point

Unlike other amateur license courses, which provide concise answers to the specific questions which the FCC includes on its published "study list" for each class of amateur license, we try to go *behind* the study-list questions and examine the principles on which each question is based.

This makes it much more difficult to simply memorize our answers before taking the exam, but we believe the advantages outweigh this disadvantage (if it really is one). If you study this course, you'll find that you can still answer the exam questions even if the Commission throws you a curve (as it has been known to do) by asking questions on the real exam which apparently were not even hinted at on the study list. The memorization method won't let you do this.

To insure that you'll get a detailed understanding of the theory behind the FCC examples, we'll rephrase these questions into new ones which can be expanded to cover an even broader area of technical knowledge.

CURRENT AND VOLTAGE

Our first question, which is ambitious enough in itself to fill a book (and many have been filled with not-too-informative answers to it), will be "What Are 'Current' and 'Voltage'?" In finding answers to that, we'll endeavor to give an accurate definition of not only current and voltage, but what fields and waves amount to and how electricity is propagated.

When we complete this seemingly impossible answer, we'll move on to ask "What about alternating current?" In the reply to this we will examine what makes ac appear to be so different.

Finally, we'll complete our basic introduction to electric circuit theory by asking "What makes a circuit?" This will give us an opportunity to look at different ways of connecting elements in a circuit, as well as at the different types of elements themselves.

Think we can do it? Let's dive in so you can see for yourself that the process really works!

What are current and voltage? To find out what current, voltage, and the like are, let's go all the way back to the beginning of man's studies of this strange thing called "electricity."

Since ancient times, men have known that some materials, such as amber and glass, have the ability to attract light objects to themselves under some conditions. The magnet's discovery is also lost in the mists of prehistory. But until the middle of the 18th century, both "electricity" and "magnetism" smacked of magic. Little was really known about either.

In 1800, though, an Italian scientist named Volta invented a chemical cell or *voltaic pile* which provided a constant source of electrical energy. Once a laboratory source of electricity was available, many investigators turned their attention to electricity and magnetism—and during the decade from 1820 to 1830 most of the basic laws of electricity were discovered.

Among these was Oersted's observation, in 1820, that any flow of electricity was accompanied by a magnetic field. This observation led the French physicist Andre Ampere to study the forces which resulted from flowing electricity, and Ampere, as a result, classified electrical effects into *electrostatic* force and *electrodynamic* motion, which he also called *electric tension* and *electric current*. Electric tension (now called voltage, for Volta, in this country, but still called "tension" in Great Britain) is a measure of electrostatic force. Electric current, measured in *amperes* (after Ampere), is a measure of electrodynamic motion.

Michael Faraday discovered, in 1831, that motion of a magnetic field would *induce* a flow of electric current in a stationary conductor. This completed the symmetry of electrical and magnetic effects, but the actual manner in which electricity and magnetism were related remained unknown. Many ideas were put forth, but none could be proved to actually be true. The most lasting, probably, of these ideas involved the *line of magnetic flux*. This line survives today as a part of basic theory as taught in engineering courses, but most authorities now believe the line to be imaginary, rather than an actual thing.

ELECTRICITY AND MAGNETISM

Since the work of Oersted, Ampere, and Faraday (as well as others less well remembered) had made the relationships between electricity and magnetism obvious, many investigators attempted to express the rules of the relationships. None succeeded until James Maxwell, in 1865, published his "Dynamical Theory of the Electromagnetic Field." While the theory itself has not yet been (and probably cannot ever be) proved beyond doubt, it *has* successfully predicted, and continues to predict, all electromagnetic actions within its intended range, and is today accepted as a true working theory.

Maxwell's equations, which define the theory mathematically, are required knowledge for any electrical engineer, and once they are understood they provide adequate explanation of all electromagnetic effects yet observed.

The actual equations are not simple. Maxwell himself required 20 separate equations in 20 variables to express them. Their reduction by later workers to the four equations taught today has added a requirement for knowledge of integral calculus, differential equations, and vector analysis, since these were the tools used to reduce the equations.

The theory, taken apart from the equations, is elegantly simple, though, and not too difficult to understand provided we don't ask a few critical questions which it cannot give us answers for. It begins with a single assumption—that the space around electric or magnetic objects contains (to quote Maxwell) "matter in motion, by which the observed electromagnetic phenomena are produced."

ELECTROMAGNETIC FIELDS

From this assumption, Maxwell went on to define the *electromagnetic field* as being "that part of space which contains and surrounds bodies in electric or magnetic conditions." He was careful to point out that this could be as complete a vacuum as could be obtained anywhere, but that there would always be "enough of matter left to receive and transmit the undulations of light and heat."

Because the transmission of light and heat was not greatly changed when transparent objects such as glass were substituted for the best available vacuum, he concluded that the "undulations" must be occurring in some form of matter which was not directly observable. He called this unknown form of matter "an ethereal substance," which later workers shortened to "the ether"; the question whether "the ether" exists is one which Maxwell's theory cannot answer, because the assumption that it *does* exist is the basis on which the theory is founded.

Later experiments by Michelson and Morley cast doubts upon the existence of the ether, and for many years now the label "ether" has been out of style. Today, we know this unknown medium as "the space/time continuum," but for our purposes the change is more one of name than of basic concept. Therefore, we will temporarily ignore the work of Michelson, Morley, and Einstein, and for the moment think only in terms of Maxwell's "ethereal substance." By doing so, we can get an adequate—and accurate—picture of how voltage, current and charge operate.

"Now the energy communicated," wrote Maxwell, "must have formerly existed in the moving medium." He based this conclusion on the observed fact that a detectable time delay exists between the energy's departure from its source and its arrival at its destination. He then borrowed knowledge from studies of the familiar motion of objects, to conclude that while the energy was in the unknown medium, it must have been "half in the form of motion of the medium, and half in the form of elastic resilience."

From this chain of reasoning, he drew the conclusion that the parts of the unknown medium "must be so connected that the motion of one part depends somehow upon the motion of all the rest," and at the same time "the connections between the parts must be capable of elastic yielding" of some sort, because otherwise the communication of motion from one part to any other would not require any time—and both heat and light, though fast, do not travel instantaneously.

ELECTROMAGNETIC ENERGY

Maxwell then reworded this conclusion into a description of the significant properties of his unknown ethereal medium; he said that it was "capable of receiving and storing up two kinds of energy, namely, the *actual* energy depending on the motion of its parts, and *potential* energy, consisting of the work which the medium will do in recovering from displacement in virtue of its elasticity."

These two kinds of energy, *actual* and *potential* correspond exactly to the "kinetic energy" and "potential energy" of ordinary objects. They also, as it happens, correspond exactly to the *electrodynamic* and the *electrostatic* effects named by Ampere.

The equations Maxwell developed provide the actual descriptions of these properties, based on the observations reported by earlier investigators. Engineers need them. They are not necessary for an understanding of "how" voltage, current, and charge work.

According to the theory as developed by Maxwell, the induction of current in a stationary conductor by a moving magnet, or in a conductor moving past a stationary magnet, is a result of the same force involved in transmission of motion from one part of the medium to another. It is, then, an effect of *actual* energy as contrasted to *potential* energy.

If the conductor does not form a complete circuit, no current can flow. Instead, a voltage or potential appears across the ends of the circuit. Should the circuit be completed, current will flow and the voltage will vanish—but so long as the circuit remains open, current will be zero and the voltage will be present.

This voltage or potential represents a transformation of energy from its *actual* state of motion into the *potential* state of storage.

This transformation of energy from actual to potential is a cornerstone of Maxwell's idea, and his summation of it is the feature which gives his theory its great power to explain and predict all electromagnetic effects.

Figure 1-1 shows how you can illustrate this principle with a row of pennies or other coins, by lining them up in a straight line so that each touches its neighbor on either side. If you strike the penny on one end with another coin, and if the motion of the "striker" coin is in line with the other coins, the penny at the other end will fly off the line but none of those in between will move visibly.

What actually happens in this demonstration is this: The energy of motion of the "striker" coin compressed the first penny by a microscopic amount at the moment of impact. As soon as the impact force had all been transformed into a compression, then nothing was holding the first penny compressed, so it sprang back into its original shape. This motion of rebound, in its turn, compressed the penny next to it, and so the motion was carried down the line of coins, one penny at a time, by an alternate series of motions and compressions.

ELECTROMAGNETIC PROPAGATION

Exactly the same sort of thing, said Maxwell, goes on to produce the propagation of electromagnetic energy. And since by his theory every "electric or magnetic body" is surrounded by his ethereal medium, and present theories hold that all physical matter is electric in nature, it follows that electromagnetic energy can be propagated in this manner throughout all parts of space which are occupied by any kind of physical matter.

Actual physical matter divides into two broad classes known as *conductors* and *insulators*; the defining property which separates these classes is that conductors permit the flow of electric current within themselves, and insulators do not.

It is generally believed that the difference depends upon molecular and crystalline structure, and that conductors contain electrons which are free to move anywhere within the boundary of the conductor without restraint, while insulators have few or no "free" electrons. It is important to keep in mind that these "free" electrons which are involved in the flow of electric current are a part of the conducting material, and are *not* associated with the "ethereal substance" which provides a means for propagation of electromagnetic energy.

What happens, according to present theories, is that the energy transformations going on in the "ethereal substance" impart some of their energy to the free electrons. The exchange is two-way in nature, so that free electrons in motion lose some of their energy to the unknown medium. We say that such energy is "lost by radiation" and the whole point of a radio transmitter is to "lose" as much energy as possible by radiation, in order to communicate over long distances.

RESISTANCE

When the free electrons of a conductor receive excess energy, they move in the direction indicated by the energy. Their motion is, as we shall soon see, by definition the flow of an electric current.

If the conductor forms a complete circuit, the free electrons can move indefinitely in the same direction as long as they maintain their energy. The energy is gradually lost, however, by being released back to the electromagnetic field and by collision with *bound* electrons and other atomic particles in the conductor.

When a free electron approaches a bound electron, the bound electron is repelled and attempts to get out of the way. The bound electron, though, is an integral part of an atom, which in its turn is fixed in place in a molecule, and it cannot get out of the way unless the whole molecule moves. To move a molecule takes energy, and molecular motion is also known by the more common name "heat."

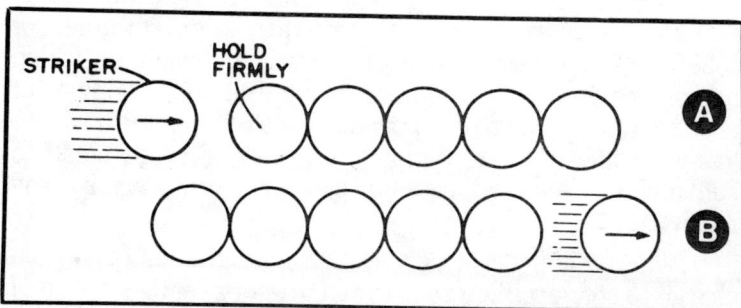

Fig. 1-1. Simple experiment with rows of coins illustrates motion-compression-motion cycle which is basis of both electric current and magnetic effects. If coins are lined up as shown at A, with one held or clasped firmly, and the line is struck end-on by another coin, the coin at far end of the line will move off of line but those in between will not move visibly. The text explains why.

15

This transformation of electrical energy of motion into the molecular motion of heat is called "resistance," and all normal conductors have resistance in varying amounts.

If the circuit is not complete, the electrons can move only a limited distance before they reach a boundary which stops their motion. The motion/compression/motion cycle must then halt in a state of compression or *storage*.

STORAGE AND CHARGE

The *storage* of energy occurs not in the conductor, but in the insulating medium which keeps the circuit open. Although the electron theory of matter was still several decades in the future when Maxwell developed his theory of fields, he considered that the electricity in each molecule of the insulator was forced to one side or *displaced* from its normal position by the *potential* energy, and remained there until either the removal of the potential energy—or the completion of a conducting circuit around the insulator—permitted the displaced electricity to resume its normal position.

This displacement of the internal structure of an insulator under the pressure of applied voltage is what we now call *charge*. Since the electron theory was developed, we consider that the displacement occurs by a bunching of the electrons within the insulating material, and, if necessary, a physical distortion of the material's molecules to align most of the electrons on one side.

We measure *charge* by a unit called the *coulomb*, which is approximately equal to the displacment of 6,280,000,000,000,000,000 electrons. An insulator which has one coulomb of charge has that astronomical number of electrons displaced to one boundary, we believe. (But no one has yet counted them, because the electron is too small to see and we cannot locate its position accurately by any means.)

When a conductor permits the potential energy of the charge to convert itself back to actual energy of motion, all of these electrons return to their normal positions, and we call the resulting flow of electrons a *current*. The unit of current is the ampere, and one definition of an ampere is a flow of one coulomb of charge in a time of one second.

One of the most important things to remember about current is that it exists *only* during the passage of time. Current is an electro*dynamic* effect, while charge can exist without motion, and so is electro*static* in nature.

Charge is a measure of quantity, similar in many ways to the "pint" or "gallon" of liquid measure. A given amount of charge may, if released from its static state, produce any amount of current, depending upon how long the flow lasts, in exactly the same way that a gallon of water may drip through a tiny hole over a period of hours or may be emptied by overturning the bucket in a matter of a fraction of a second.

The other fundamental electric unit, the volt, is a measure of potential, tension, or pressure. Though the volt is a static unit, it is like the ampere in that one coulomb of charge may be stored at any voltage.

What determines the voltage for any given quantity of charge is the electrical capacity of the insulator in which the charge is stored. *Electrical* capacity is not quite like *physical* capacity; the thicker the insulator, all other things remaining equal, the smaller its electrical capacity, but the greater the surface area the larger the electrical capacity will be. This comes about because charge is stored *only* at the boundaries between the insulating material and the conductors. The farther apart these boundaries are, the more pressure is required to keep all electrons displaced; the larger the area of each boundary, the more room for charge.

CAPACITANCE

To distinguish between physical capacity and electrical capacity, we call electrical capacity by the special name *capacitance*. The voltage associated with any particular amount of charge depends not only on the amount of charge, but upon the capacitance in which the charge is stored. The unit of capacitance is the *farad* (for Faraday), and one definition of a farad is *that capacitance which will store one coulomb of charge at a potential of one volt.*

The units of *charge* and *potential*, then, as well as that of *capacitance*, describe electrostatic effects or static *electricity*, while the unit of *current* describes an electrodynamic effect. When a circuit includes either a chemical cell or a generator which converts motion to electrical energy, the electrostatic units of potential and capacitance may still take part in the circuit, but the flow of current is the essential item.

ELECTRIC FIELDS

In today's world, most authorities don't use Maxwell's own names of *actual energy* and "potential energy" to describe the motion/storage/motion cycles. Instead, the component which Maxwell called "actual" and which is the "motion" part of the cycle is called the *magnetic field* and the component which Maxwell called "potential" and which is the "storage" or "displacement" part of the cycle is known as the *electric field*.

Either of these fields can, apparently, exist apart from the other so long as no motion is introduced. A permanent magnet has its own magnetic field, but if it is at rest no electric field surrounds it. Similarly, a charged capacitor contains an electric field, but if no current is permitted to flow, no magnetic field is associated with it.

Introduction of motion, whether it is the physical motion of the field with respect to its surroundings, the physical motion of the surroundings with respect to the field, or the electromagnetic motion produced by providing a complete circuit and thus permitting discharge of an electric field, results in the presence of both kinds of fields. Whether the motion causes the field to exist, or the field causes the motion, is not really a matter worth worrying about. Sometimes it seems to be one way, sometimes the other, and in many cases nobody can determine which is cause and which is effect.

While we still have no knowledge of the nature of that "ethereal medium" which is known as the "the electromagnetic field"—in fact, no one has much more knowledge of its nature than that set forth by Maxwell in his original description, when he wrote that the electromagnetic field is "that part of space which contains and surrounds bodies in electric or magnetic conditions"—we *do* have an explanation of a possible means by which voltage and current are produced and propagated. And, as it turns out, that's all we really need in order to do anything we desire in electronics.

Now that we know more than we really wanted to about charge, voltage, current, and capacitance, let's look at some other factors involved in the flow of charge, or *electricity*.

OHM'S LAW

The unit of resistance is the *ohm*, for Georg Simon Ohm, who first formulated the law which relates voltage, current, and resistance. Ohm's law is of such fundamental importance to all electrical work that it comes as something of a surprise to discover that Ohm formulated the rule simply as a byproduct of his studies into the effects of heat upon resistance, and in his original description did not express the law in the formula which today bears his name. The familiar formula is the invention of Maxwell, who included the law as one of his 20 equations.

We saw, when looking at the qualities which separate conducting materials from insulators, that conduction is accomplished by "free" electrons, which can migrate from place to place (as contrasted to "bound" electrons which are held firmly in place in an atomic structure). We also saw in that study that free electrons always lose some of their energy as heat, because of interaction between free and bound electrons.

This interaction within any normal conductor is the major factor limiting current flow within that material. It, in turn, depends upon the number of free electrons within the material, which is determined by both the material itself and by the physical size of the conductor. Different materials have different ratios of free to bound electrons, and the larger the conductor the more free electrons will be available within any one cross section of it.

The more free electrons a material contains, the better that material is as a conductor. We say, it has low resistance. Silver has the lowest resistance of any normal conductor, but copper is almost as good a conductor and costs considerably less. For this reason, copper is the most widely used conductor.

Some materials have few free electrons, although still enough to be classed as conductors rather than as insulators. The resistance of such materials is high. One such material is carbon. Another is the nickel-chromium alloy known as *nichrome*.

True resistance, or "pure" resistance (to use a jargon term,) is the conversion of electrical energy to heat. Many effects other than true resistance can also cause energy to be "lost" from a circuit, though, and most usually these effects are also known as *resistances*. One example is the radiation of energy from a circuit such as a transmitting antenna.

The difference between true resistance and these other effects which may also be labeled as resistances is that true resistance produces heat while "imitation" resistance does not. The *radiation resistance* of an antenna, for instance, has no heat associated with it; the energy is radiated instead.

In a complete circuit with steady current flow—a *direct current, or dc,* circuit—the potential, current and resistance are all related by Ohm's law, which states that the potential in volts is equal to the product of the current in amperes multiplied by the resistance in ohms. Stated in its most familiar form. $E = IR$, where E is voltage, I is current in amperes, and R is resistance in ohms.

If we know any two of these three quantities, the third is determined by Ohm's law. Using high school algebra we can rearrange the formula to find resistance if voltage and current are known ($R = E/I$), or current if voltage and resistance are known ($I = E/R$).

A number of memory aids have been developed to help students remember Ohm's law. One that is particularly often used is the simple word "ear," which is what comes out if you try to pronounce "$E = IR$" and consider the "=" to be silent. Another way is to simply write the three letters from left to right in alphabetical order placing the "equal" sign at the appropriate spot. If you're handy at algebra this is all you need, because it's not difficult to turn the formula around into the other two versions.

Figure 1-2 shows still another, which uses a circle divided into one half and two quarters to provide the same illustration.

Although these simple versions of Ohm's law are those most widely used, it's necessary to keep in mind that they apply only to dc circuits. If the current is changing, modifications become necessary. A chemical battery provides dc. A simple generator, however, produces current of quite another kind which we will take up a little later.

With voltage, current charge, resistance, and capacitance all defined, we have completed answering our question. There's still one more major electromagnetic effect, but it does not apply when current flow is steady. So we will wait a bit to study it.

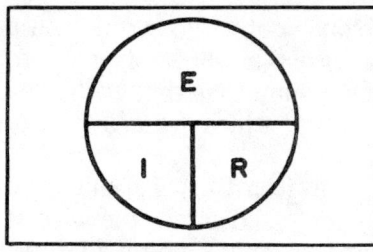

Fig. 1-2. This simple picture may help you remember Ohm's law. E stands voltage, I for current, and R for resistance. To get the formula for calculating any one if you know the other two, cover the unknown and the picture shows you the right formula. For instance, covering R gives E/I.

ALTERNATING CURRENT

We just mentioned that while a chemical battery provides dc, a simple generator gives current of quite another kind. Let's look into this a little deeper.

Assuming that the magnetic field is of constant strength, then the faster the motion the greater the current. If the polarity of the field is reversed the direction of current flow will also reverse.

There facts provide a method for converting mechanical motion into electrical energy, by using a magnet and a conductor, one of which remains stationary while the other is moved.

Figure 1-3 shows an impractical example to illustrate the idea, before we go on to more practical techniques. Let's

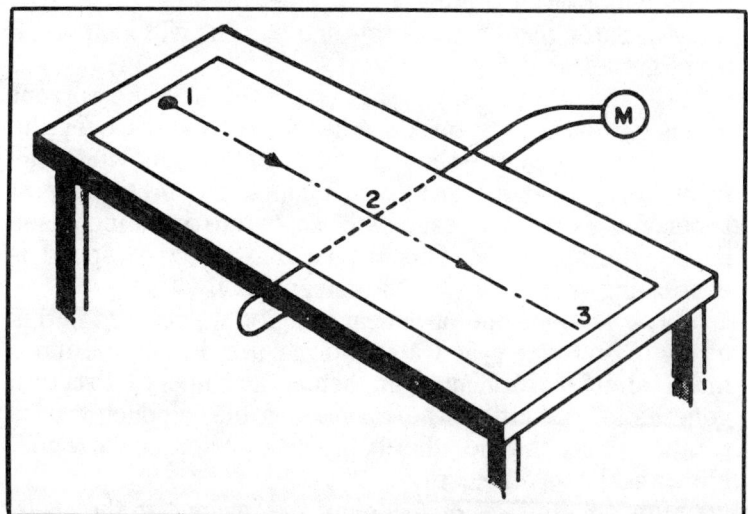

Fig. 1-3. Principles of converting motion to electric current are shown here. Magnet at 1 is moved along the line shown crossing the wire. Motion of the magnet induces a flow of current in the wire. See text for more detail.

assume that we have a wire conductor stretched flat on a large tabletop and covered with a thin sheet of paper to provide a smooth surface. Let's assume also that we have a small, powerful permanent magnet, which we are free to move across the table.

Let's start at point 1, well away from the wire, and move the magnet at a constant speed up to and across the wire at point 2, continuing on away from it to point 3, and at all times measuring current in the wire.

As we do so, we find that the instant the magnet begins to move, a small amount of current flows. At the beginning, the flow increases rather slowly, because during any tiny fraction of a second the distance from magnet to wire undergoes little change so long as the distance is large.

For example, if point 1 is 100 inches from the wire, then if the magnet moves at a steady rate of one inch per second, it will move the 0.1 inch from 100 to 99.9 inches in the first 0.1 second. The change in distance in only about 0.1%. Since "speed" is a measure of distance moved in a given time, the effective speed is low.

But when the magnet is only 10 inches from the wire, then in the next 0.1 second it will move the 0.1 inch from 10.0 to 9.9 inches, which is 1% of the distance or 10 times as much relative motion as at point 1. The effective speed is thus 10 times greater than at the start, and current will also be 10 times greater.

Current flow, likewise, is great. In fact, the current intensity reaches its peak as the magnet passes over the wire. Once the magnet has passed point 2, it's moving away from the wire rather than toward it, and so the *effective motion* is becoming smaller at exactly the same rate that it increased during the approach. This means that effective speed is decreasing, and so must be the current flow.

The straight-line or "linear" motion we've used in this example isn't very practical for normal use. Rotary motion is much simpler to handle in actual machinery. Practical generators put either the magnet or the conductor on a rotating shaft, and the other half goes on a stationary frame, known as the *field*, close by.

For our illustration, let's put the magnet on the shaft, and keep a single wire just outside the circle described by the rotating tip of the magnet. We'll use a bar magnet, with its two opposite poles, to make the illustration of bit more practical. Figure 1-4 shows the arrangement.

When we change from *linear* to *rotary* motion, the way in which effective motion varies with time undergoes a few changes of its own. Let's try to see just what happens; to do so, we'll have to use a little geometry, but very little. All we have to accept from the geometers is the rule for determining the length of one side of a triangle if we know the other two sides.

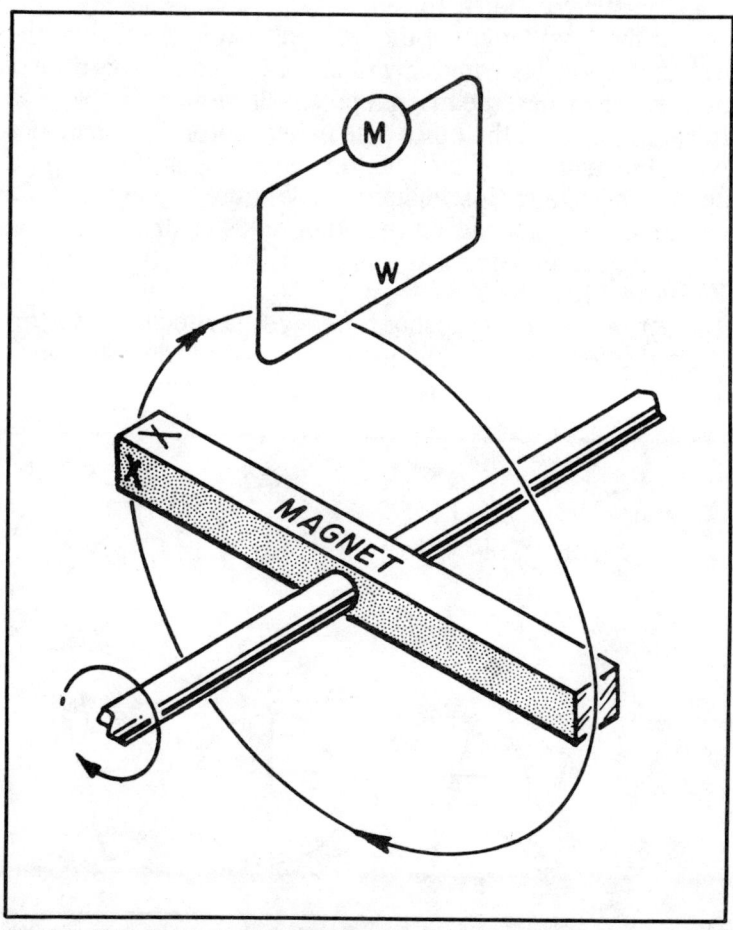

Fig. 1-4. This sketch shows the simplified generator we used to illustrate the nature of ac and how it gets that way. It's just a bar magnet on a rotating shaft, with one wire just outside the path of the magnet poles. As the magnet rotates, currents are induced in the wire. Intensity of current depends upon shaft position at each instant, as discussed in the text.

Figure 1-5 shows several successive "snapshot" sketches of our generator, seen end-on, as the shaft turns through a little more than a fourth of a full revolution. Starting at a point which we arbitrarily called zero degrees, the "snap-shots" show the relation between magnet and wire at 15-degree intervals as the shaft rotates.

You can see that the distance XW from magnet to wire is the *effective distance*, and the *effective motion* during any period of time must be the change in this distance. *Effective speed* must be related to the effective motion.

If the total length of the magnet is 20 inches, then the radius of its circle rotation will be 10 inches. At the time of our first snapshot, the magnet pole is as far from the wire as it can get without the other pole being nearer. This happens when the magnet and the line from wire to shaft form a right angle. Because of this, distance XW is the hypotenuse of a right triangle, and each of the other sides is 10 inches long. Applying the geometry at this point, we find that distance XW is 14.142 inches.

At the second snapshot time, WX has become shorter. It's down to about 12.1752 inches. By the third, it's down to 10 inches, and by the fourth, to about 7.65 inches.

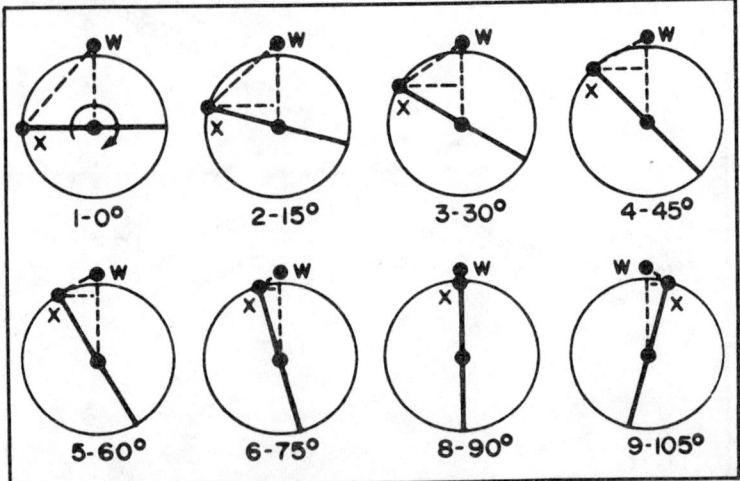

Fig. 1-5. "Snapshot" sketches of our simplified generator, seen end-on, as it would appear at eight points from 0° to 105° of rotation. All rotation here is clockwise. Key factor is the distance from the nearer pole tip to our wire, indicated as distance XW. Triangles (dotted lines) help us to calculate this distance for each position of the magnet.

Table 1-1. Calculations of Distance XW, Effective Motion, and Change in Speed, as Produced by Computer.

Angle	Distance XW	Effective Motion	Speed Change
*0	14.1421	0.000000	0.000000
5	13.5118	0.630331	0.630331
10	12.8558	0.656051	0.025720
*15	12.1752	0.680523	0.024472
20	11.4715	0.703699	0.023176
25	10.7460	0.725536	0.021837
*30	10.0000	0.745991	0.020456
35	9.2350	0.765027	0.019036
40	8.4524	0.782606	0.017579
*45	7.6537	0.798696	0.016090
50	6.8404	0.813265	0.014569
55	6.0141	0.826286	0.013021
*60	5.1764	0.837734	0.011448
65	4.3288	0.847588	0.009854
70	3.4730	0.855828	0.008240
*75	2.6105	0.862439	0.006611
80	1.7431	0.867408	0.004969
85	0.8724	0.870726	0.003318
*90	0.00001**	0.872387	0.001661
95	0.8724	0.872360**	0.000000**
100	1.7431	0.870726	0.001634**
*105	2.6105	0.867408	0.003318

*— Points used for "snapshot" calculations (Fig. 1—5)

**— Results affected by accumulated tiny error in computer.

That is, during the time between the first and the second points the effective motion was 1.97 inches, between the second and the third the motion was 2.18 inches, between the third and fourth 2.35 inches, and so on. Just as in the case of linear motion, the motion during a fixed period of time becomes larger as points X and W approach each other.

The change in effective *speed* is also important, though, and this is where much of the difference comes in. Unlike the change in effective motion, the change in speed is greatest when X and W are farthest apart.

The calculations to determine this are much too much arithmetic to go into here (in fact, they are so much arithmetic that we fudged and gave the problem to a computer. Table 1-1 is the table of results it gave us back. But when we do them we find that the relative change in speed between snapshots 2 and 3, for instance, is 0.2083, while from 3 to 4 it is only 0.1711 and from 4 to 5 it's down to 0.1310. The change becomes smaller until X passes W, then gets larger again.

In Table 1-1, the column headed *Speed Change* shows the continual change of effective speed as the shaft is rotated.

Students who have been exposed to trigonometry may recognize the triangle dotted into Fig. 1-5 as bearing some resemblance to the "unit circles" used to define and illustrate the various trig functions.

As it happens, the resemblance is so great that it turns out that effective speed is related to the sine function of the angle through which the shaft is turned.

Now that we've seen how the effective motion changes with changes in shaft position, let's perch ourselves right on that wire at point W and watch the magnet poles as we rotate the shaft.

When the shaft rotates, each pole in turn "rises" over the "horizon" formed by the rim of the circle, as in snapshot 1 of Fig. 1-5. It moves rather slowly at first, but picks up speed rapidly.

As the pole approaches us, its effective speed continues to increase, but the acceleration slows down. During the final part of the approach, its speed is great but almost constant. It passes us at a rather rapid rate, and immediately begins to slow down as it curves away.

The deceleration is slow at first, but as speed becomes less and distance increases, the rate of deceleration increases in an exact mirror of the acceleration rate during the approach. Finally this pole disappears over the "far horizon" just as the opposite pole "rises." This succession of pole passages continues so long as the shaft is rotating.

The result, in the circuit of which our wire is a part, is a continual reversal of current flow as one pole "sets" and the other "rises." At these instants, current flow in the wire is zero because the opposing magnetic fields of each pole cancel. In between these zero points, the nearer pole's field takes over, and current flow either rises to a positive peak or falls to a negative peak, as each pole in turn flashes past the wire.

Because the effective speed of the magnetic field (and the resulting intensity of current flow), is associated with the sine of the angle through which the shaft has turned we call the resulting current a "sine wave."

Because the current alternates in direction twice during each full revolution of the shaft, as the two opposite poles pass, we also call it *alternating current* or ac.

Early experimenters had no practical use for ac; the reversals of polarity canceled out many of the effects in which they were most interested. To get rid of it, they put a rotating switch called a *commutator* on the shafts of their generators to reverse the connections between the conductor and the outside circuit whenever the current direction reversed and thus cancel the effect of the current reversal, effectively *rectifying* the ac into dc.

They also used many windings, rather than just one conductor, so that at any one instant they were taking current only from the conductors which were at that time generating peak current.

Only recently has the *alternator*, or simple generator such as we've described here, come into popularity. In a practical alternator, the magnetic field is produced by a field coil rather than by a permanent magnet, and a multiturn output winding rather than a single conductor is used, but the principles remain the same.

One important factor necessary to talk about ac, which was not required with dc, is a measure of the number of times each second that the current changes direction. The older standard for this was the *cycle per second* with one cycle being the result of one full revolution of the shaft in our simplified generator. That is, one cycle includes a positive peak, a negative peak, and two zero crossings (one in each direction). A cycle may begin at any point on the waveform, and when that point has been reached the next time from the same direction, the cycle is complete.

Several years ago, the U.S. adopted an international standard of frequency, called *hertz* (in honor of Heinrich Hertz, who was the first person to test Maxwell's theory by transmitting and detecting radio waves). One hertz is the same as one cycle per second.

Whether measured in hertz, or in cycles per second, the quantity so measured is called *frequency* because it is a count of the frequency with which the current alternates.

WAVELENGTH

Closely allied to frequency, but not identical, is the term *wavelength*. A *wavelength* is the distance traveled by the energy during the course of one cycle, to put it roughly.

If the current flow moves at the speed of light, then wavelength in meters is equal to approximately 300,000,000/frequency in hertz. This comes about because the accepted (though not exact) figure for the velocity of light is 300,000,000 meters per second. If, for example, an ac wave is alternating at a frequency of 50,000,000 hertz, then in one second it will undergo 50,000,000 cycles, and these 50 million cycles will be able to travel for 300 million meters in that time. Each of them, then, must stretch over a distance of 300/50, or 6 meters, so the length of each wave, or wavelength, is 6 meters.

If frequency is expressed in megahertz or megacycles, the conversion is simplified and becomes wavelength in meters = 300/f, where f is frequency in megahertz.

INDUCTANCE

In dealing with any ac circuit, it's essential to remember that what makes it ac is the fact that the current's direction reverses as time passes (which requires current intensity to change with time, as well). This fact makes ac behave very differently from dc in many instances. One of the differences shows up when we try to measure ac voltage or current, and we'll look at that in a little while. Another difference brings us back to that one remaining major electromagnetic effect, which we postponed looking at in the previous discussion. It's time now.

We have seen, with the help of Maxwell's theory, how a flow of current is always associated with a magnetic field, and have also learned that a magnetic field in motion is always associated with either a current (if a complete circuit is present) or a charge (if no complete circuit is available). It follows, with a little thought, that if a current is flowing in a circuit and anything changes the flow of the current the associated magnetic field must change—and this resulting change must, in and of itself, be associated with a *secondary* or *self-induced* current which is completely different from the original current.

As it happens, this self-induced current is of such a polarity as to oppose the change of original current which called it into being in the first place, which means that electric current flow has a property very similar to that called *inertia* in moving objects. This property tends to keep any current flow constant, and opposes any change.

Because it is produced by self-induced currents, this inertia-like property is called *inductance*, and it is measured in a unit known as the *henry*. The usual definition of inductance involves the *line of magnetic flux* which, as noted earlier, is probably imaginary rather than real. By this conventional definition, inductance is a measure of the number of lines of flux which encircle the total current ("linkages"), per ampere of current. The henry is defined in this manner as the number of linkages divided by current in hundred-millionths of an ampere.

The *true* inductance of any specific inductor depends not only upon the material, shape, and size of the inductor, but also upon the amount of current flowing in it, the rapidity with which current flow changes, the number, size, conductivity, shape, and proximity of all surrounding objects, and several other factors.

The most important fact to keep in mind about inductance is that it represents an effect which impedes any change of current flow in a circuit. The more rapid the change, the greater the effect. Similarly, the greater the absolute flow, the greater the effect. Once a current flow becomes steady, however, its associated magnetic field is fixed and does not move. As a result, the self-induced current disappears, and inductance is no longer a factor.

Thus inductance applies to dc circuits only during the small periods of time when current flow is changing. Since current flow in an ac circuit is continually in a state of change, inductance is an important factor in the functioning of ac circuits.

Now let's look at some of the differences in measurement between dc and ac.

MEASURING AC AND DC

In most dc circuits, measuring voltage or current is relatively simple. Both voltage and current are steady as time passes, and all we must do is determine the intensity. But in an ac circuit, current is constantly reversing its direction and voltage follows right along (though in some ac circuits there may be a time delay between current and voltage). If we want to determine intensity, we must also determine just when to take the measurement.

Because the current is zero at two points in the cycle, at a positive peak at one point, at a corresponding negative peak at another point, and at all other points is balanced by a corresponding value of opposite polarity, if we simply take the average intensity over a complete cycle, both current and voltage will average out to zero regardless of the peak values of either.

This is exactly what happens when we try to measure ac with a dc meter. The meter faithfully averages values over complete cycles and indicates zero, even though the actual energy may be enough to burn the meter out.

One way of measuring ac might be to look only at half of the cycle, and take the average over that half-cycle. Then we could multiply the reading by two to account for the other half-cycle, or simply ignore it. This kind of measurement is sometimes made, and it's called the *average value*.

Unfortunately, the voltage and current readings we get by "average" measurement of ac don't correspond very well in practice to dc readings. We expect a 110-volt light bulb to have the same brightness on 110 volts of ac that it does on 110 volts of dc. If our 110 volts of ac is based on an "average" reading, the bulb will be much brighter on the ac.

It would be nicer to have a method of measuring ac which would give readings directly comparable to dc values. Such a method exists; we can measure the amount of heat generated by ac in a resistor, and find out how much dc is necessary to produce the same amount of heat.

Such a value is known as the *effective* value of the ac, and also as the *rms* value. The *rms* stands for *root mean square*, and refers to a mathematical technique to convert other kinds of measurements to effective values.

When ac voltage or current is not labeled as being in some other method, the effective or rms value is understood. Thus the 115 volts of the ordinary wall plug is 115 volts rms; the average voltage of this same plug is just under 103.

Sometimes the easiest way to measure ac is to measure intensity from zero to either peak. This is known as *peak* voltage, and its value 1.414 times the rms value for the same signal. Peak voltage of 115-volt household power is about 163 volts, which is why some simple power supplies produce about 150 volts of dc from a 115-volt input. It's peak voltage that you feel when you touch a defective appliance and get shocked.

In a few cases the most meaningful measurement is from one peak to the other, or "peak-to-peak" readings. With a sine wave signal, this is 2.83 times the rms value (or twice the 1.414 peak reading); with other ac waveforms there may be no way to relate readings. Peak-to-peak values are important because they represent the maximum pressure or potential impressed by the signal upon any insulators.

In an ac circuit, voltage, current, and resistance are not related so simply as they are in a dc circuit. This means that in order to handle ac, we must make some minor modifications in Ohm's Law.

CIRCUITS

All the way through this chapter, we've been taking the word *circuit* for granted in order to build some basic ideas. Now let's find out just what makes a circuit a circuit.

For this chapter, let's look only at dc circuits. We'll assume that a battery is our source of electric current during the rest of this discussion.

The simplest circuit, using our definition rather than the textbook version, would be merely a length of wire connected from one pole of the battery to the other. Current would flow, but not much else would happen. In order to do anything useful, we must include *components* in our circuit, and that brings us to the textbook's definition.

The most common qualities affecting simple electric circuits are resistance, capacitance, and inductance. Of these, capacitance and inductance are effective only when current flow is changing, so we'll consider only components which produce resistance when they're included in our circuit. Such components are called resistors.

Each component in a circuit is known as a circuit *element*. The connecting wire is not usually considered to be an element if the circuit contains anything else.

Circuits are divided into three major categories, known as *simple*, *series*, and *parallel* circuits. *Simple* circuits are those which contain only one circuit element, such as the one shown as A in Fig. 1-6.

Series circuits contain two or more elements connected end to end, so that the current will pass first through one element, then through the next, and so forth until the current has passed all the way around the circuit. A series circuit of two resistors is shown as B in Fig. 1-6. Elements of this type of circuit are said to be connected *in series*.

Parallel circuits contain two or more elements, none of which are connected in series. Each element is connected directly to the power source, so that the current which flows through one element does not flow through any of the others. A parallel circuit of two resistors is shown as C in Fig. 1-6. Elements of this type of circuit are said to be connected *in parallel*.

The major difference between series and parallel circuits is that in a series circuit, all the circuit current flows through each element in turn, while in a parallel circuit, that current which flows through one element returns directly to the source and does not flow through any other elements in the circuit.

From this difference, we can determine how resistors combine in both series and parallel connections to give total resistance. If two resistors are connected in series, all the circuit current must flow through each. In the first, a part of the energy will be converted to heat, leaving less for the second. In the second, the remaining energy will be converted to heat, so that none remains to return to the source.

Let's use the circuit of Fig. 1-6B, and assume that one of the resistors has a resistance of 5 ohms and the other is a 10-ohm unit. Let's assume also that we measure total circuit current and find it to be 2 amperes.

By the rule set forth in Ohm's law, the voltage across the first resistor must be 2 x 5 or 10V (E = IR). Across the second, it's 2 x 10 or 20V. Total voltage of the battery, then, must be 10 + 20 or 30V.

Now we apply Ohm's law again, using the 30V value we just calculated for battery voltage, and the 2-ampere measurement of circuit current. Resistance must be 30/2 or 15-ohms (R = E/I). In a series circuit of resistors, then, total resistance must be the sum of the individual resistances.

Let's try it again with the same two resistors connected in parallel (Fig. 1-6C) and using the same battery. We know now that the battery's voltage is 30. We can use Ohm's law to calculate current through each resistor individually. Through the 5-ohm resistor, it's 30/5 or 6 amperes (I = E/R), and through the 10-ohm unit it's 30/10 or 3 amperes. Since none of the 6 amperes going through the 5-ohm resistor flows through the 10-ohm unit, the total circuit current must be the sum of 6 and 3 or 9 amperes.

With a 30V source and 9 amperes circuit current, we return to Ohm's law and calculate total resistance as 30/9 or 3.333-ohm (R = E/I)—less than that of either resistor alone.

Effective values of resistors in parallel can be calculated in other ways, too, and some are much simpler. The way we've done it here, though, shows *why* resistance is lower, and is the basis for all the others. We'll meet the other ways next time around when we go into the manner in which other circuit elements combine in series and in parallel.

Incomplete circuits, which are complete except for their power source, can themselves be considered as a sort

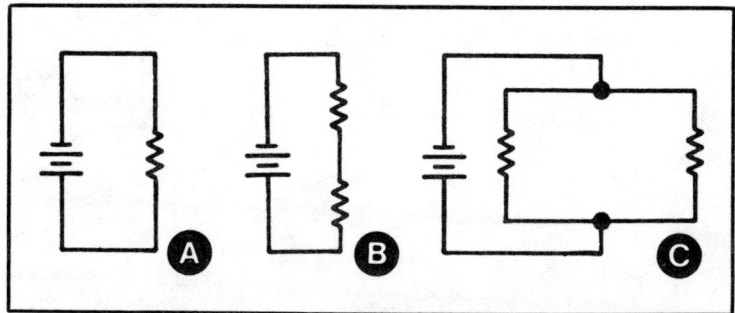

Fig. 1-6. These are, respectively, "simple" circuit (A), "series" circuit (B), and "parallel" circuit (C). Number of circuit elements (resistors here) and division of current flow are factors determining which type is which.

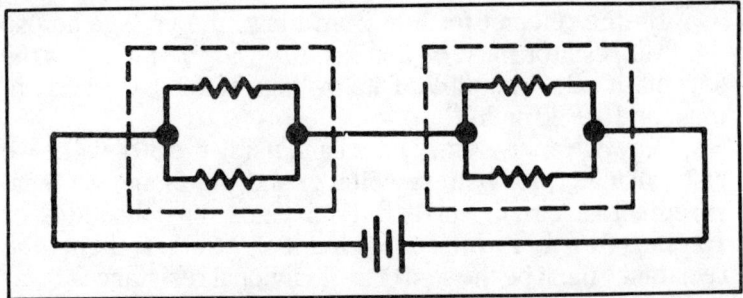

Fig. 1-7. Not all circuits are as simple as plain "series" or "parallel" connections. Subcircuits shown in dotted boxes are parallel connections, but subcircuits themselves are in series. Total circuit is called "series-parallel" but it could be reversed to "parallel-series" without changing components or connections simply by re-drawing dotted boxes. Most real circuits are like this, only more so.

of circuit element in a larger circuit. Figure 1-7 shows a larger circuit with two such "subcircuits" in it, each outlined by dotted lines. Each subcircuit is a parallel circuit, but the two are connected in series. Such an arrangement is known as a *series-parallel* circuit.

Basics of AC Theory

The General class license examination covers many fine points of radio theory and practice—and many basic points as well. Since the basics are necessary as a foundation upon which to build the fine points, we began this study course by accounting for voltage, current, resistance, magnetism, and the propagation of electromagnetic energy. This chapter covers the basics of ac circuitry.

In this license study course, we're working from the official FCC study list of "typical questions"—but we're not answering the Commission's own questions. Instead, we're expanding their range to cover a broader area, and examining that broad area in as great detail as we find necessary.

Because we know from our previous study what inductance, capacitance, volts, ohms, and amperes are, as well as series and parallel circuit connections, but have not yet met "reactance" or "resonant frequency," let's begin by asking "How does ac behave differently from dc?" This will permit us to explore not only reactance and skin effect, but a more basic concept called impedance. From there, we will try to determine "What is resonance?" This discussion sets the stage for much of the later "practical work" on tuning of both receivers and transmitters. The "Q" factor is closely allied to "resonance," but is enough different that the question "What is Q?" deserves its own discussion.

By the time we explore these three general questions, we should have a solid foundation in both ac and dc circuit theory, and be ready to move on to some advanced theory. Let's get started, then.

CURRENT AND VOLTAGE RELATIONSHIP

We saw in the last chapter that ac must be measured differently from dc, because of the fact that ac is continually changing not only its intensity but its direction of current flow. We said that voltage, current, and resistance were not so simply related in an ac circuit as they are with dc, as well, but we didn't dig in any deeper.

In a capacitor, for instance, a steady or dc potential will cause an accumulation of charge across the insulator or dielectric. If the potential is then increased, the charge which already exists will be of the same polarity. Since like poles repel each other, the charge already existing will tend to oppose the increase.

However, because of the *brute force* provided by the increased potential, electrons *will* move within the dielectric to establish a new state of charge and bring the capacitor's charge into balance with the applied voltage. Only after this movement has occurred will the full applied voltage exist across the insulation. The time involved is exceedingly small, but the net result is that a capacitor delays the voltage until current flow has stabilized.

If we apply ac rather than changing the voltage of applied dc, the same basic rules apply—except now, the current flow never stabilizes. It just keeps on changing, and the voltage never quite catches up to it. The net result is that the voltage cannot reach a peak value while current is flowing, and when current is at a peak of intensity no voltage exists across the capacitor. In between these times, current will decrease from its peak as voltage increases from zero. We express this by saying that the voltage *lags* behind the current.

Most usually, a simple statement that the voltage is lagging doesn't describe things accurately enough. We need to know just how far behind the current peak the voltage peak is located.

We could express the time in fractions of a second, but then we would also have to specify the frequency of the ac involved. To avoid this entanglement, we use the ac cycle itself as our time scale. Recall our simplified ac generator in the previous chapter, which generated one full cycle of ac during one revolution of its shaft through 360 degrees of arc. We use this fact, and say that a full cycle of any ac sine wave occupies 360 degrees of phase; that's our way of measuring time in ac circuits.

Since a full cycle is 360 degrees, let's arbitrarily set our zero-degree point on the instant when the current is crossing zero in a positive-going direction. The current's positive peak will then occur one quarter cycle later, or 90 degrees. The next zero crossing will be another one quarter cycle after that, or at 180 degrees, and the negative peak will be still another one quarter cycle later, or at 270 degrees. At 360 degrees, we will complete the cycle and be back at the zero point of the next cycle.

To apply this to the voltage lag in a capacitor, let's see what happens to the voltage at these same four points in the current waveform.

When the current is at zero degrees, with the waveform crossing zero and going positive, the voltage is at its negative peak. This is the 270-degree point on the previous cycle, or 90 degrees behind the current.

When the current is at 90 degrees, at its positive peak, the voltage is just crossing zero going positive. This is the 0-degree mark of the current cycle, the same intensity which the current had 90 degrees earlier. Voltage is still 90 degrees behind current.

Note that there's a very major difference here from what happened in a dc circuit. With dc, the capacitor blocked all current flow. With ac, current flow continued right on through the capacitor, but the lock-step relationships between current and voltage (which existed when the ac left its generator) has been disturbed, so that voltage now lags by 90 degrees.

CAPACITIVE REACTANCE

Since we're assuming that there was no resistance anywhere in the circuit, no energy was turned into heat. However, with the current and the voltage waveforms out of step they are no longer capable of producing the same amount of power as they were before meeting the capacitor. The energy has undergone an apparent loss, but actually it's just out of time.

Because this apparent loss is due to a reaction brought about by the capacitor, we call it an effect of *reactance*. Because the apparent loss is real enough to affect any circuit connected to it, we measure reactance in ohms just as we do resistance. The major difference between reactance and resistance, at this point of the game, is that resistance loses the energy permanently by turning it into heat, while reactance merely locks it up and makes it unavailable for us to use.

Going back now to the reason why voltage and current get out of step in a capacitor—the fact that the existing charge at any instant opposes any increase in charge—it shouldn't be extremely difficult to see that the more rapid the attempted increase, the more rapidly the change will occur. Similarly, the greater the capacitance involved, the more thinly spread will be the existing charge and the less it will be able to oppose the changes.

These two observations lead us to the point that the *greater* the frequency of the ac signal, and the *greater* the capacitance in question, the *smaller* will be the reactance.

Going in the other direction, as the frequency becomes ever lower it gets closer and closer to being dc, and since a capacitor is an open circuit for dc it must be a very high value of reactance for very low frequency ac. There's a formula for calculating reactance in ohms of any capacitor if you know its capacitance and the frequency of the ac signal involved, but we won't go into that just now. First, let's look at another form of reactance.

INDUCTIVE REACTANCE

We have just seen that a capacitor introduces reactance into an ac circuit because its charge opposes any change in voltage across itself. We also know that an inductor opposes any change in current through itself. This characteristic is similar enough to that of a capacitor and voltage that we might suspect that inductors also have reactance—and we would be correct! The inductor's opposition to change, though, is not identical to that of the capacitor. Where the capacitor opposes changes in voltage, the inductor opposes changes in current.

Because of this, the reactance of an inductor does not cause the voltage to lag behind the current as does that of a capacitor. Instead, with an inductor, the current lags behind the voltage.

The reasons are exactly parallel, however, to those which apply in the case of the capacitor. When the current flow is zero, there's no opposition (for a brief instant) and voltage intensity is at its peak. When current flow is at a peak, opposition is at a maximum and voltage hits zero. Just as with the capacitor, the inductor's reactance causes voltage peaks to coincide with current zero-crossings, and voltage zero-crossings to coincide with current peaks.

This means, automatically, that a 90-degree phase difference is introduced between the current and voltage waveforms by pure inductive reactance. Since it's the current which is delayed, this time the voltage is 90 degrees ahead of the current rather than 90 degrees behind.

Figure 2-1 illustrates both of these conditions, as well as the normal nonreactive relationship between voltage and current in a resistive ac circuit.

Current lag instead of voltage lag is not the only difference in reactance between capacitors and inductors. They behave a bit differently with respect to inductance values and to frequency, also.

The more rapid the attempted change of current in the inductor, the more opposition is developed. Going the other

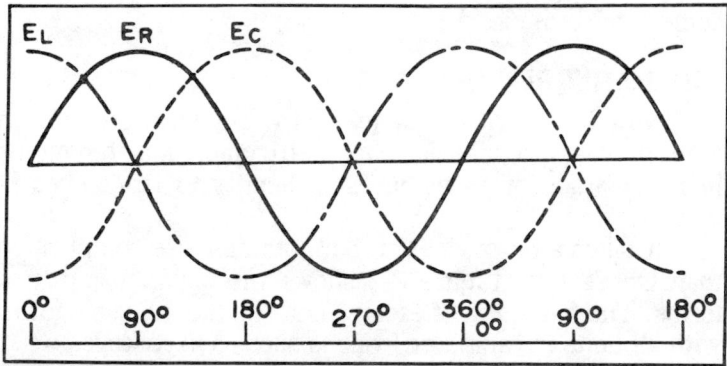

Fig. 2-1. Effects of capacitance, resistance, and inductance on phase relationship between voltage and the current are shown here. In all three cases, the current phase is the same as waveform E_R. Capacitance causes the voltage to lag 90° behind current. The resistor has no effect and voltage and current remain in phase. The inductor causes current to lag 90° behind the voltage, which is the same as causing the voltage to lead the current by 90°. The phase scale at the bottom refers to the phase of the current waveform.

way, the nearer the signal frequency gets to dc, the less effective the inductance is, since inductance has no effect at all on steady-state dc.

This means that while the reactance of a given capacitor decreases as frequency goes up, the reactance of a given inductor goes up as frequency increases. The higher the frequency, the greater the reactance.

Similarly, the greater the inductance at any one frequency, the more the opposition. Thus the reactance increases as the inductance value climbs, while capacitive reactance decreases as capacitance is increased.

Since capacitive reactance and inductive reactance behave so differently in these three different areas, we must have some way of telling them apart in our equations and theories. The behavior is so different that they almost appear to be opposites. One causes current to lead; the other causes it to lag. One increases with frequency; the other decreases. One decreases as its corresponding unit increases; the other increases right along. This "oppositeness" of characteristics is used to tell them apart in our symbols. We use the "+" and "−" signs of arithmetic to designate which kind of reactance we mean. If the reactance is capacitive, we tag it as "−"; if it's inductive, we call it "+."

This would indicate, and Fig. 2-1 would back us up on it, that we could use some of one kind of reactance to cancel out some of the other kind.

CALCULATING REACTANCE

Now let's turn the tables just a bit, having met both kinds of reactance, and find out how to relate capacitor values in microfarads or inductances in henrys to reactances in ohms.

It takes a bit of algebra, but nothing more complicated than Ohm's law. Figure 2-2 shows the equations, in two forms. The first form of each equation is the full basic form, with all units in "standard" values such as hertz (cycles per second), farads, henrys, and ohms. The second form of each is the "practical" one for radio work, with units in more convenient form much as mF, kHz, microhenrys, etc. The answers, however, still come out in ohms.

The physical facts which these equations represent are those which we've already met in our introduction to

$$X_L = 2\pi f L$$

INDUCTIVE REACTANCE
(full formula, ohms, hertz, henries)

$$X_C = \frac{1}{2\pi f C}$$

CAPACITIVE REACTANCE

X in ohms	X in ohms
f in megahertz	f in kilohertz
L in microhenrys	C in microfarads

Fig. 2-2. Reactance equations. Versions at the top are full basic formulas, with all quantities in basic units. Those at the bottom are practical versions. See text for details on their use.

reactance. A reactance in a circuit which contains no other elements will always have applied to it the maximum voltage which the source can provide. The reactive element's opposition to change—that is, to change in voltage in the case of a capacitor or to change in current in the case of an inductor—will act to limit the flow of current in the circuit to just that which is necessary to keep the circuit electrically balanced at all times.

We already know that a capacitor, for instance, offers less reactance to a high-frequency signal than to one of low frequency. The more rapid change of applied voltage brought about by the higher frequency signal requires a large current flow to keep the circuit in balance or equilibrium. Similarly, for two capacitors of different capacitance but with the same frequency of applied ac, the larger capacitor will require more current to keep the charge at just the level which will satisfy the applied voltage.

The reactance equations simply express this relationship between voltage and current, in terms of capacitance (or inductance) and frequency.

Now that we are able to express a capacitor's or an inductor's value as "ohms of reactance" at any specific frequency, let's see how ohms of reactance combine with ohms of resistance. Cast an eye upon Fig. 2-3, which shows us a resistance-capacitance series circuit.

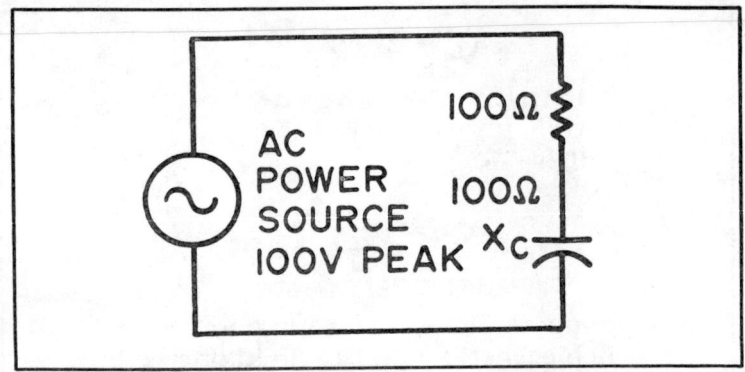

Fig. 2-3. Series R-C circuit illustrates how the reactance and resistance combine. Effects of each influence the action of the total circuit, and the result is neither purely resistive nor purely reactive. Inductance behaves similarly.

Since the circuit is a series circuit, by the definition we learned in the previous chapter, all of the circuit current must flow through each element. With dc applied, there will be no current after the initial transients die out, because the capacitor interrupts its path. But with ac, current will flow at all times, as the capacitor charge follows the applied voltage during each cycle.

Were the capacitor alone in the circuit, the voltage would be varying up to the full peak value provided by the power source, and the current would be that value defined by the capacitive reactance formula; with a 100-ohm reactance, it would be 1/100 the voltage value. In addition, the current waveform and the voltage waveform would be 90 degrees out of phase with each other.

Were the resistor alone in the circuit, the voltage would also be varying up to the full peak value provided by the power source, and the current would be that value defined by Ohm's law; with a 100-ohm resistance, it would be 1/100 of the voltage value just as for the reactance we just looked at. In the resistor, however, current and voltage waveforms would be *in* phase with each other.

What happens when both are in the circuit? We know that any current which flows through one must flow through the other. Let's assume that our power source is putting out 100V peak (200V peak to peak, or 70.7V rms) and that both the resistance and the reactance have values of 100-ohms each (at 0° and −90° phase angles, respectively). That would indicate that one ampere peak would pass through either element alone. Let's see what happens when we assume that one ampere is passing through *both* in series.

We would, by Ohm's law, find 100V peak, in phase with the current, across the resistor—and by similar reasoning, we would find 100V peak across the capacitor also, but this would be 90° out of phase with the current and so would also be 90° out of phase with the voltage across the resistor as well.

What happens when we have two voltages out of phase with each other in the same cicuit? Figure 2-4 shows some waveforms which may help us to see the result. When one of the voltages is at its peak, the other is just crossing zero. When the second reaches peak, the first is crossing zero. For a part of the cycle, both have the same polarity, and for a part of the cycle, they have opposite polarity.

But we cannot have *two* voltages in one circuit, not across the outside terminals of the circuit. At any one measuring point, we can have only *one* observable voltage and *one* observable current. Since we can imagine each of our

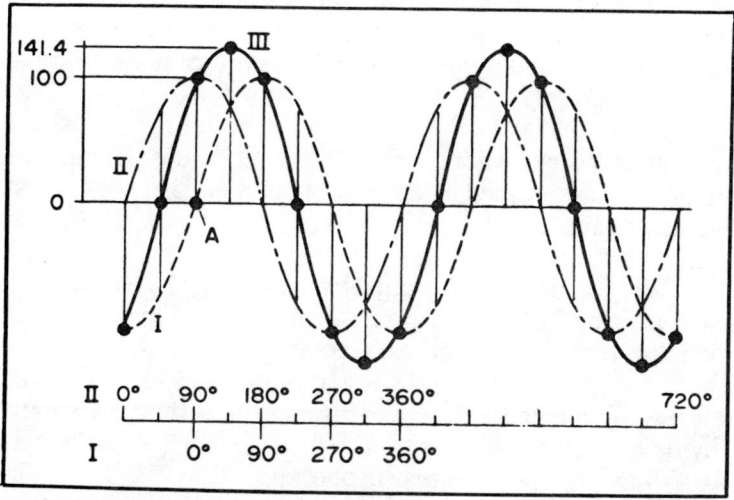

Fig. 2-4. Waveforms of voltages in the circuit of Fig. 2-3 show what happens in a combined circuit. Voltage I represents the voltage across the resistor, Voltage II is that across the capacitor, and Voltage III is that across the total circuit. Voltage III is determined by adding together the values of I and II at each point along the time scale. Note that the final voltage is halfway between the original voltages in phase, and is greater than either alone but not twice as great as either.

voltages separately, let's stretch our imaginations a bit more and see what happens when we merge them together into a single waveform.

At point A, voltage I is zero and voltage II is at positive peak. The merged result (III) must be simply whatever value voltage II has right then. A little later, the two original voltages cross each other, one on the way up and other coming down. Voltage III at this point must be equal to the sum of both, or twice as great as either.

In fact, all the way through the cycle voltage III must be equal to the sum of voltage I and voltage II. Where the two are of opposite polarity and tend to cancel out, read *difference* (most texts use the phrase "algebraic sum" to indicate that cancellation or reinforcement either can occur, depending on relative polarity).

And voila! Voltage III is still a sine wave, just like the two waveforms of which it is composed—but it's not in phase with either of them. It does, in fact, split the phase difference, so that it is 45° out of phase with the current.

We still have one small insect in the unguent. Our original 100-ohm values were based on a 100V-peak source and one amp current. Voltage III has a peak value considerably greater than 100V. It works out to be 141.4V, if you care to measure it (and if our illustrator has drawn his sine wave curves accurately).

However, in a series circuit, the total voltage across the outside terminals where we're observing Voltage III must be equal to that provided by the power source. That means that we can't have 141.4V here; we are limited to 100V for the total.

We compensate for this by assuming a reduction in current, and this is only reasonable since whatever the resistor turns into heat isn't going to be available to the capacitor. To cut the total voltage from 141.4 to 100 we can merely divide it by 1.414. This indicates that we should divide our current figures by 1.414 also, so that we have only a 0.707 ampere peak instead of one amp.

With a 0.707 ampere current through the resistor, the voltage across it becomes 70.7V peak. That same 0.707 ampere through the capacitor sustains a 70.7V peak potential across it, and if both Voltage I and Voltage II are 70.7V peak, then their combination into Voltage III will produce 100V peak, and everything balances properly.

IMPEDANCE

We have seen, through some illustrations and a lot of specific numbers, that 100-ohms of capacitive reactance combine with 100-ohms of resistance to produce a result different from either reactance or resistance. What is this result called?

Its name is *impedance*, and like resistance and reactance it is measured in ohms. The name impedance, comes about because impedance is the combination of resistance and reactance, both of which impede the transfer of electrical energy.

How many ohms of impedance does our sample circuit have? We could take a stab at figuring it out by applying Ohm's law to the figures we finally worked out of 100 volts and 0.707 amperes; this would give us a figure of 141.4 ohms for R—but it's impedance rather than resistance we're figuring, and R is the abbreviation for resistance. Impedance is always abbreviated as Z, while reactance is X with a C or L subscript to identify it as capacitive or inductive.

That stab is exactly accurate, because Ohm's law does work for ac circuits just as well as for dc if all the R factors are replaced by Zs. However, there are simpler ways of expressing impedance than by having to figure out all the currents and voltage drops in a circuit.

One of the simplest ways to express impedance, and one which is in wide use by engineers and sufficiently general use in the ham field to make it necessary that you know it, is not to combine the R and X at all into a single figure, but to express impedance as a sum, $R \pm jX$. The sign is plus if X is inductive, and minus if X is capacitive (that is, the sign in the sum follows the sign of the reactance). This is known as the *complex* expression of impedance, because it's the form used in mathematics to express complex numbers, but it's actually less complex than the arithmetic you have to go through to combine the R and the X into a single value of Z.

The impedance of our sample circuit, expressed this way, would be: $100 - j100$ ohms. The "j" is an indicator telling us that the number it accompanies is a reactance rather than a resistance.

For many purposes, this is the clearest way to put it. One of the major ham applications of complex impedance values is in rating feedpoint impedance of antennas and

45

transmission lines. The object in this case is to get the "j" value to zero, which we can do by adding reactance of the opposite sign, and the R ± jX way of expressing the value tells us directly how much and what kind of tuning reactance is necessary.

The other way of expressing impedance reduces the sum to a single number, such as the 141.4 ohms we came up with a few paragraphs back. To be accurate about it, such a number must always be accompanied by its phase angle, so that the impedance of our sample circuit is actually 141.4 − 45° ohms. To show why the angle is necessary, and to show how to figure the number and angle, we're going to have to put it in some more pictures.

If you've been exposed to the "new math" either as a student or as a sidelines observer, you have probably met a "number line" already. If not, one awaits in Fig. 2-5. This is just a line divided into regular spaces, like a ruler, with zero in its middle. Each space to the right of the zero point is one more positive number, and each to the left is a negative number, so that the spaces run −4, −3, −2, −1, 0, 1, 2, 3, 4 ... The idea is that this "number line" offers a picture of the way we count things, and we can use a number line to illustrate anything that is countable, which includes the number of ohms of impedance, resistance, or reactance in a circuit.

We can cross two number lines as shown in Fig. 2-6, crossing them at their zero points and placing them at right angles to each other, to show how resistance and reactance combine. It doesn't really show *how*, but as we shall see everything fits properly, so the idea is workable (the real *how* is the combining of voltages which we've already suffered through back in Figs. 2-3 and 2-4). We shall call the horizontal scale the resistance or R scale and the vertical scale the reactance or X scale. *Negative resistance* is hard to find, so most of the time we'll use only the right-hand half of the resistance scale, but both halves of the reactance scale come into play because of our convention that capacitive reactance is negative while inductive reactance is positive.

If we plot our 100 ohms of resistance from Fig. 2-3 and the −100 ohms of capacitive reactance on the graph of Fig. 2-6, we come out as shown in Fig. 2-7 with a pair of dashed lines. The point at which these lines meet represents the

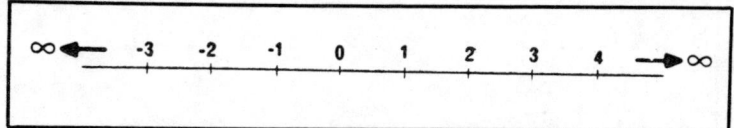

Fig. 2-5. This "number line" is the basis of the graphical method of determining impedance values. It provides conversion from picture to actual values.

combination of 100 ohms resistance and −100 ohms reactance.

We can now draw a straight line from this point back to the common zero point of both lines, and this single straight line represents the combined impedance of the circuit.

If we measure it, we will find (assuming that our illustrator remains accurate in his work) that it comes out to the same 141.4 ohms that we found by arithmetic before, and that it is at a −45 degree angle from the resistance scale.

Rather than drawing the graph and measuring the impedance line and angle, the normal practice is to calculate them from the relationships of the right triangle. The impedance line is the hypotenuse of the triangle, while the resistance and the reactance are its legs, so that Z^2 must equal $R^2 + X^2$. The trig functions of angles save us the need of working out all of this, because the phase angle can be determined by using the ratio X/R and finding the angle which has this ratio for its tangent. This angle is the phase angle of the impedance; the cosine of this angle is equal to the ratio R/Z, while the sine is equal to the ratio X/Z, so that

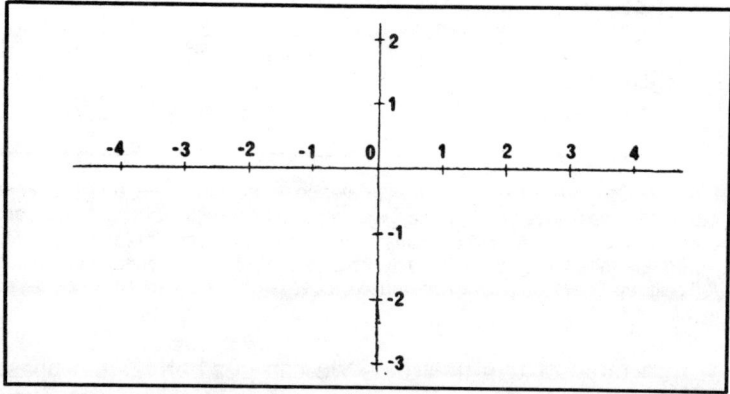

Fig. 2-6. Crossing two number lines at right angles gives us a "number plane" on which we may draw a graph. The two number lines provide scales for graph.

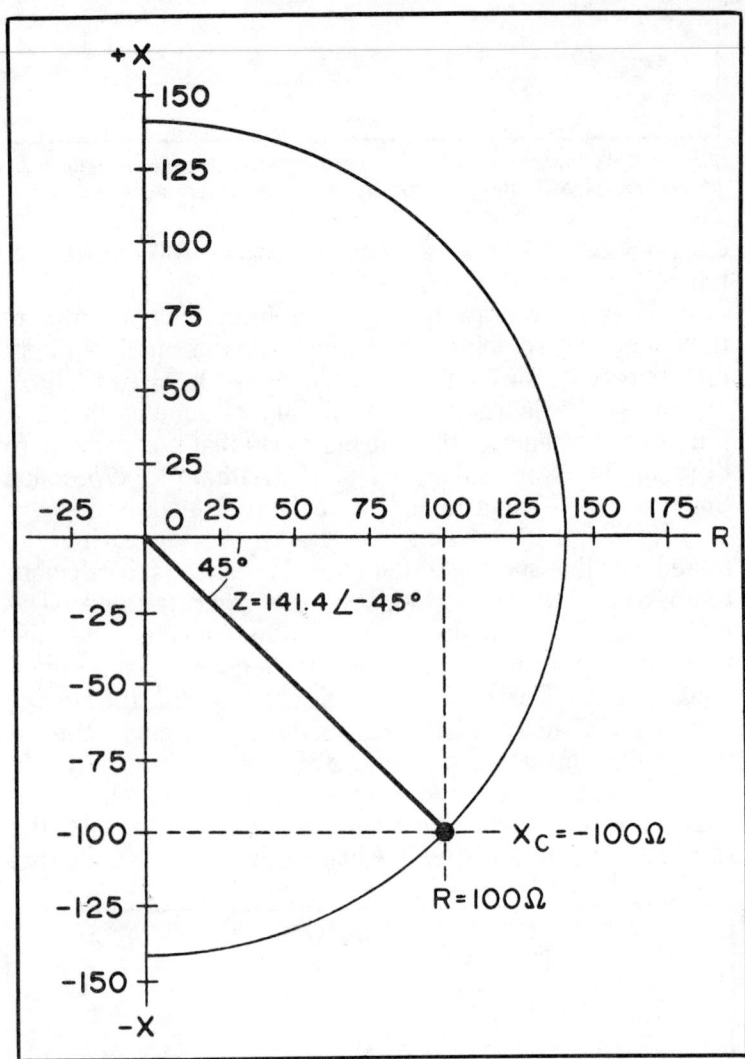

Fig. 2-7. Combination of 100 ohms resistance (horizontal scale) and 100 ohms capacitive reactance (vertical scale) gives us an impedance of 141.4 ohms at a phase angle of −45°. Text explains in detail how the number plane is used to determine these values. Only the right half of the plane is needed because the left half represents *negative resistance* values, which are seldom encountered in practice.

with a table of trig functions we can get both Z and phase angle if we know R and X, or can get R and X if we know Z and phase angle. Figure 2-8 shows all the appropriate equations for converting from one way of expressing impedance to the other.

Before we leave Fig. 2-7, though, we should pause to see why it's so important to always include the phase angle when we're talking about impedance if we don't use the complex method.

We found, using Fig. 2-7, that the impedance of our sample circuit was 141.4 ohms. However, 141.4 ohms of impedance might be 141.4 ohms of pure resistance, 141.4 ohms of pure reactance of either sign, or *any* combination of R and X which would bring us to a point lying on the dotted circle in Fig. 2-7, because every point on this circle is 141.4 ohms away from the zero point.

To make sure that the *one* point representing the combination 100 − j100 ohms is understood, we must add the −45° phase angle information. This angle defines a line which starts at the zero point and goes on indefinitely at a 45-degree angle. However, the line and the 141.4-ohm circle can cross each other only once, at a single point, and so the two together make our impedance specification exact.

COMBINING REACTANCES

Now that we're on reasonably solid ground with resistance, reactance, and impedance, we can turn our attention

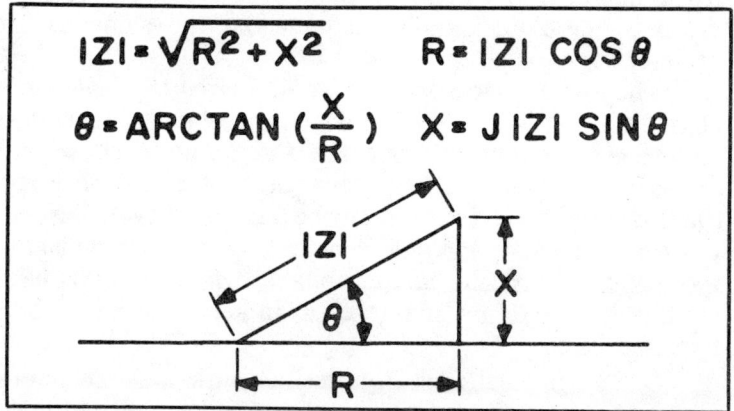

Fig. 2-8. These are the four equations which are used to convert from a polar plot impedance expression (Z ohms at phase angle φ) to a *rectangular plot* or complex expression (R + jX). Vertical bars on either side of Z in the equations mean absolute value without regard to sign. Sketch identifies variables with respect to Fig. 2-7. *Arctan* means: the angle whose tangent is.

to the question of how like reactances combine in series circuits. So long as it's the same kind of reactance, and assuming (it's important to note that this assumption can hardly ever be achieved in practice) that each reactance is completely independent of every other one in the circuit, they will simply add together like resistors would. What they will add, however, is their *reactance* rather than their capacitance or their inductance. Two capacitors, each having 100 ohms reactance by itself, would produce 200 ohms reactance when connected in series. A quick look at the reactance equations will show that this is the same as one capacitor of *half* the capacitance.

Most textbooks have a string of equations for you to memorize in this respect, in order to calculate the total effective capacitance of any number of capacitors connected in series. We're showing you the equations as Fig. 2-9, but you don't really need them. All you need do is figure out the effective reactance of each capacitor at some arbitrary frequency such as 159 kHz (which will cancel out all conversion factors if you use it), then total up all the reactances and convert the resulting total reactance back to a single value of capacitance by reversing the conversion formula.

The same trick works with inductors—but if you use a frequency of 1/159 MHz for them, you'll find that it boils down to simply adding up inductances the same as for resistors.

The same trick works for parallel circuits, if we use a couple of other characteristics which are related to reactance and resistance but are not the same thing. These are *susceptance*, which is the reciprocal of reactance, and *conductance*, which is the reciprocal of resistance. Susceptance is abbreviated B, and conductance as G. Impedance, too, has a reciprocal—*admittance* abbreviated Y. In parallel circuits, just total up admittances; in series circuits, total impedances.

Since like reactances in series behave just like resistors in series so far as total effective value is concerned, and since Ohm's law holds for ac as well as dc when we substitute Z for R in the equations, it's only natural to expect that ac voltage across a series string of like reactances would divide just as dc voltages do across a string of resistors.

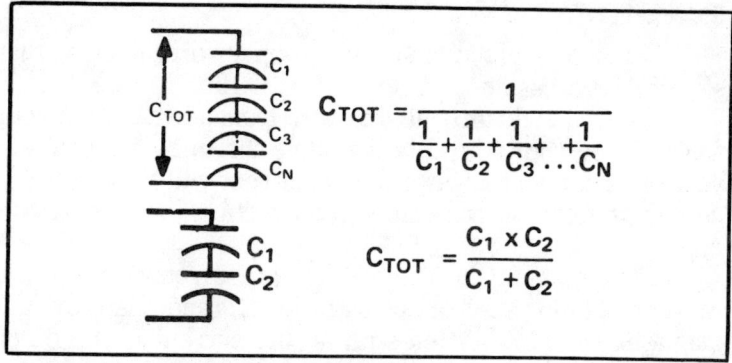

Fig. 2-9. General equation for determining the effective value of several capacitors connected in series is shown at the top. If just two capacitors are involved, the simplified equation at the bottom may be used. Since both these amount to an addition of susceptances (see text) and conversion back to reactance, they also apply to parallel resistors and to parallel inductances, by substituting R or L for C in the appropriate formula.

The expectation is correct. In similar fashion, the current through parallel reactances of like sign will divide just as dc divides among parallel resistors.

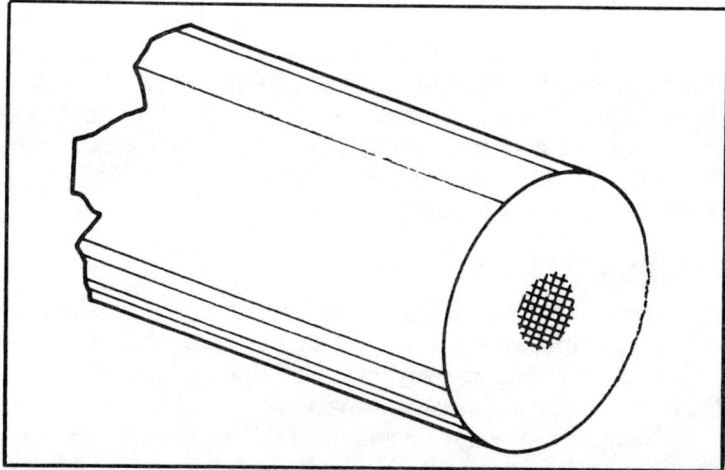

Fig. 2-10. Cross-section view of copper wire shows how the center portion of the wire is surrounded completely by the conductor, while the surface region has conducting material only on one side of it. This causes the center to have more inductance than the surface, which will in turn force ac current to seek the surface and avoid the center. The higher the frequency of the ac, the less effective will be the interior of the wire. At a sufficiently high frequency, all current travels on the "skin" and almost none flows in the interior. This is called *skin effect*.

SKIN EFFECT

Consider a piece of straight copper wire, such as that sketched in cross-section view in Fig. 2-10.

If we push dc through this wire from a battery, current flow will be approximately the same throughout the cross-section of the wire. Just as much current will flow near the center as flows in the same cross-sectional area near the edge.

That current in the center is surrounded by the magnetic field associated with itself, and also by the magnetic field associated with all the current in the outer parts of the wire, while the current on the wire's surface is surrounded by magnetic field from only itself and adjacent currents.

This means that there will be, inherently, more inductance in the center of the wire than on its surface. With dc, it makes no difference, because inductance is a factor only when current flow is changing.

With ac, the resulting inductive reactance means that the wire's impedance is lower on the surface than it is in the center. So long as signal frequency is low enough, the effects are not noticeable. At radio frequencies, though, the effect becomes appreciable. Virtually all the current is flowing near the surface of the wire, and the center might as well not be there. This variation of impedance between the surface of the conductor and its interior is what is known as *skin effect* because the rf current seems to flow on the "skin" of the conductor and avoid the interior.

RESONANCE

Back in the beginning of our preceding discussion, we mentioned in passing that it was possible to use a little reactance of one kind—say inductive—to cancel out some of the other kind—in this case, capacitive.

Now let's take full advantage of this fact, with the series circuit shown in Fig. 2-11 which contains an inductor, a capacitor, and a resistor, all in series with each other.

We know already that the resistance of the resistor will be the same at all frequencies, and that the reactances will change as the frequency changes. Let's plug in some specific figures to see how this circuit can be expected to behave. For instance, let's make the resistor 10 ohms. Reactance of the

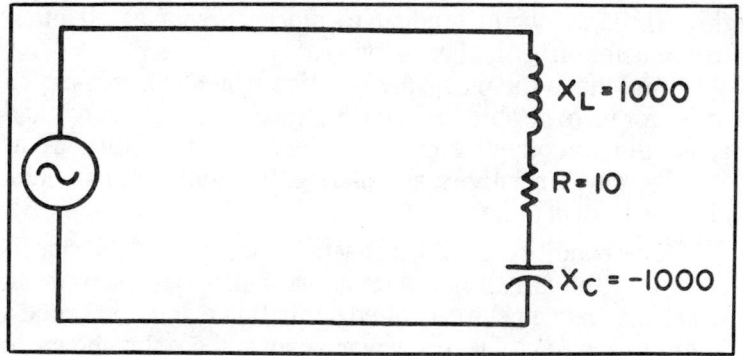

Fig. 2-11. Series L-C-R circuit illustrates the effects of resonance, which is possibly the most important single phenomenon aside from energy propagation itself so far as radio operation is concerned. Resonance permits "tuning" of equipment to select one frequency and reject all others, which in turn makes it possible for more than one transmitter to be used at one time. Without tuning, radio as we know it would not be possible.

inductor will vary with frequency, being much less than 10 ohms at frequencies near zero Hz, and being in the megohm region at extremely high frequencies. Similarly, the capacitor's reactance will vary from a very high value at low frequencies, to only a few ohms at high frequencies.

Notice that where the inductor's reactance is small at low frequencies, and that of the capacitor is large, and vice versa. It stands to reason, then, that at *some* frequency within the infinite range possible, their reactances must be equal.

Going in the other direction, at the frequency 10 times that at which the values cancel, the inductance is up to $+10,000$ ohms and the capacitor is down to -100 ohms, which gives a circuit impedance of $10 + j9900$ ohms or, in single values, 9900.5 ohms at +89.5 degrees.

If our circuit is fed by an ac source which provides the same voltage—say 99V—at all frequencies, then at either the high or the low frequency only about 1/100 ampere of current will flow through the circuit. But at the single frequency where the reactances cancel, current will be limited only by the resistance and in this case the current flow will be 9.9 amperes. That's 990 times as much as at either the high or the low frequency.

In this manner, the series L-C-R circuit of Fig. 2-11 selects current at a single specific frequency and permits it to flow through, while tending to block current at all other frequencies either higher or lower.

And that's the particular function which is necessary in order for us to be able to choose a signal at one frequency and reject those at other frequencies. For this reason, this circuit and its close relatives are among the most fundamental circuits in all of radio.

The condition in which reactance is completely cancelled out of the circuit is known as *resonance*, and a circuit in which all reactance is cancelled out in this manner is called a *resonant circuit*. The particular resonant circuit shown in Fig. 2-11 is known as a *series* resonant circuit, because its driving source is in series with all circuit elements.

The impedance of a resonant circuit depends upon a number of factors, but they are lumped into two general headings. One is the frequency of the applied signal, and the other is the Q factor which we will be examining shortly. Figure 2-12 is a graph of a "universal" resonance response curve; the solid line can be thought of as representing current flow through the circuit of Fig. 2-11 as compared to maximum current flow at resonance.

In addition to the series resonant circuit, we also make use of *parallel* resonant circuits. Parallel resonance was at one time called *antiresonance*, but this word is rather rapidly fading from use. Figure 2-13 shows a parallel resonant circuit.

If we ignore the power-source connections in Fig. 2-13 we will see that the reactances and the resistance are still in series with each other. The difference between parallel resonant circuits and series resonant circuits depends on the way in which the power or driving source is connected. If the source is in series with the reactances, it's series resonant. If the source is in parallel, it's parallel resonance.

The impedance characteristics of a parallel-resonant circuit are markedly different from those of series resonance. In series resonance, circuit impedance is high except at the resonant frequency. In parallel resonance, though, the reactances are in parallel so far as the power source is concerned, and this means that rather than cancelling out reactances, we must combine *susceptances*. High reactance means low susceptance, and vice versa.

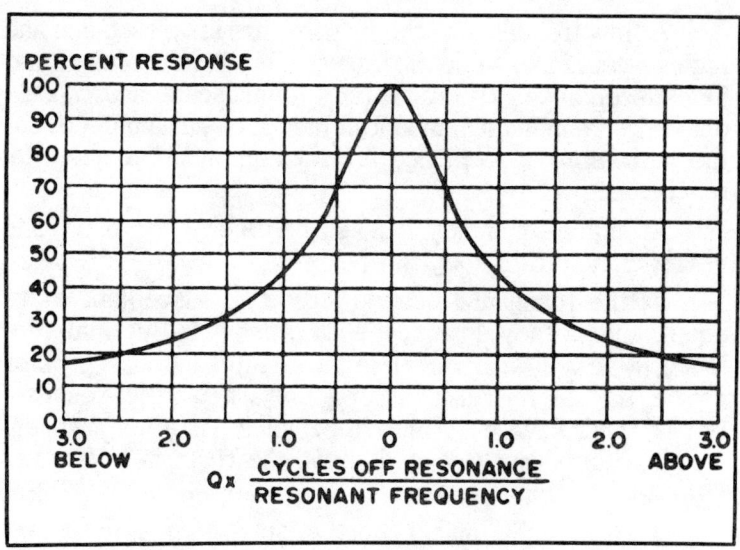

Fig. 2-12. Response curve of any resonant circuit is shown here. Horizontal scale is dependent upon two factors, Q and *resonant frequency*, which makes this graph universal. For example, a resonant circuit with a Q of 100 and a resonant frequency of 10 kHz would make the horizontal scale come out to be 100/10000 times cycles off resonance, or 0.01 times cycles. At a frequency of 10.01 kHz, 100 cycles above resonance, the value to use on the horizontal scale would be 0.01 times 100 or 1.0, and response would be 45% of that at resonance.

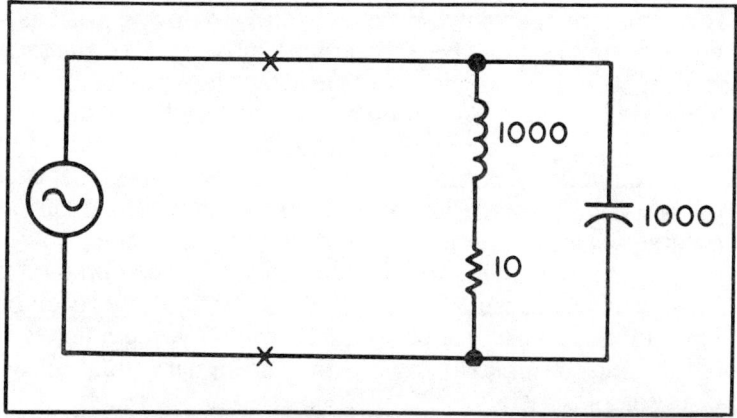

Fig. 2-13. Schematic of a typical parallel-resonant circuit. If the power source is disconnected at points marked X and the circuit is then opened at any point, it becomes the same as the series-resonant circuit of Fig. 2-11. Resistance is shown here in series with the inductor because in practice, inductors usually have more stray resistance than do capacitors, and a circuit behaves as if all resistance were in the inductive leg. See text for details of circuit action.

At low frequencies, the inductor has low reactance and high susceptance, while the capacitor has high reactance and low susceptance. The capacitor's low susceptance cancels out its corresponding amount of inductive susceptance, but the remaining susceptance is still high which makes net reactance low.

At high frequencies, it's just the other way around, and net reactance is still low.

At the resonant frequency, the two susceptances are equal, and so they do cancel each other out. But a susceptance of zero is equivalent to an infinite impedance. This means that at its resonant frequency, a parallel-resonant circuit must have very high impedance. At other frequencies, the impedance depends upon the reactance left over after cancellations.

The "cancellation" is effective only so far as the power source is concerned. In each reactance, its own current is flowing. However, this current cannot get out of the circuit—and so it circulates between inductor and capacitor. It's known as *circulating current*, and is very real indeed as anyone who has watched the output coils of his transmitter melt from its effects can testify!

Because the circulating current *is* larger, a closed resonant circuit in which circulating current flows can act as a voltage amplifier. The circulating current through each reactance will produce a voltage drop determined by the amount of current circulating and by the reactance value. If the driving power is coupled into the circuit by means of a tap on the inductor, it can force large circulating currents, and an output voltage much larger than the input voltage can be obtained. This isn't something for nothing, because no power gain is achieved. The higher voltage can only be obtained if little or no current is taken from the circuit. Many receiver input circuits make use of this fact to help overcome effects of tube and circuit noise, by stepping up signal voltage ahead of any amplifier stages.

To determine the exact frequency at which any specific pair of inductance and capacitance values will be resonant, we just combine the inductive-reactance and the capacitive-reactance formulas and come up with the equations shown in Fig. 2-14. As before, one is exact and the other is more

$$F = \frac{1}{2\pi \sqrt{LC}} \qquad F^2 = \frac{25{,}330}{LC}$$

Fig. 2-14. Equations for resonant circuits. Version at the left is the basic formula, with frequency in hertz, inductance in henrys, and capacitance in farads. That at the right is the practical version with frequency in megahertz, inductance in microhenrys, and capacitance in picofarads. For any value of inductor, some capacitor exists to make it resonant at any desired frequency. In practice, choices are also limited by the effect upon circuit Q and the feasibility of actually obtaining desired values.

practical, having all the conversion factors and the "pi" constant built into a single magic number. While any specific pair of values will have only one resonant frequency, in theory any inductor can be resonated at any desired frequency by simply choosing the proper capacitor value, and vice versa. In practice, Q is also a factor to be reckoned with and tends to limit the choice of L and C values.

CIRCUIT Q

What is *Q?* Q is a symbol which shows up in radio theory just about as often as do the R which stands for resistance, and the E for voltage, the I for current, the L for inductance, or the C of capacitance. You'll meet Q as part of the description of nearly any coil, many capacitors, and all tuned circuits. It is used to describe the characteristics of quartz crystals, and of antennas. But what is it?

The symbol, Q, originally stood for *quality factor*, and almost all textbooks define it in the same way—as the ratio of inductive reactance to total resistance in a circuit (X_L/R). But very few texts bother to show why the ratio of reactance to resistance should be of much importance, and only a couple of the more than a dozen we studied in preparing this discussion bothered to give any other definitions of Q.

Nevertheless, Q *is* important, and it has many definitions. All are equally true in every case, but for any special case, some are easier to apply than others.

So let's try to see first why the ratio X_L/R affects behavior of a resonant circuit, and then explore some alternate definitions of Q. We'll begin by returning to our series-resonant circuit shown in Fig. 2-11, with the same component values we used before. That is, 10 ohms resistance and 1000 ohms reactance for each reactance.

What we failed to say before is the fact that we can determine the voltage across either the coil or the capacitor alone, either by Ohm's law or by actual measurement with a good vtvm. Here, we'll do it with Ohm's law in the form $E = IZ$.

Since it's a series circuit, I is the same for all three elements, or 9.9 amperes with our 99V source. Impedance of the coil is 1000 ohms at 90°, and so the voltage across the coil must be 9.9 times 1000, or 9,900V at 90°. Impedance of the capacitor is also 1000 ohms at $-90°$, so its voltage must be 9,900V also, but at $-90°$. Since these two high voltages are 180° out of phase with each other, they cancel each other and cannot appear to the external circuit—but inside the circuit, measured across each element by itself, they're present.

Some authorities call this action the *voltage magnification* of the resonant circuit, and speak of the 9900/99 ratio between reactive voltage and applied voltage as the magnification factor. Others call the ratio the quality factor or merely Q.

The relation between this magnification factor and the definition of Q as being X_L/R is not overly obvious. It comes about because each of the voltages—that across the reactance and that across the total circuit—is related to the same series current by Ohm's law. In both cases, $E = IZ$, where Z is the impedance and I is the current.

For the reactance, Z is equal to X_L or X_C, and at resonance they're the same absolute value. For the total circuit, both reactances cancel out and the impedance is simply R. The voltage ratio E_x/E_t then becomes IX_L/IR; the I cancels out of the calculation since I/I is always one, leaving us only X_L/R to define the magnification factor or Q.

One of the most important uses of Q is as a measure of the effectiveness of a resonant circuit. That is, it measures how effectively the circuit can separate two signals of different frequency. We can demonstrate the relation between Q and selectivity in two ways, one of which is easy to see and the other of which, though much more accurate, requires more thought. Let's try both, the simpler one first.

Recall that the selection capability of a series resonant circuit comes about because at the resonant frequency, current flow is limited only by the R of the circuit, while at frequencies far from resonance, the X is the limiting factor and the R has very little effect.

If we increase the resistance and leave reactance alone, it will have little effect on current at the extreme frequencies, but will reduce the current flow at resonance. Thus the ratio between current at resonance and current far from resonance will become smaller, and the circuit is therefore less selective.

Since Q goes down as resistance goes up, if X is constant, this means that a high-Q circuit must be more selective than one with low Q.

While this explanation shows how circuit Q affects selectivity in general, it unfortunately does not tell us much about what happens at frequencies near resonance. To find out how Q works in this region—which, after all, is the one in which we are most often really interested—we must go to the more detailed explanation. It takes a little arithmetic to illustrate.

We defined Q by looking at the magnification factor— the ratio between the voltage across either reactance, and the total voltage applied to the circuit.

At frequencies near resonance, the reactances will not vary greatly from their values at resonance. If our inductor has 1000 ohms reactance at 10,000 Hz, then at 10,010 Hz its reactance will be 1001 ohms, and at 9,090 Hz it will be 999 ohms. The capacitor will behave similarly, with reactance values of -999 ohms at 10,010 Hz and -1001 ohms at 9,090 Hz.

Because of the mutual cancellation of opposite types of reactance, the net reactance of the circuit will be the difference in reactance values. At 10,010 Hz, net circuit reactance would be $1001 - 999$, or 2 ohms inductive.

If our circuit had a Q of 100, this would mean a resistance of 10 ohms (and with a 100-volt-peak source, 10 amperes current flow at resonance). Impedance of the circuit at 10,010 Hz, then, would be $10 + j2$ ohms, or about 10.19 ohms at a very small phase angle. Current flow at this frequency (E/Z) would be 100/10.19 or about 9.84 amperes, nearly as much as the 10 amperes at resonance.

If we leave the reactances alone, but reduce the resistance to only 2 ohms, we will raise circuit Q to 500

(1000/2). Current flow at resonance would be five times as great, or 50 amperes with a 100V source. At 10,010 Hz, however, circuit impedance would now be 2 + j2 ohms, or 2.83 ohms at 45°. Current at this frequency would be 100/2.83 or 35.4 amperes. This is only 35.4/50 or 70.8% of the current at resonance, while with a Q of 100 the currents at the two frequencies were almost equal.

The same thing happens at all other frequencies, whether near to or far from the resonance point of the circuit. The result is that a high-Q circuit is "sharper" than one of low Q. Figure 2-15 shows this by comparing relative currents for circuits with Q's ranging from 10 to 100, for frequencies from 20 percent below resonance to 20 percent above.

Circuit Q is sometimes defined as being the inverse (or reciprocal) or the *power factor* of a component or circuit. The pwoer factor is the ratio of power consumed to apparent power, which boils down to being a ratio of resistance to reactance. Since this is the inverse of the ratio which defines Q in the most common definition, defining Q as the inverse of power factor is simply a second-hand way of providing the same definition.

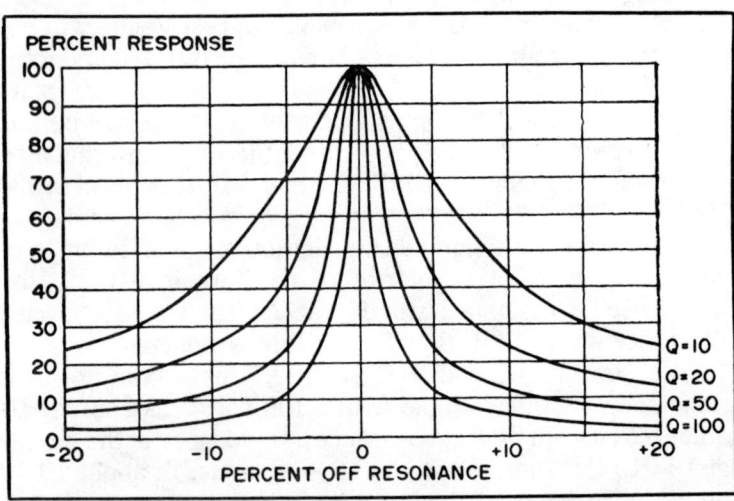

Fig. 2-15. Effect of Q upon circuit selectivity is shown here. This is similar to Fig. 2-12, except that factor Q has been removed from the horizontal scale and different curves have been drawn for different Q values. The greater the Q, the sharper the selectivity. Notice that effects are greatest in regions near resonance, but very little difference exists at frequencies almost exactly on the resonant of frequency.

All of our discussion of Q so far has been in terms of series resonant circuits. It applies equally, though, to parallel-resonant circuits by making proper substitutions. The net result after these substitutions are made is that the circuit impedance, rather than voltage, is multiplied by Q. That is, in a parallel-resonant circuit each of whose reactances is 1000 ohms at resonance and having a Q of 100, the impedance at resonance will be 100,000 ohms at 0°.

If such a circuit is then "loaded" by connecting a perfect 100,000-ohm resistor across it, the net impedance will be reduced to 50,000 ohms. The same result would be produced by raising the internal resistance from 10 to 20 ohms and so lowering Q to 50. In most radio theory the effects of such loading are accounted for by assuming that the load is transformed into the equivalent series resistance. In this case, the 100,000-ohm resistor's value would be transformed to 10 ohms, and its effective connection point would be changed from being in parallel with the resonant circuit to being in series with the inductor inside the circuit.

This is what is meant by the *impedance transformation* effects of a tuned circuit, and transmitters make use of tuned circuits in this way to a large degree.

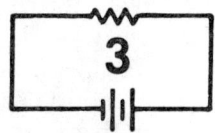

Power, Decibels and Harmonic Frequencies

It may seem that we're putting too much emphasis, in this study course for the General license examination, on the ultrabasic fundamentals of electricity and electronics. After all, nearly three-quarters of all the questions on the FCC's official "study list" deal with matters much more complex than those we have been examining since we began.

However, our stress on the ultrabasics is deliberate. With a strong understanding of the major principles of electrical physics as a foundation, the more specialized matters met on the actual examination will be much more solidly seen. In other words, we're going a bit slowly at the beginning, so that we can do a better job later on and make it all appear to be as simple as it really is.

POWER

Power is one of those words we all *think* we know the meaning of—and because of this, it's extremely important that we be sure right at the start of our studies of electrical power that we're all using the *same* meaning for the word.

So let's ignore what we think we know about the meaning of the word, and develop a brand new definition to use during our examination of this subject. Let's begin by going back to what we learned in the first chapter of this course, about current and voltage.

Use of Energy

There, you may remember, we found out what current and voltage were by relating them to the two kinds of energy involved in the magnetic field and the electric field. We found that either of these two kinds of energy (actual energy and potential energy) could apparently exist indefinitely by itself so long as nothing moved, but that as soon as motion came into the situation the total energy present began changing from one form to the other, and that this change of form of the energy was what we know as electric and magnetic effects.

While we didn't bring it out in so many words at that time, motion itself involved the use of energy—which might mean that putting motion into our balanced situation simply unbalances it by adding additional energy. This is one of the areas in which physicists are still uncertain as to exactly what goes on, but the assumption appears reasonable in view of the current belief that everything in existence is either energy, or equivalent to energy, and that the total amount of energy in the universe cannot increase or decrease but must remain forever unchanging.

What all of this means to us at this point in our studies is simply that any use of electricity involves a use of energy.

We also found, earlier, that resistance in a circuit causes some of the electrical energy present to be dissipated in the form of heat—which, again, is another kind of energy. If it won't cause too much confusion, we can consider that electrical energy which is changed to heat energy has been "used up"; we do not mean by this that it's no longer energy, but merely that it's no longer in the right form to be used electrically. In contrast, electrical energy which has been "locked up" by a change of phase relationships in an ac signal is not used up, because we can easily get to it again by changing the phase relationships back again.

When we do anything at all to electrical energy by means of our circuits, we must "use up" at least a part of that energy. Some of the energy we "use up" is wasted—that is, it doesn't do anything effective for us. If the circuit is operating properly, though, most of the energy "used up" will be used in doing what we want the circuit to do.

For example, the whole purpose of a radio transmitter is to transmit information to one or more other locations by radiating radio energy from the antenna. The energy which we radiate is "used up" so far as the transmitter is concerned; we don't get it back to use again, but must continue to supply more energy in the form of the operating voltages which keep the rig going.

Now we get back to our immediate subject, power. Power, as it applies to electricity, is simply the *rate* at which electrical energy is used. This is *not* the same as the amount of energy used, any more than miles per hour is the same as distance. The power measures only the *rate* of energy use.

Back in the first chapter, we learned that voltage is a measure of electrical pressure or potential, while *charge* is a measure of the actual amount of energy at any instant. Current is a measure of the *rate* of charge, and so power, which is the rate of energy usage, must be related to both voltage and current.

And the formula for determining power is simple: power equals voltage times current. If potential is in volts and current is in amperes, the resulting power will be in the unit of power, *watts* (for James Watt, inventor of the steam engine).

Notice that because power is a measurement of rate, like miles per hour, rather than an absolute measurement such as distance, it tells us nothing about the *total* amount of energy used. Just as the question "Which goes farther, a bicycle moving at 10 mph or an airplane doing 500 mph?" cannot be answered until the time of travel of each is known (if the bicycle goes for 72 hours while the airplane goes for only half an hour, the bicycle will go nearly three times as far), neither can the question "Which uses more energy, a 5W transmitter or a 50kW transmitter?" be answered until the time during which each is operated is known. The 5W transmitter must operate 10,000 times as long as the 50 kW one to use the same amount of energy—but if it operates more than 10,000 times as long, it will use more!

To measure total energy consumption, we use such units as the kilowatt-hour or the watt-second. One kilowatt-hour is the amount of energy used by a 1 kW device in one hour, or a 1W device in 1000 hours, or any other combination which multiplies out the same way when power in kilowatts (thousands of watts) is multiplied by time in hours. Similarly, a watt-second is the amount of energy used by a 1W load in one second, or simply power in watts multiplied by time in seconds. The watt-second is also known as the *joule,* after the physicist who first stated many of the fundamental ideas about energy and power, and is the standard unit of electrical energy usage. In practice, power companies charge for their product by the kilowatt-hour while photographers rate their strobe-light energy storage capabilities in watt-seconds or joules, and radio operators seldom worry about either since they are more concerned with the rate of energy use than with the total amount used.

Power Measurement

We've just met one way in which power is measured—in watts. And we have learned that 1W is the rate at which energy is used in a circuit which operates at a potential of 1V and requires an energy flow of 1A. We also met the kilowatt, which is 1000W. Now let's look at some of the other measurements of power. Suppose for a start that we wanted to know how much power was required by a circuit operating at 100V dc, but that for some reason we had no ammeter with which to measure its current. We do know, however, that its resistance is 250 ohms.

We can apply Ohm's law using the known values of voltage and resistance to determine how much current flows. This tells us that current equals voltage divided by resistance, so we know that the current is 100/120 or 0.4A. We can then plug this 0.4A figure into the power equation $P = EI$ to get power 100V times 0.4A, or 40W.

For a shortcut, we can combine the equation of Ohm's law that tells us $I = E/R$ with the power equation $P = EI$ by substituting the entire right-hand half of the Ohm's law equation into the power formula in place of the I. This gives us a new power formula, $P = EE/R$. Since E times E is usually written as "E squared" (E^2), this gives us $P = E^2/R$, which defines power in watts as the square of the applied voltage, divided by the circuit resistance.

The same thing can be done if we know current and resistance but do not know voltage. In this case, we use Ohm's law to determine voltage ($E = IR$) and substitute the IR from that equation in place of the E in the power formula to get $P = IIR$, or $P = I^2R$. That is, power dissipation in a circuit is equal to the square of the current, multiplied by the resistance.

The relation between power dissipation, current flow, and resistance which we have just developed is extremely important throughout electricity and electronics. It's the reason, for instance, that the power distribution circuits which carry ac power around the country from generating plants to the ultimate users operate at such high voltages. By using high voltage, current can be kept comparatively low while transmitting the same amount of power. Keeping current low reduces the I^2R loss in the power lines, so more of the energy gets to the ultimate user and less goes to provide a foot-warmer for stray birds.

It's also the reason why a high-Q tank circuit in a ham transmitter may get hot enough to melt its plastic coil supports, even when there's apparently very little power going anywhere. A tank circuit with high Q has high circulating current, which may be measured in the hundreds of amperes, and the wire of the coil has a definite minimum resistance. If circulating current is 100 amperes and coil wire resistance is only 0.01 ohm, the I^2R power dissipated in the wire is still 100 x 100 x 0.01 or 100W. And a hundred watts worth of heat is quite a bit; if you don't think so, touch a 100W light bulb after it's been on for an hour, or maybe the tip of a 100W soldering iron.

The three major ways of determining power in watts, then, are by multiplying voltage and current ($P = EI$), by relating voltage and resistance ($P = E^2/R$), and by relating current and resistance ($P = I^2R$). These equations, like Ohm's law, work for all dc circuits, and for resistive ac circuits with certain restrictions.

The major restriction upon use of these equations with ac circuits involves the *units* of measurement to be used. With dc, a watt equals the power of a volt at a current of one ampere. When ac is introduced, the voltage and current are always changing—and this means that the watt becomes open to confusion unless we restrict things.

Even if we follow the conventional engineering standard and ignore all ac except the kind which has a sine waveform, both the current and the voltage are continually varying from a negative peak value through zero to a positive peak value and back again. To get a power calculation, we must choose which of these unlimited voltage and current levels to use.

Peak Power

While we might at first be tempted to use the peak levels, if we did we would find that volts times amperes would give us a figure that represented more power than volts times amperes in a dc circuit does. In fact, the figure would be just twice as great as in a dc circuit.

This, of course, is not a happy state of affairs. A watt should be a watt, regardless of whether it's dissipated in ac or dc circuits.

If, instead of the peak levels, we use levels which will produce a watts figure just half as large, everything will come out as it should. But we cannot do this by using half the peak voltage and half the peak current, for that would give us half of a half, or one fourth, the power.

What we have to do is find a voltage level and a current level which multiply out properly. If we divide the peak value in each case by the square root of two, we will have such levels.

What's more, these levels turn out to be the *rms* or *effective* levels we have already met. In fact, the rms levels were chosen as the normal values for ac voltage and current specifically because they do provide power figures which are equal to those obtained with dc. Whenever ac power is mentioned, unless it's qualified with some sort of modifying word, what is meant is the power determined by multiplying rms voltage by rms current.

Another restriction on the determination of power in ac circuits comes from the fact that power can only be dissipated when voltage and current are in phase. That is, nothing but *resistance* can dissipate power. The ac circuits, however, normally contain reactance as well.

Reactance acts just as does resistance to limit voltage or current in the circuit—but this control is not achieved by dissipating the power into heat energy. Instead, the phase is altered. Since no power is dissipated by doing this, the effect cannot be expressed in watts or any other unit of power.

To cover such situations, the unit known as the voltampere was invented. Like power, it's obtained by multiplying rms volts times rms amperes, but it doesn't necessarily indicate any use of energy. If the circuit is composed of resistance only, its voltampere figure will be the same as its power figure; if reactance is present, there will be more voltamperes than watts, and the difference between them indicates the amount of reactance. The ratio of voltamperes to power is known as the power factor, and you may see it in the ratings for transformers or electric motors.

Average Power

Now that we've seen how ac power is calculated and why the rms values of voltage and current are used, let's look at some of the modifiers often attached to the phrase—for instance, *average* power.

This is a term most frequently used with respect to the power of an amplitude-modulated radio signal, and has nothing to do with average voltage or average current. Instead, it's the average of the signal's rms power level, taken over several cycles of the modulating signal.

For this to make any sense, we must jump a little ahead of ourselves at this point and take a quick look at amplitude modulation (more exact discussions of modulation will come quite a bit later in this course). Figure 3-1 shows the *envelope* waveforms of both a modulated and an unmodulated AM signal. You can see that the modulated waveform rises to higher peaks than does the unmodulated or *carrier* signal.

If the modulating signal is a sine wave, and if the maximum amount of modulation is applied as it is in Fig. 3-1, the total power of the modulated signal (averaged over several cycles of modulation) will be half again larger than that of the carrier. The average power, then is 1.5 times greater with such modulation than without.

Instantaneous Power

Instantaneous power, on the other hand, refers to the power figure that results from taking one specific instant during a cycle and multiplying the voltage at that instant by the current at the same instant. These are *not* rms values— the *instantaneous* values are used instead. If we choose the instant when both voltage and current are peak, the instantaneous power comes out to twice the average power—and in the modulated signal with its higher peaks, the maximum instantaneous power is four times as great as the carrier's average power.

The maximum instantaneous power level possible for a modulated signal—that is, the instantaneous power at the

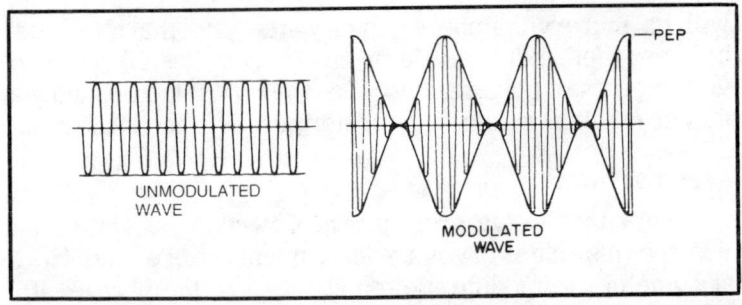

Fig. 3-1. Waveforms of an unmodulated or *carrier* signal, and the complete modulation envelope for 100% amplitude modulation are necessary to illustrate some of the various power measurements often encountered. Peak envelope power, for instance, is the power calculated from voltage and current at the modulation peak, as indicated at right.

peak of the modulation cycle—is called the peak envelope power or *PEP*. This figure is often used to rate output power of SSB transmitters, and to compare the power levels of various types of modulation.

When we attempt to measure the power involved in an ac signal which is *not* a sine wave, we have a problem. While it's theoretically possible (as we shall soon see) to analyze *any* ac waveform into a combination of sine waves, this is not a practical approach to power measurements.

The most accurate way to measure the power content of a non-sine-wave ac signal is to dissipate the power into heat by means of a suitable resistor, and use a thermometer to measure how much heat it produces. From this, we can determine the power figure by converting degrees-of-heat-rise to calories, and calories to watts.

WHAT IS A DECIBEL?

Contrary to what you may believe or have heard, the decibel (abbreviated dB and often pronounced "dee bee") was *not* invented solely for the purpose of making an already complex situation more confusing. In fact, the whole idea behind the decibel was to make things simpler!

Most textbooks approach decibels from a mathematical viewpoint which isn't such a bad idea, except that you don't really have to comprehend the math behind them in order to make good use of them. We'll try another approach.

In many situations involving a comparison between two values of the same quantity—such as the amount of light present in a room, or the intensity of sound produced by a PA system—it's not so much the *absolute* difference between the values that's important, as the *ratio* between them.

That is, you wouldn't compare the brightness of two different light bulbs by saying that one was 25 watts more powerful than the other. Instead, you would say that it was twice as bright.

If one bulb was a 25-watter and the other rated at 50 watts, you might expect twice the brightness from the doubling of power, but you wouldn't expect to keep getting twice the brightness every time you went up another 25W in power.

Actually, the relation between brightness and wattage of the bulb isn't really all that simple and we don't mean to imply that it is. Our purpose is to bring out the difference between the *ratio* of twice or double, and the *absolute* increase of 25W.

Now we bring decibels back into the picture. All that the decibel amounts to is a unit of the measurement of power *ratio* rather than any absolute value. Any time power is doubled, it has been increased by 3 dB, no matter what the starting point. If the power is cut in half, it has been decreased by 3 dB.

While the decibel is the unit normally used in all technical work, it is not the basic unit. The basic unit of power ratio is the *bel*, named for Alexander Graham Bell of AT&T renown, and it's defined as being a power ratio of 10:1. The decibel is one-tenth of a bel—but it's not a 1:1 ratio (the 1:1 ratio is zero bels or 0 dB). And right here is one of those porcupine quills.

The problem is that *ratio* itself involves multiplication or division. A 2:1 ratio implies multiplication or division by 2, a 10:1 ratio by 10, and so forth. The *units* of ratio, though, are not multiplied or divided, but instead are added or subtracted, a 10:1 ratio is 10 dB. A 100:1 ratio is not 10 *times* 10 dB or 100 dB, but rather is 10 *plus* 10 dB or 20 dB.

If you're familiar with the mathematics of a log table or of exponents, you have probably realized by this point that the decibel behaves like the log of the true ratio. The bel, in fact, *is* the log of the power ratio, and since 10 dB equals 1 bel, the formula for calculating decibels becomes merely 10 times the log of the power ratio. If you have a log table handy, the ratio itself can be calculated by logs—but that's not required in order to make use of decibels, as we shall see shortly.

Strictly speaking, the bel and decibel are units for the measurement of *power* ratios and cannot be used for measuring anything else. This might appear to be rather confining until you recall that power is simply a measurement of the rate at which energy is used, so that whenever energy is actually being dissipated you have power whether you realize it or not.

The situation isn't helped any, though, by the fact that many people speak (incorrectly, as it happens) of "voltage decibels" and "current decibels."

These voltage and current decibles came into vogue because it's possible to calculate power when only voltage or current and circuit resistance are known. We saw this in the previous chapter, and developed a pair of shortcut equations by combining the power formula with the appropriate form of Ohm's law.

Since decibels are calculated by first getting the ratio between two power figures, then taking the log of that ratio and multiplying by 10, it's possible to plug the shortcut power equations in place of the actual power figures, and come up with a corresponding pair of decibel equations which apparently compare voltages or currents. Both the shortcut power equations involve squaring the voltage or current figure, and when we move this into the decibel equation we do the squaring by multiplying by 20 rather than 10.

But what we have is still a *power* ratio, and in addition we have introduced a restriction that both voltage figure (or current) must be in circuits of identical resistance.

Since a 2:1 power ratio comes out to 3 dB and a 2:1 voltage ratio comes out as a 6 dB, people began calling the decibel values obtained from the composite equations voltage or current decibels and the belief grew that they were somehow different from power dB.

Actually, there's no difference between one set of dB and any other. The apparent difference is actually the difference between a power ratio and a voltage ratio.

If we double the voltage present across a resistor without changing the resistance value, the current through it must also double (Ohm's law proves this). That means that when we figure the power, we must multiply twice the voltage times twice the current, and we come out with a power ratio of 4:1.

A power ratio of 4:1 comes out to be 6 dB, and a voltage ratio of 2:1 is also 6 dB. Since a 2:1 voltage ratio across the same resistance (which you recall is a requirement for using the "voltage decibel" equation) is the same as a 4:1 power ratio, you can see that there's no difference in the decibels themselves whether we get them by voltage/current calculations, or by power calculations.

If you're going to use log tables to deal with decibels, here are the two equations:

$$\text{decibels} = 10(\log P_1 - \log P_2)$$
$$\text{decibels} = 20(\log V_1 - \log V_2)$$

For current, use the voltage equation but substitute current in place of voltage. For either equation, the larger of the two quantities is usually used as the first, so that the result will usually be positive.

But you don't need to use a log table or a slide rule—or even know what they're all about—to figure out decibels accurately enough for all practical purposes. All you need to do is memorize a couple of facts and four specific decibel values.

We've already met the facts—they are that decibels add together, so that if you know a 10:1 ratio is 10 dB, you can figure out a 100:1 ratio as being 20 dB and 30 dB as being 1000:1. The process can continue as long as need be. Adding dB multiplies the ratio, and subtracting dB divides the ratio. If 100:1 is 20 dB and 2:1 is 3 dB, then 50:1 is 20−3 or 17 dB.

Now for the values. One decibel is 1.25:1, or 5:4. Three dB is 2:1, 5 dB is 3.16:1, and 10 dB is 10:1. From these, we can get any other values we may need by adding and subtracting. For instance, 9 dB is 3+3+3 dB, or 2x2x2 or 8:1.

From just these values, you can find out how many decibels any specific ratio amounts to, or conversely what ratio is meant if you have a known number of decibels, because you can build up any number you want by proper addition or subtraction of 1, 3, 5, and 10.

For instance, how many decibels stronger than a 25W transmitter is a 500W one? First, we convert the absolute power figures of 500W and 25W to a ratio by dividing 500 by

25, and discover that the ratio is 20:1. Next, we divide the 20 by 10 and note down the 10 dB effect of that division; the resulting ratio is 2:1. But a 2:1 ratio is 3 dB, so we add 3 dB to the 10 we already have, and the answer is 13 dB.

HARMONICS

Back in the first chapter of this course we made the acquaintance of ac, and met one of the major characteristics of any ac signal—its frequency.

An ac signal may have any frequency greater than zero, although most common ac signals lie in the range from around 10 or 15 Hz up to several thousand MHz. Physicists consider light itself as an ac signal of exceptionally high frequency (so high, in fact, that it's not normally measured in hertz at all), and for some purposes it turns out to be convenient to treat dc as being an ac signal with a frequency of zero.

Since an ac signal may have any frequency at all within this unimaginably wide range, it's only natural that some signals must have frequencies which are exact multiples of those of other signals. For instance, a 24 Hz signal is at two times the frequency of a 12 Hz one, or 3 times that of an 8 Hz signal, or 4 times a 6 Hz signal, and so forth.

We can go the other way, too. For instance, if we start with an 8 Hz signal, we could have signals at 2 times 8 or 16 Hz, 3 times 8 or 24 and so forth all the way up until we run out of numbers.

The point is that no matter what the frequency of the signal with which we start, other signals may exist at frequencies which are exact multiples of our original. As it happens, these other signals bear some special relationships to our original signal, and for this reason are known as the *harmonics* of the original (which is itself known as the fundamental frequency signal or merely the fundamental).

Harmonic Phase Relationships

One of the special relationships between a fundamental and its harmonics involves their mutual timing. If we start all the signals from zero at the same instant, then whenever the fundamental's voltage (or current) waveform crosses zero, so will that of each harmonic.

This is always true, because of the definition of a harmonic as an exact multiple of the fundamental frequency.

If the frequency is two times that of the fundamental, then during the time that the fundamental goes through one half of its cycle and gets back to zero, the harmonic will complete a full cycle and reach zero at the same instant. When the fundamental has completed one full cycle, the harmonic has completed two, and again they reach zero together. The harmonic also goes through zero when the fundamental is at either peak—but we don't care about that right now.

If the harmonic frequency is three times that of the fundamental rather than twice, then when the fundamental is half-way through its cycle and reaches zero, the harmonic will have completed one full cycle and will be half-way through another one. At the end of a full cycle of the fundamental, the harmonic will have completed three cycles.

Harmonics are identified as second, third, and so forth, according to the multiplying factor involved. The second harmonic is at a frequency two times that of the fundamental, the third at three times fundamental frequency, and so on. The *order* of a harmonic usually refers to its separation from the fundamental; second, third, and fourth harmonics are often called *low-order*, while fifth and higher are known as *high-order*. The dividing line between low and high harmonics, though, depends largely on the particular problem under discussion.

The same arguments which we just applied to the second and third harmonics apply to all of higher order as well. Those which are at even multiples (even harmonics) of the fundamental will be completing a cycle when the fundamental reaches the midpoint of its cycle, and odd harmonics (those at odd-numbered multiples) will be in the middle of a cycle when the fundamental is at midpoint.

Because the timing relationship between a signal and its harmonics is made constant in this manner by the definition of harmonic, we can talk about *relative phase* between a signal and its harmonics although in general, phase and refer only to timing differences between two signals of identical frequency.

That is, an odd harmonic which is in phase with its fundamental crosses zero at the same time as the fundamental, and is going in the same direction (whether positive or negative).

The same signal 180 degrees out of phase with its fundamental will still cross zero at the same instant as the

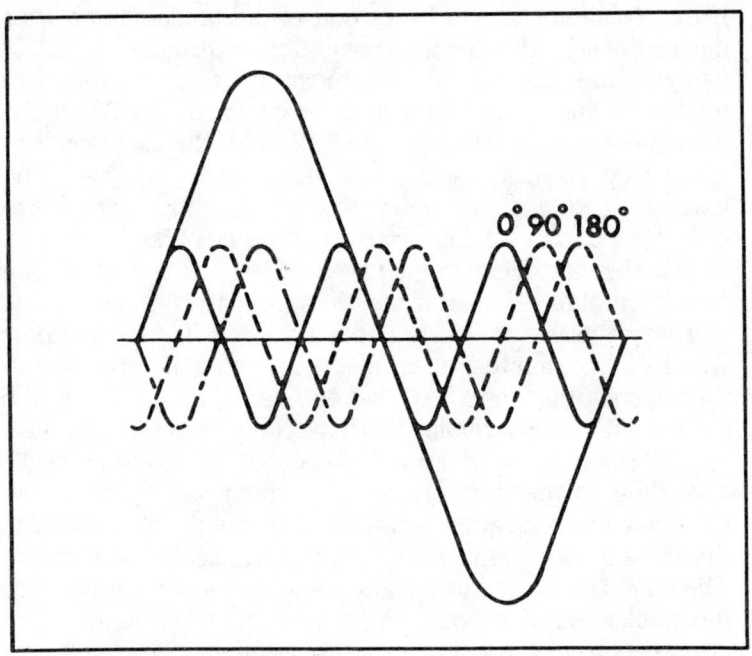

Fig. 3-2. Phase relationships between fundamental and odd harmonics are illustrated here using the third harmonic. At 0 degrees, or in-phase condition, harmonic and fundamental both cross the zero line at the same direction. At 180 degrees, or opposite-phase condition, they cross in opposite directions. The 90-degree phase relationship is midway between these (dotted line).

fundamental every time, but will be always going toward opposite polarity.

If the two are 90 degrees out of phase with each other, the harmonic will be at a peak value whenever the fundamental crosses zero. (See Fig. 3-2 for an illustration of these phase relationships between a signal and its third harmonic.)

The exact description of phase relationships between a signal and its harmonics depends to a large degree upon the order of the harmonic. Whether an even harmonic is in phase or 180 degrees out of phase depends upon which half-cycle of fundamental you start from, as shown in Fig. 3-3 for the second harmonic.

Another special relationship between a signal and its harmonics involves the way in which they combine to produce signals which are not sine waves at all. Take, for example, a combination of fundamental and third harmonic such as that shown in Fig. 3-4.

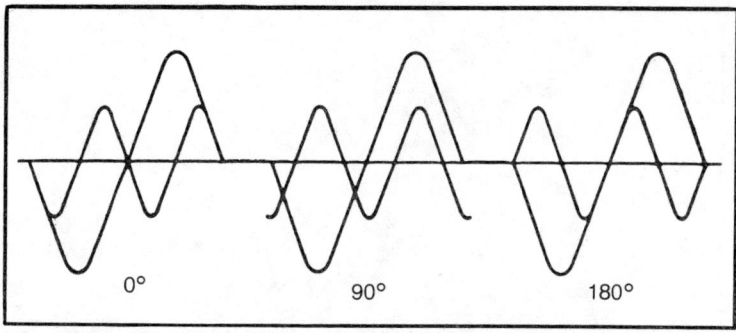

Fig. 3-3. Phase relation between fundamental and even harmonics (second is shown here) is a bit more confusing, since both the 0– (in-phase) and 180-degree relationships appear very similar.

When both these signals are present in a circuit which permits them to combine, the net energy at each instant is the total of the energy in both signals. At certain moments they add together, and at others they cancel; the result is the waveform shown by the heavy line in Fig. 3-4.

If we change the phase relationship between the two signals, the waveform changes, because the addition or cancellation occurs at different times.

Similarly, if we introduce still more harmonics into the mixture, we can produce far different waveforms.

When enough different harmonics are present, and with the phase of each being independent of all the rest, we can reproduce any ac signal at all. This is another reason why engineers deal only in sine waves, although sine waveforms are rare in nature; any other signal may be treated as merely a combination of sine waves, each of appropriate phase and strength.

Even such decidedly non-sine waveforms as the square wave (Fig. 3-5C) can be built up from sine waves. Figure 3-4 shows the result of combining a fundamental and its third harmonic. Figure 3-5A shows what happens to this when the proper proportion of fifth harmonic is added, and Fig. 3-5B shows what happens when the first 100 odd harmonics are included; it's difficult to distinguish between the waveforms of Fig. 3-5B and Fig. 3-5C.

Harmonic Suppression

So now we have met the family of harmonics, and have seen that they are essential to most types of actual ac signals.

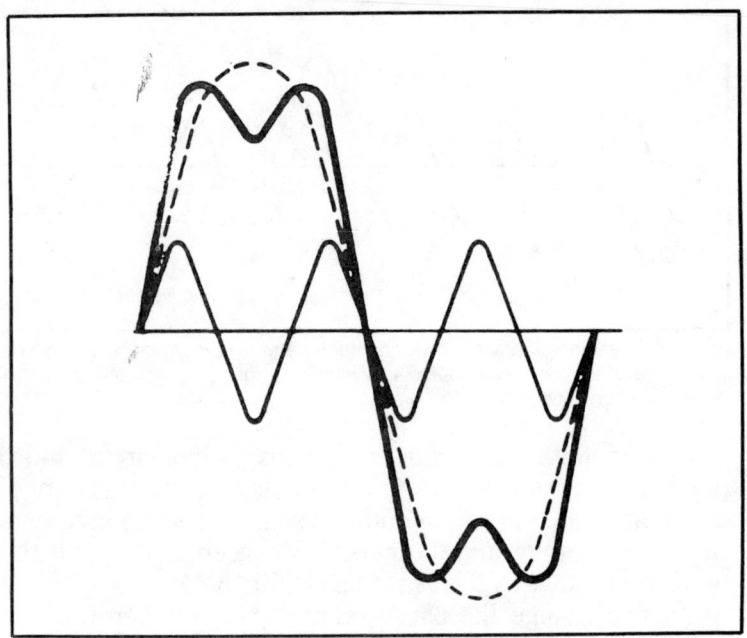

Fig. 3-4. Sine waves can be combined to produce any ac waveform. Combination of fundamental and third harmonic, shown here, produces a tooth-shaped waveform (heavy line) which is the first step toward a square wave.

Yet most discussion of harmonics centers upon methods of suppressing them, or preventing them in the first place. Why?

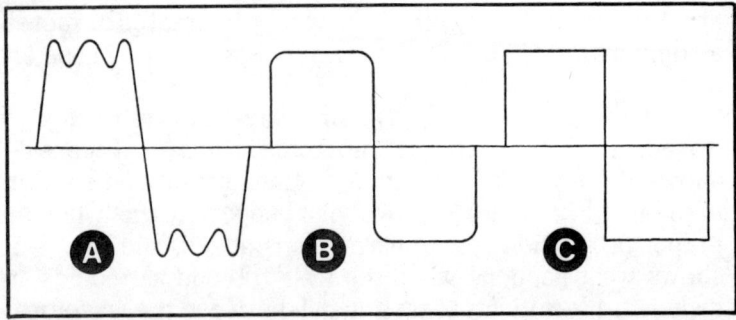

Fig. 3-5. Square wave can be considered as a sine wave together with all its odd harmonics in proper proportions. At left (A) is the result when the fundamental, third, and fifth harmonics are mixed. Center (B) shows the waveform when the first 100 odd harmonics are included. Idealized square wave (C) is difficult to distinguish from the version produced by the first 100 odd harmonics; few scopes would be capable of telling which is which.

A harmonic is good only if you want it. Like anything else, an unwanted harmonic or one which you cannot control is to be avoided whenever possible. FCC regulations prohibit transmission of excessive amounts of harmonic energy, and almost anything you can detect is considered to be excessive. Yet we do make extensive use of harmonics in ham transmitters—always attempting to keep them under control.

Harmonics may be generated either deliberately or by accident, whenever a sine-wave signal is distorted by any means. If the signal was sine to begin with, and is not now, something must have been either added or taken away to make it different. As it turns out, what was added was harmonic energy.

The distortion need not occur in an amplifier. *Any* action which distorts the signal will suffice. Most harmonic generation in ham transmitters, though, occurs in the amplifiers, because the normal rf power amplifier circuit distorts the signal greatly in the interests of handling high power with comparatively small tubes.

In general, harmonics can be controlled by either or both of two major routes. The first concentrates upon preventing their existence, while the second places the emphasis upon containing them so that they cannot escape.

Most practical situations require both types of control. The first, alone, is impractical and probably impossible as well, because all the efficient rf amplifier circuits capable of handling high power also generate harmonics prolifically. Without the first, the second becomes an unimaginably large job. Both together, though, hold down the strength of the harmonics which are generated, and make them easier to bottle up.

Deliberate harmonic generation is usually accomplished by circuits known as *frequency multipliers*, which look like any other rf amplifiers so far as the schematics are concerned. What makes one generate harmonics for us is the specific manner in which it is operated.

To get a good harmonic generator, we bias the tube's grid much more strongly than for normal amplification, so that its switching action becomes more abrupt; then we apply much more driving power than normal to the stage, again to make the switching as rapid as possible and the output waveform as close to square as we can get.

When we do this, the tube's output signal has a high percentage of harmonic energy, and it's a simple matter to pick off the one we want with a selective (high-Q) resonant circuit. Usually, we take either the second or the third, because it's difficult to get enough power at higher multiples in a single stage. VHF transmitters often use an 8.3 MHz crystal followed by a tripler to 25 MHz and a doubler to 50 MHz, before beginning real power amplification.

Now if we want to generate as *few* harmonics as possible, all we need do is reverse this philosophy. We take special care to bias the tube only enough to get the job done, and to keep driving power at the lowest level capable of producing proper output. The output will then be as free of harmonics as we can get—but they're still present, so our job is not yet done.

If we increase the Q of our resonance output circuit, we will make it more selective and it can reject the harmonics more efficiently. This will reduce the harmonic level at the antenna. In an rf circuit, we can raise Q by increasing the capacitance and then reducing inductance to keep the circuit resonant. Most of the effective resistance of the circuit is in the coil, and so when we reduce inductance we reduce resistance as well. The resistance decreases more rapidly than the reactance, and Q goes up accordingly.

When we have reduced harmonics to the practical minimum by applying these ideas to all straight-through (input and output at same frequency) amplifier stages in a transmitter, we have almost exhausted the possibilities of the first method of harmonic control and it's time to turn our attention to the second—that of bottling up the remaining harmonic energy so that it cannot escape to cause trouble elsewhere.

Before we do so, though, one point needs special emphasis and a couple of tricks with harmonic generators deserve mention.

The point which we must emphasize is that a harmonic cannot be *generated* by anything which does not introduce distortion into the signal, but that the *radiation* of a harmonic can be suppressed by nondistorting devices.

In particular, neither a feedline nor an antenna can generate harmonics if it is in decent working condition; about the only thing in either which could *generate* a harmonic would be a corroded connection. However, if a harmonic is generated back in the transmitter, the feedline and the antenna can easily radiate it to the world—or just as easily can help you keep it contained.

Now for the harmonic-generator tricks. When we use a harmonic generator to double or triple a frequency, we want a single specific harmonic and we don't want any others if we can help it.

As it happens, there's a special type of circuit which can generate even harmonics but suppresses odd ones, and another circuit which generates odd harmonics and cancels out the even ones. Use of these circuits makes life much easier, because the selectivity of the output circuits doesn't have to do so much of the selection job.

The even-harmonic generator is usually called a *push-push* doubler and the schematic appears as Fig. 3-6. Here's how it works:

The centertapped grid circuit divides the input signal into two independent signals, identical in strength but opposite in phase, and applies each to its own amplifier tube. Operating conditions for both amplifier tubes are made identical by the common grid resistance. Outputs at the plates of these tubes will be 180 degrees out of phase with each other at the fundamental frequency, but the bias is chosen so that each output also contains an appreciable amount of second-harmonic energy. These second-harmonic signals are in phase with each other (Fig. 3-7) despite the out-of-phase condition of the two halves of the fundamental signal.

When both plates are connected, in parallel, to the same tuned circuit, the fundamental portion of the output signal cancels itself out. No such cancellation occurs, however, for the second (or any even) harmonic, so by tuning the output circuit to the desired frequency we can recover the even-harmonic energy and couple it out to a later stage.

All odd harmonics are canceled, just as was the fundamental, because they are effectively 180 degrees out of phase with each other at the output.

If we want the odd rather than the even harmonics, a minor modification to the circuit as shown in Fig. 3-8 does the trick.

The input and amplification action is the same as in the push-push circuit. However, rather than the plates being

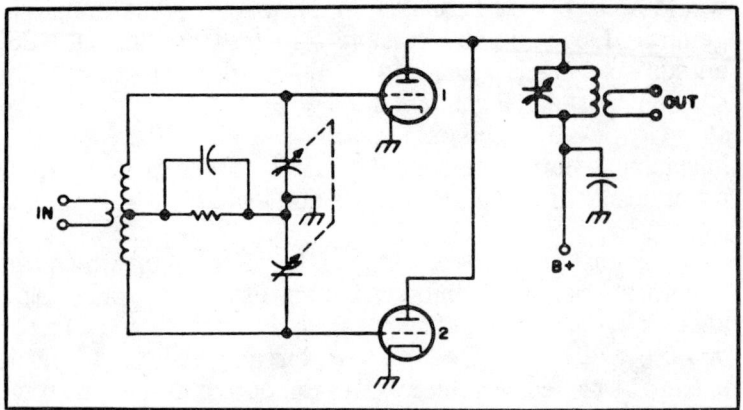

Fig. 3-6. This circuit, known as a push-push multiplier, generates only even harmonics of the input signal and cancels out the fundamental and all odd harmonics.

connected in parallel, they are connected to the opposite ends of a *split* circuit which reverses the action of the input circuit, to combine the out-of-phase halves of the signal into a single, in-phase, output.

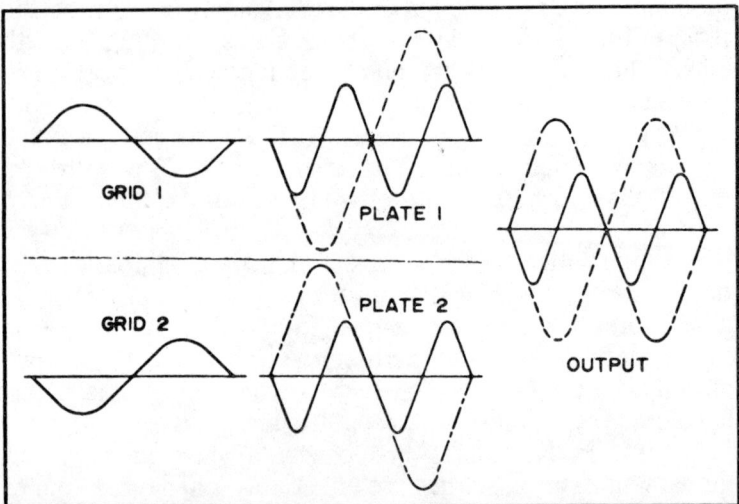

Fig. 3-7. Input and output waveforms show how a push-push doubler works. Grid waveforms, at the fundamental frequency, are equal in strength but opposite in phase. Same waveform, only amplified and reversed in phase, appears at the plates of the tubes—together with the second-harmonic component introduced by the tube's operating conditions. In the output circuit, the two fundamental waveforms cancel each other out, leaving only the second harmonic.

The recombination is accomplished by effectively reversing the phase of signals from one side, and leaving those from the other alone. This means that the previously in-phase even harmonics are now out of phase with each other, while the out-of-phase fundamental and odd harmonics are pulled into phase.

This circuit is known as the *push-pull* amplifier, because the phase relationships make it appear that half the circuit is "pushing" the signal while the other half is "pulling" it and vice versa. It finds side use as a straight-through amplifier as well as a tripler, and always cancels out even harmonics without effect upon the odd harmonics or fundamental.

When we're trying to reduce harmonic generation to a minimum, use of either a push-push doubler or a push-pull tripler will help; either eliminates at least half the harmonics which might otherwise be generated in a frequency-multiplier stage.

Now let's move on to the other half of the harmonic-control situation and see how to bottle up or contain harmonics which we cannot avoid generating.

The starting point in containment of harmonics is to make it as difficult as possible for them to follow the normal signal-energy route. This means using enough tuned circuits in the normal path, and having adequate selectivity in each, so that harmonics will find it rough going.

Antenna Choice

The proper antenna choice helps minimize harmonics. The popularity of *all-band* antennas helps us forget that any wire which works on 40 meters as well as it does on 80 cannot possibly help you reject the second harmonic of an 80-meter signal. Most *single-band* antennas, on the other hand, have such high Q that they refuse to radiate signals at even harmonics of their band. They may perform nicely at odd harmonics, but the antenna tuner can easily take care of that by preventing the third harmonic from getting to the feedline.

When all these points have been covered, the only ways left for harmonic energy to get from the transmitter to the outside are by direct radiation from circuit components, and by conduction through wires which are not carrying the output signal (such as power leads, audio cables, etc.).

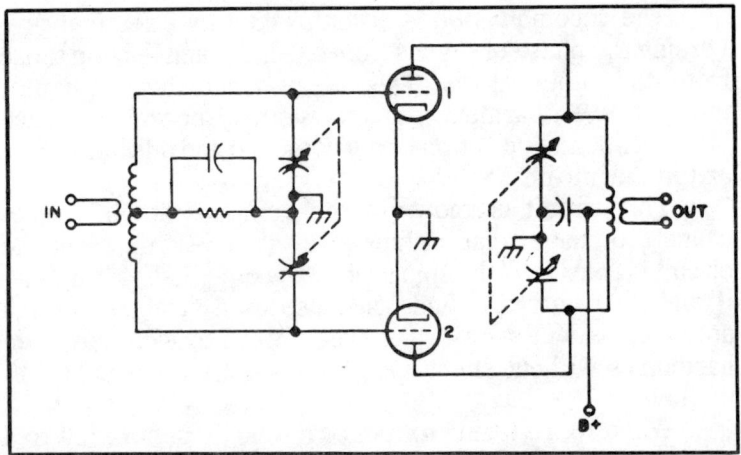

Fig. 3-8. Push-pull amplifier reverses the action of the push-push doubler to cancel out all even harmonics while amplifying the fundamental or odd harmonics. If the plate circuit is tuned to the harmonic, it's a frequency multiplier, while if plate circuit is tuned to the fundamental, it operates straight through.

Direct radiation can be controlled by completely shielding the transmitter. If adequately done, this will make it impossible for rf to escape except by conduction on wires passing through the shield.

If we then add filters to each wire, to make certain that no energy travels in or on it except that which is supposed to be there, we can be fairly sure all harmonics are under control.

The filters may range from simple bypass capacitors on power lines, to complex arrangements of rf chokes, audio inductors, and capacitors, for lines which carry ac. Bypass capacitors often suffice.

In fact, the normal legal requirements for control of harmonics are usually met without needing to shield and filter the equipment and its connections, but problems with television interference may make some shielding and filtering necessary.

We just mentioned filters—and that was getting ahead of ourselves by a fraction. In the next chapter, we'll go into filters and their action rather deeply, along with impedance matching and solid-state devices.

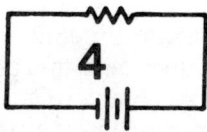

4
Matching, Filtering and Amplifying

In this chapter we're concentrating on three major subjects: impedance matching, how filters operate, and transistors. For starters, we'll attempt to discover why and how impedances are matched. This exploration may shock a few oldtimers as much as it shocked us during the research stages—because the "why" is, it turns out, rather different from the reasons most popularly believed.

Then we'll turn our attention to filters, and describe what they are, which should bring some order into the names applied to filters as well as uncovering their various uses (and many disguises). Then, we'll examine the means by which the more common circuits accomplish their purposes.

With matching and filters accounted for, we can take up transistors—but to provide some groundwork in this area, we'll first explore amplification and bring all kinds of amplifiers into our picture of circuit theory. With that established, we will be in much better position to understand how transistors amplify. This will take care of the other transistor details involved in the FCC questions as well.

IMPEDANCE MATCHING

In previous chapters, we have put together a number of different types of electrical circuits, building each circuit up from different kinds of components. In every circuit, we have had a "generator" or power source of some sort, together with the passive components which make up the rest of the circuit.

In examining these circuits, we have seen that the action of the entire circuit depends upon the relations between the various components which make up the circuit. If any single component is changed, the circuit's action changes. In other words, for any circuit to act as its designer expects it to, every part of the circuit must be what it was intended to be.

Now we can take any single component, or a group of several components which go together to make up a subcircuit, and draw an imaginary line separating that part of the complete circuit from the rest. Having done this, we can name the part of the complete circuit which contains the power source as the *source* and the other part of the circuit as the *load*. Doing this does not change the fact that it is one complete circuit, so that every part must still be what it was intended to be for the expected circuit actions to occur.

In particular, any variation of component values or types in the *load* will affect operation of the *source* just as it would when we looked at the whole thing as being a single circuit.

In practice, almost every electrical or electronic device is an *incomplete* circuit. Only the simple circuits which we study to learn how things work are complete and self-contained in themselves. A factory-built transmitter, for instance, does not come from the store complete with its own 117V ac generator, nor is the antenna a part of the rig as you purchase it. For the transmitter to operate, you must connect it to a power source, and you must provide an antenna as the rig's load.

The importance of impedance matching stems directly from the fact that all practical circuits are incomplete, and must be completed by the user in order to operate. To complete the circuit, the items connected by the user must *match* the equipment—and yet the equipment designer cannot possibly predict exactly how every possible user may connect his product.

The situation is saved by the fact that impedance can be considered to be a constant factor. That is, any circuit which exhibits a specific impedance characteristic can be substituted for any other circuit which has the same impedance characteristics. This is sometimes called the *black box* theory; the name comes from the fact that if you conceal a circuit inside a black box which cannot be opened, the only method anyone has of determining what is inside the box is to measure its performance at the terminals which connect it to the outside world—and all circuits which show the same characteristics in such measurements can be considered to be identical to each other no matter what is actually inside the box.

By specifying the required impedance levels for the power source and the load, then, the equipment designer can assure the user of his product that it will perform as intended.

In its most-used sense, the phrase *impedance matching* means simply providing the required impedance level, and says nothing about what this impedance level may be. In some cases this may mean providing a high impedance compared to the source, and in some cases it may mean a low impedance. The important thing is that the *required* impedance is provided, and it's up to the equipment designer to say what this required impedance may be.

However, impedance matching may also mean other things, and that is where the confusion comes in. One of the most respected ham radio reference books puts it—incorrectly—this way: "It is possible to show that any source of power will deliver its maximum possible output when the impedance of the load is equal to the internal impedance of the source. The impedance of the source is said to be *matched* under this condition."

The error is easy to see if we take the circuit of Fig. 4-1, in which the impedance of the source is 50+j50 ohms (inductive reactance as well as resistance). If we match this with a 50+j50 ohm load, the total impedance applied to the generator will be 100+j100 ohms. The 100 ohm resistance is accompanied by another 100 ohm of inductive reactance, and if the generator supplies 100V rms we will get a current through the resistance of only about 700 mA. This 700 mA through the 50 ohm resistance of the load will develop a potential of 35V and the power dissipated in the load resistance is 700 mA (0.7A) times 35V, or 24.5W.

If we change the load impedance to 50−j50 ohms, though, to make the reactance of the load equal in amount but

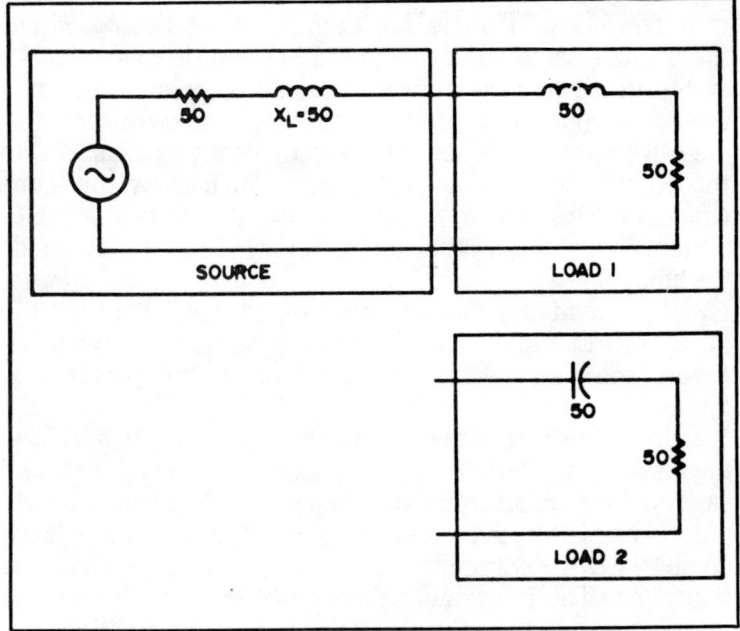

Fig. 4-1. Differences between *image impedance* matching and *maximum power transfer* matching are shown here and discussed in the text. Using load 1 follows the image-impedance rule, in which load's impedance is an exact duplicate of the source impedance. You can see that reactance is present, reducing power from the maximum possible. Use of load 2 follows the maximum power transfer matching rule in which resistances match but reactances are of opposite polarity. Reactance then cancels, and power is at maximum.

opposite in sign to that of the load, the reactances will cancel each other. The circuit is now series resonant. Net impedance seen by the generator is $100 \pm j0$ ohms, and 1.0A flows. This 1A develops 50V across the resistance in the load, or a 50W dissipation.

An important point to keep in mind during this discussion of matching is that we are limiting ourselves to making changes in the value of the load impedance only, and are leaving source impedance strictly alone. Only with this restriction do our rules and definitions apply.

For instance, let's take the circuit of Fig. 4-2, which consists of a battery, its internal resistance (drawn separately), and a load resistor. If we could vary *both* resistances at will, we would find that the maximum power in the load resistor would be achieved not when both were of equal value, but when the internal resistance was much smaller than the load resistor. With the internal resistance fixed, however, (say at 6 ohms) we find that the maximum power is developed with the same value in the load. With a 12V source and a 6 ohm load, we get 6W dissipated in the load. Reducing the load resistor to 5 ohms makes our current 12/11A and our dissipated power 720/121W ($I^2 R$), or 5.95W. Increasing load resistance to 7 ohms makes the current 12/13A, and the power becomes 1008/169 or about 5.98W.

Before confusion becomes complete, let's review the key points we've encountered. Impedance matching has at least three meanings. The most common is *providing some specified, required value of impedance as a device's load*. The second is identified as *maximum-power-transfer matching*, and means to make the load impedance equal in resistance to the source impedance, and equal-magnitude-but-opposite-sign in reactance. The third, called *image-impedance matching*, means making the load impedance the exact duplicate of the source impedance. Finally, in case both source and load impedance are free of reactance and involve only resistance, the maximum-power-transfer and the image-impedance matches turn out to be identical to each other.

Before we move on to see how such matching is performed, let's look at some examples to illustrate the various meanings.

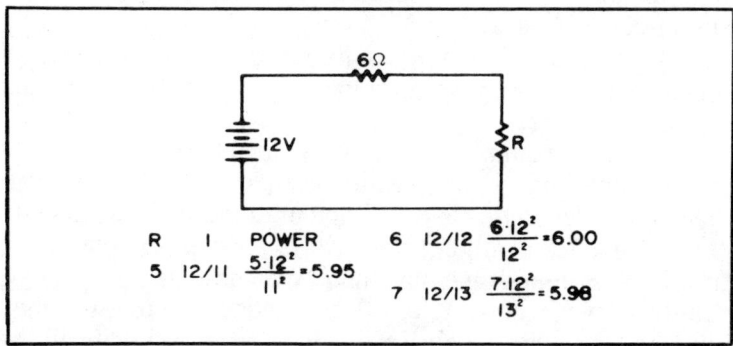

Fig. 4-2. When impedances are purely resistive, maximum power is always taken by the load which is neither higher nor lower in resistance than the power supply's own internal impedance. If the internal impedance is fixed at a 6 ohms, as shown here, then the load resistance of 5 ohms, draws more current, but the smaller resistance results in less power being taken. A load resistance of 7 ohms draws less current, and again power goes down. Only when source and load impedances are equal does power taken by the load reach the maximum 6W level.

One ready example not directly connected with ham radio is the matching between amplifier and loudspeaker in a top-quality hi-fi or stereo system. Most of the tube-type amplifiers are rated to operate into a 16-ohm load impedance, although they likely have output connectors marked as 8 ohms and 4 ohms. The speakers, similarly, are rated at either 4, 8, or 16 ohms. We simply connect the speaker to the amplifier terminals which are marked with the same impedance value.

In actual fact, the impedance rating is somewhat inaccurate since it includes neither reactance nor phase angle, which we saw a while back, is required for accurate specification of impedance. In addition, the true impedance of any speaker varies widely over its operating frequency range, and few if any come anywhere within 20% of their rated impedance at more than three or four spot frequencies.

These points make very little difference, because the actual source impedance presented by the amplifier is almost always far smaller than the rated impedance.

What does make all the difference is that the 4, 8, and 16-ohm ratings provide a means for specifying the "black-box" characteristics of both amplifier and speaker. Any amplifier with a 16-ohm output terminal can be expected to operate with any speaker rated at 16 ohms impedance, regardless of actual impedance characteristics of either amplifier or speaker.

This, then, is an example of the first meaning of impedance matching; we provide a specified, required load impedance value.

Another example is offered by the 117V power outlets in almost every home. While we usually talk about our electric power in terms of voltage or sometimes wattage (a 100W bulb, for example) this still is a close relative of the first kind of impedance matching. We know that any device rated to operate on 117V, 60Hz power will operate when plugged into a normal outlet; we don't have to specify all the electrical characteristics of the power company's generators and transformers on each electric appliance.

Maximum-power-transfer matching is the usual rule for connecting transmission lines and antennas to radio transmitters. The objective is to radiate as much of the transmitter's output as possible, which is just another way of saying maximum power. Antenna and feedline matching is often thought of as though it were done on an image-impedance basis—but when the normal practice of tuning the feedline to eliminate reactance is looked at in a slightly unusual light, it can be seen as a way to provide reactance of equal value but opposite sign and so to achieve max-power matching.

In fact, the whole art of matching transmission lines to antennas and transmitters *is* the art of maximum-power-transfer matching, and we will get into this type of impedance matching much more deeply when we get around to looking at antenna theory.

Transformer Matching

Now that we've defined impedance matching in its three variants, let's see how it is accomplished:

The most common method of matching impedances is by the use of a transformer. A transformer consists essentially of two inductors coupled to each other so closely that the magnetic field associated with one of them coincides with the magnetic field of the other. When this is the case, a change in current flow through the first inductor causes variations in the common magnetic field, and these variations in the magnetic field induce current flow in the second inductor. Energy is thus transferred or coupled from one circuit to another with no conductive connections; the common magnetic field does the "connecting" replacing the usual wires.

Because a change in the magnetic field is necessary for a transformer to operate, the device is effective only with ac. In a dc cirucit, a transformer produces one spike of energy in the output when power is applied and the initial current flow begins, and another spike of opposite polarity when power is removed and current flow ceases. In between, nothing happens.

Because the transformer does its job by means of the magnetic coupling between two inductors, it introduces a 90-degree phase shift just as would any other inductor. This is usually ignored.

The transformer's ability to serve as an impedance-matching device depends upon the relationships that exist between current intensity, magnetic field strength, and inductance. We've gone through most of the necessary background when we met inductance to begin with. All we need to point out now is that the strength of the magnetic field depends upon the number of turns in the inductor, and similarly the intensity of the induced current also depends upon the number of turns, but in opposite ways.

This means that if one of our inductors contained 10 turns it might develop a magnetic field strength of 10 units with one ampere of current. If the other inductor also contained 10 turns, the resulting magnetic field would induce an identical current intensity (assuming that the coupling is perfect and none of the magnetic energy goes to waste). However, if the second inductor contained 20 turns, the current intensity would be only half an ampere because each unit of magnetic field strength would have to release its energy to twice as much wire and could only give each turn half as much. And if the second inductor were only five turns, then each unit of magnetic field would affect half as many turns, and so would produce twice the current intensity. See Fig. 4-3.

Keeping in mind that we're talking about a "perfect" transformer which loses no energy, all of the energy put into the circuit must appear in the output. If the input energy comes out to be a 10W rate, with 1A flowing, it must be at 10V. We get that same 10W out; if current intensity is only 0.5A, the voltage must have been increased to 20V, and if the current intensity is raised to 2A the voltage must have been reduced to only 5V. Figure 4-3 summarizes these points.

By adjusting the ratio of the number of turns on the input inductor (usually called the primary winding) to those on the output inductor (secondary winding), any desired ratio of impedance transformation can be obtained. The rule is simple; the ratio of impedances is equal to the *square* of the number of turns in a transformer with tight coupling between winding. The square gets into the act because the ratio of voltages is the same as that of the number of turns, while the ratio of currents is the inverse of the number of turns (because current goes up as turns ratio goes down). Impedance ratio equals the voltage ratio divided by the current ratio, and dividing any number by its inverse is the same thing as squaring the number.

Most transformers you can buy are for use at audio or i-f frequencies, but the transformer principle is also used in rf circuits. Almost all transmitters use transformer coupling between stages, although not by that name.

Special amplifier circuits, known as *cathode followers* and *anode followers* when vacuum tubes are used and as *emitter followers* or *common-collector* circuits when transistors are employed, also have an ability to transform impedances, but they do so in a much different fashion. In general, transformer matching cannot always be replaced by special amplifiers, although in special cases they can.

Matching Pads

One method of impedance matching sometimes used in audio and communications work is by means of matching *pads*, which are circuits of series and parallel resistors so arranged that from one side of the circuit it presents one impedance level, and from the other side, another. The simplest matching pad is the resistive voltage divider, which is merely two resistors in series.

The advantage offered by the matching pad as compared to the transformer is that it is much less expensive—but this is offset in most cases by the fact that a matching pad must waste more than half the energy present in the circuit, and the greater the ratio of impedances to be matched the greater the loss in a matching pad. For an 8:1 impedance ratio, the loss is 35 dB—more than 99.9% of the energy lost! The transformer, while it costs more, rarely has more than 1 dB loss.

FILTERS

In any electrical circuit, it's sometimes necessary to select specific kinds of signals and separate them from other kinds also present in the circuit. In general, the devices or circuits used to do this job are known as filters. The name comes from the similarity to a physical filter such as the oil filter in an automobile, which separates grit from the oil.

Any device or circuit which separates signals into two or more groups can be rightfully called a filter. Since a filter is anything which does this job, we must have various categories into which to classify filters.

Active and Passive Filters

One major classification method is to divide filters into *active* and *passive* filters. Active filters are circuits which

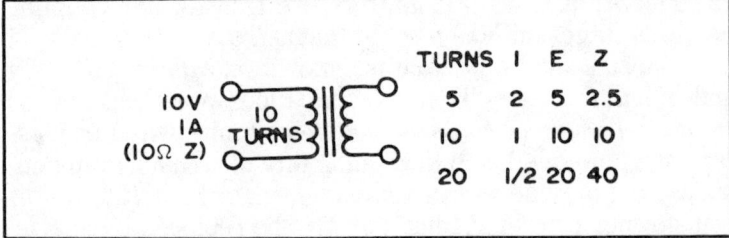

Fig. 4-3. Transformer operates by converting current changes in the primary winding into a changing magnetic field, and reconverting the magnetic field changes which result into new current changes in the secondary. Voltage, current, and impedance levels are all transformed from original values to new values which depend on the ratio of the number of turns in each winding. Chart shows results for three sizes of secondary winding, if the primary is fixed at 10 turns, and takes 10W (1A) from a 10V source for an impedance level of 10 ohms.

make use of amplifiers as a part of the filter itself. They are relatively recent developments and are seldom encountered in ham radio; about the only devices used by amateurs which may be termed active filters are the Q-multiplier and the select-o-ject circuit. For now, we won't pursue active filters further.

Passive filters are circuits which do not contain amplifiers within the filter circuit. Most filters are of the

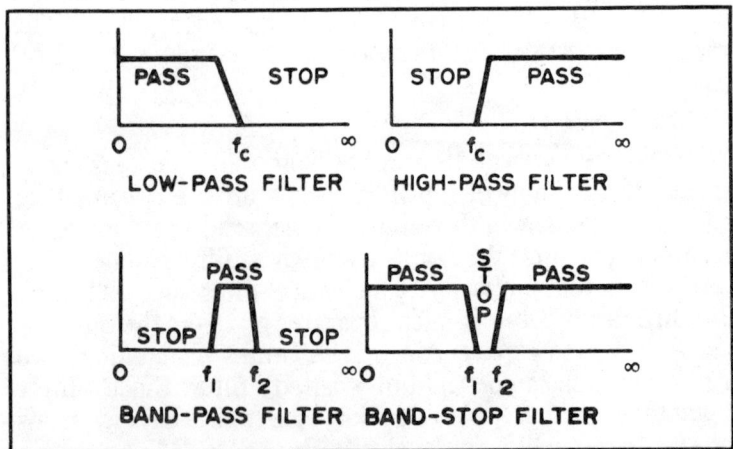

Fig. 4-4. Idealized frequency-response charts illustrate principal characteristics of low-pass, high-pass, band-pass, and band-stop filters. Each separates the applied input signals into groups which are either passed or stopped. Frequencies which mark the boundaries between these bands are known as cutoff frequencies.

passive sort, and throughout this chapter when we use the term filter we will be speaking of the passive type.

Filter Performance

Another way to classify filters is by the type of action they perform (Fig. 4-4). All filters separate signals into two or more groups, but they do so in different manners. This classification scheme puts any filter into one of four groups—low-pass, high-pass, band-pass, and band-stop.

The separation action of a filter can be described as letting one group of signals (passband) go through, and blocking all other signals (rejection band or stop band). The frequencies at which the passband and the rejection band meet are known as cutoff frequencies.

A low-pass filter passes all signals below its cutoff frequency and rejects those above; that is, its passband goes from 0 Hz up to cutoff, and its stop band goes from cutoff on to infinitely high frequency.

A high-pass filter rejects all signals below cutoff, and passes all those above. Its stop band corresponds to the low-pass unit's passband, and vice versa.

A band-pass filter has one passband and two stop bands. The first stop band runs from zero frequency up to the lower cutoff frequency, and the second stop band runs from the upper cutoff frequency to infinity. The passband is the region between lower and upper cutoff frequencies. A resonant circuit, by virtue of its selectivity, is a band-pass filter.

A band-stop filter is the opposite of a band-pass filter; it has two passbands and one stop band, and passes all signals except those between its lower and upper cutoff frequencies.

Filter Design

Performance is not the only classification criterion, however. Filters may also be grouped according to the techniques used in their design. This involves two major techniques, each of which has a number of subcategories.

The major design techniques are known as *image impedance design* and *modern network theory* methods. Most filters in the past 50 years have been designed using image impedance methods, although this technique requires the use of physically impossible components and so cannot produce accurate results. Within the last decade or so, the modern network theory method has been brought to prominence to provide more accurate filter action.

Filters designed by image impedance techniques fall into two subcategories, known as *constant-k* and *m-derived*. Both these labels are just names, and bear no relation to the performance characteristics of the filter. The k and m refer to mathematical variables used in the design equations.

Filters designed by modern network theory methods bear names associated with the mathematicians who studied the various types of equations involved in the theory. The most widely known of these filter types is the *Tchebychev*.

To complicate the question a little more, filters always include both series and parallel components, at least one of which (and often both) is reactive. This means that the most primitive building block possible in a filter is the combination of one series element and one parallel element. Such a circuit (Fig. 4-5) is known as a half-section.

Half-sections can be connected together so that the two series elements meet, or so that the parallel elements meet.

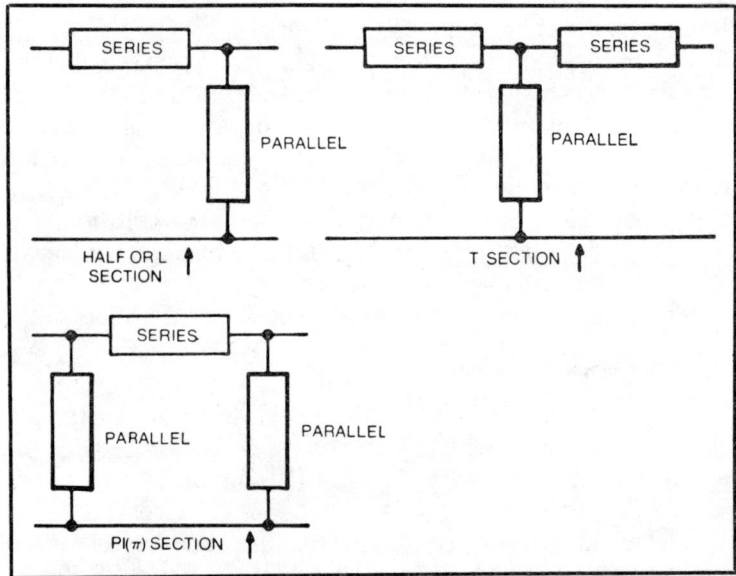

Fig. 4-5. Different arrangements of components in filter circuits bear names suggested by resemblances to various letters of the alphabet. Pi-section resembles no letter in our own alphabet, but is a dead ringer for π of the Greek alphabet. Elements represented by boxes here may be resistors, capacitors, inductors, or combinations of any or all of these.

In either case, the two elements which meet can be combined into a single element; the resulting three-element circuit is known as a section. If it has two series elements but only one in parallel, it's called a T-section (from its resemblance to the letter). If there's only one series element and two in parallel, it's a pi-section, from the resemblance to the Greek letter π. And if we don't combine at all but use only a half-section, it's sometimes called and L-section.

That's not all the possibilities, but we have enough on hand for the moment.

The functional description—high-pass, low-pass, band-pass—is most generally used, unless it's necessary to be more explicit about the particular filter circuit involved. Even then, the functional description is usually retained, and the added classification simply tacked onto it, to give us a "low-pass constant-k filter using π-sections" or a "high-pass m-derived filter using T-sections."

Operation

We have just seen that filters act to separate signals into groups by permitting signals in the passband to go through, and rejecting signals in the stop band(s). How is this accomplished?

The clue is hidden in our statement that every half-section of any filter contains at least one and often two reactances. A reactance is frequency-sensitive. That is, its characteristics depend upon signal frequency. By combining reactances of proper types and values, we can get almost any kind of filtering action we may desire.

To show how this works, let's begin with the simplest possible filter arrangement (Fig. 4-6) containing only a single half-section and a single-reactance. We already looked at this circuit, but not in this context. It contains a resistor and a capacitor, with input applied across both and output taken across only one.

This circuit can act as either a low-pass filter or as a high-pass unit, depending upon which element the output signal is taken across. To see how this works, let's assume that the resistor's value is 1k, and that the capacitor's reactance is 1 k at a frequency of 1kHz.

For an input signal of 10V at 1 kHz, then, since resistance and reactance are equal at this frequency, we will

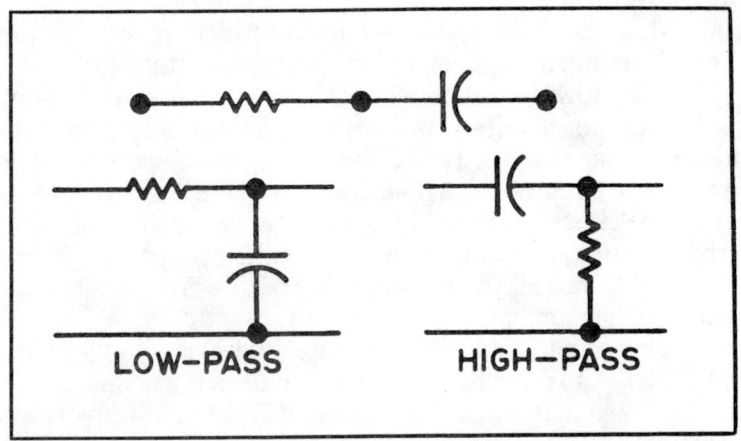

Fig. 4-6. Simplest filter circuit consists of one resistor and one capacitor in series, as shown at the top. This forms on RC L-section, which may be used for either low-pass or high-pass purposes as shown in the two lower schematics. Method by which the filter operates is explained in the text; all filters work on this basic principle, but more complicated circuits modify it in various ways because they contain not just one but several reactances, each of which behaves differently.

get the same voltage across either the resistance or the reactance. Because of phase shifting, we will actually find 7V across either; we went into the reasons for this before. With 10V input and 7V output, we have 3 dB loss.

If signal frequency goes up to 2 kHz, the resistance will not change but the reactance will decrease to 500 ohms. We now have less voltage developed across the capacitor than across the resistor. As signal frequency is increased still more, reactance keeps going down; when we reach a signal frequency of 10 kHz, the reactance is only 100 ohms, and at 100 kHz the reactance is down to 100 ohms. Resistance remains unchanged, so that the voltage division between resistor and capacitor reduces capacitor voltage and leaves most of the voltage across the resistor.

If our output voltage is taken across the capacitor, we have a low-pass filter. As signal frequency goes up, output votlage goes down. If, on the other hand, we take the output from the resistor, our filter is a high-pass unit, as output voltage increases with frequency.

At dc or zero frequency, all the input voltage appears across the capacitor. At the frequency at which resistance and reactance are equal, 70% of the voltage (half the power) appears across the capacitor. At infinite frequency (a purely theoretical idea, granted) no voltage at all appears across the capacitor. In between, whenever signal frequency is doubled, the output voltage drops to half what it was before, because the reactance drops to half its previous value.

This somewhat idealized description leads to the idea of a 6 dB per octave cutoff rate. Cutting voltage in half is introducing a 6 dB loss, and doubling the frequency is equal to going up an octave in the musical scale. The theoretical limit of the effect of a single reactance is 6 dB per octave.

An actual RC filter like the one we've been examining does not reach this rate in practice. Its cutoff frequency is defined as being the frequency at which X and R are equal (1000 Hz in our example) and in that region the slope is only 3 dB per octave.

Improving Filter Performance

We can improve its performance in several ways. One is to cascade additional half-sections. This modifies its action in many ways, since the first reactance now has another half-section in parallel with it and no longer behaves as a pure reactance. Another (Fig. 4-7) is to replace the resistor with an inductor, thus putting two reactances in rather than one.

When we do either of these things, we immediately find ourselves involved in the intricacies of filter design. With both inductive and capacitive reactance in the circuit, we have the certainty of resonance at *some* frequency, and this may or may not be good for our purposes. This is not the place to delve deeply into filter design theory; all we can do is skim the surface enough so that we are able to identify the various parts of any kind of filter.

The constant-k filter design procedure amounts to a method for replacing the resistor with an inductor, together with adding additional half-sections, to produce a filter with the desired cutoff frequency and cutoff rate. The resulting filter's input and output impedances will vary in a definite manner which the designer cannot control, so that this type of filter can be used only when its impedance characteristics are satisfactory for the purpose. A typical power-supply filter amounts to a constant-k low-pass filter; the *choke-input* filter uses a T-section followed by an L-section (with two of the three inductors combined into one so that it appears to be

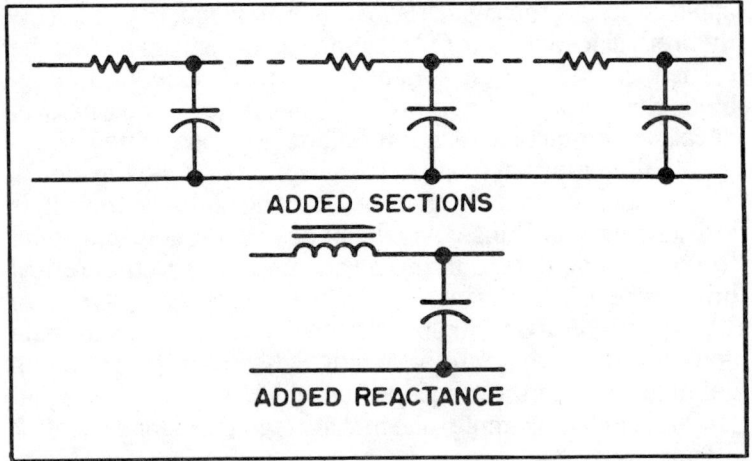

Fig. 4-7. When performance of the simple filter of Fig. 4-6. is not adequate, either or both of two routes may be taken to improve it. Sections may be added (left) or additional reactances may be included (right). Usual filters encountered in practice do both; the design technique is rather complicated but results justify it.

two L-sections), and the *capacitor-input* filter is a π-section standing alone.

The m-derived filter design procedure offers the designer another level of control. He can achieve the cutoff frequency and rate desired, just as in the constant-k procedure, and in addition he has the option of controlling either the impedances or the attenuation characteristics (but not both). To do this, a part of the circuit resulting from the constant-k design is converted into a resonant circuit. The *ratio* between *cutoff frequency* and the *resonant frequency* of this converted portion yields the variable named m, which gives the procedure its label.

The distinction between a constant-k filter and an m-derived filter (Fig. 4-8) is simply the presence of a resonant circuit where one would expect to find only a single reactance. If such a circuit is present, the filter is m-derived.

Filters designed according to modern network theory techniques may look the same as either type of image-impedance circuit. In these designs, the engineer trades off the ability to control certain characteristics in order to gain control over others. The most popular Tchebychev designs, for example, permit some degree of ripple (uneven response) in the passband to achieve a cutoff rate greater than the theoretical 6 dB/octave/reactance limit for image-impedance designs.

These trades are made by designing each separate section of the filter for a different cutoff frequency; the complex math involved in the design procedure specifies the particular cutoff frequencies to be used.

Balanced vs Unbalanced Operation

Regardless of the design procedure, a filter may be built to operate in either a balanced or an unbalanced condition.

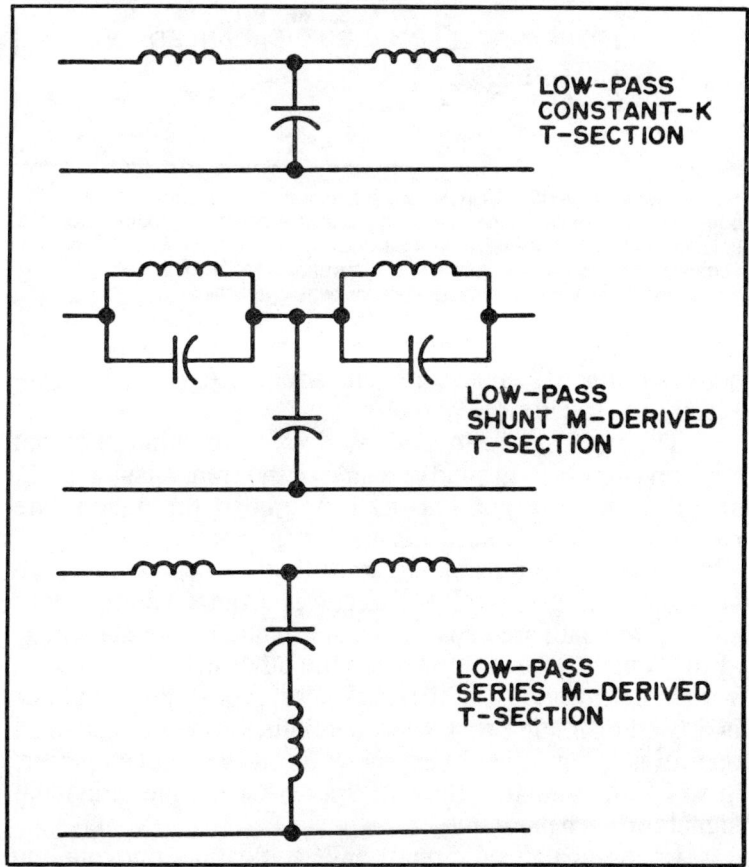

Fig. 4-8. Differences between *constant-k* and *m-derived* filters are apparent when the schematic diagrams are examined. Any m-derived filter contains resonant circuits replacing simple reactances. Series and shunt portions refer to the type of resonant circuit, either series-resonant or parallel (shunt).

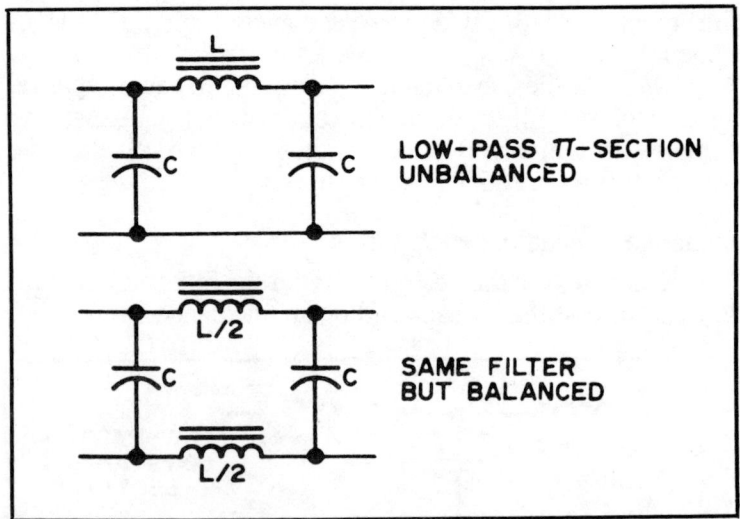

Fig. 4-9. Most conventional circuits are unbalanced in nature, as at the top, with both input and output circuits sharing a single common ground line. If a balanced circuit (bottom) is in use, input and output circuits have no terminals in common. Any design may be converted from unbalanced to balanced operation by splitting the series elements between the two signal lines.

These terms are not limited to filters—they apply too all signals transmission circuits, and affect filters only because filters are used in signal circuits.

The most conventional practice for any kind of signal transmission is to ground one side of the transmission path, and then work a hot line against ground throughout the resulting system. This is unbalanced operation.

It's equally feasible to keep both sides of the path separate from ground. If the circuit is arranged so that both sides of the path are separate from ground, but so that one is always above ground level when the other is below and vice versa, the circuit is said to be *balanced*. A push-pull amplifier is a good example of a balanced circuit. So is the industrial operational amplifier circuit known as a *differential* amplifier, in which the signal is the *difference* between the individual signals on each input line.

In an unbalanced circuit, any accidental stray coupling between circuits cannot be readily corrected since there is no way to distinguish between desired and undesired signals. Most of the undesired signals are inherently unbalanced in nature.

The balanced circuit permits such a distinction, because it is possible to reject any unbalanced signals appearing in a balanced path. Commercial telephone circuits are all balanced for precisely this reason; it permits the reduction of crosstalk to acceptable levels.

A filter for use in an unbalanced line must be itself unbalanced. The sample circuits we've been examining are all unbalanced; the series element appears in only one side of the filter, and the input and output terminals have a common or ground reference.

To convert such a circuit to a balanced filter for use in a balanced line, the series elements are divided between the two signal paths (Fig. 4-9). Then half of each series element appears in each side of the line. The total series impedance is unaffected by this change, but there is no longer a common terminal between input and output.

AMPLIFICATION

In all the circuits we've looked at so far, we have had only one generator or power source and all the rest of the circuit has been composed of passive components that dissipate energy or convert it from one form to another, but which have not added any additional energy.

In practice, radio deals with extremely small amounts of energy. It's necessary to boost the energy levels while retaining the major characteristics of the original energy. Any device which does this job is known as an amplifier.

An amplifier, then, is a device for adding energy while controlling its characteristics—and amplification is the process of doing this.

One way of adding energy is to introduce a generator into the circuit. Another way is to add a battery. Unfortunately, while the energy from a battery is plentiful enough, it's difficult to control.

However, we have been making much use of the voltage-division capability of series circuits composed of resistors. If we have two resistors in series with our battery, one of them fixed in value and the other variable, we can control the added energy by varying the resistance of the variable resistor.

We now have the basics of a rudimentary amplifier—a source of energy to be added, and a means for controlling the

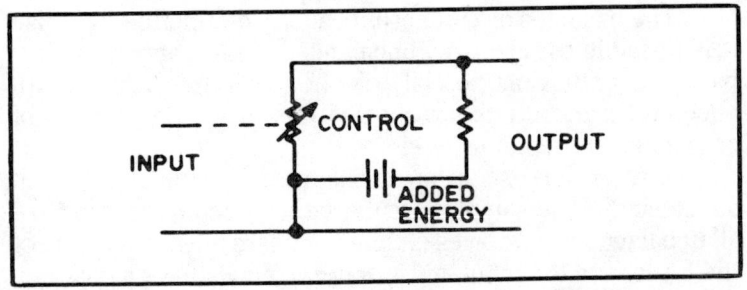

Fig. 4-10. Basic amplifier circuit consists of a source of added energy and variable resistances controlled by an input signal which together permit the input signal to control the amount of energy added. Schematic diagram shows the principle involved, but no real amplifier looks much like this. Control function is usually provided by either a transistor or a vacuum tube, which is acting as an electronically variable resistance.

energy. It may come as no surprise to find that almost all practical amplifiers boil down to this simple basis—a battery and a variable resistor. (See Fig. 4-10).

The variable resistor, though, seldom looks much like what you would expect. Conventional variable resistors achieve their variations by mechanical means, and cannot exert control rapidly enough to follow all the gyrations of an ac signal. The variable resistors we find in practical amplifiers to control the added energy achieve their variations by *electronic* means rather than mechanical. We call them vacuum tubes and transistors.

We'll examine the actual methods by which vacuum tubes and transistors achieve variable resistance later; each does it differently. The important point at this stage is that so far as the electrical circuit itself is concerned, they do act as variable resistances, whose resistance depends in some way upon the input signal. The British term for the vacuum tube—valve—is highly descriptive of the function. A small input signal controls the resistance which the tube offers in the energy-adding circuit, and thus produces an output signal which matches the input in every desired way.

By proper circuit design, we can make the output signal's characteristics match all of those of the input signal, or we can select only a few features of the input signal and eliminate the rest from the output. The ability to do this makes it possible for us to modify signals in just about any way we like. For instance, we can preserve the frequency characteristic while eliminating any changes in strength, and this is what makes FM so much more free of atmospheric noise than AM. We can preserve the variations in strength while changing the frequency, and this makes today's sensitive receivers possible by permitting the *superhet* receiver design. Or we can preserve all the original characteristics, boosting only the strength, and we have full fidelity sound systems. All of this depends upon the control inherent in the amplification process. More about how transistors and tubes operate in the next chapter.

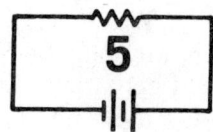

Tubes and Transistors

For nearly 50 years, the vacuum tube reigned supreme and virtually unchallenged as *the* amplifier for electronics. From the passing of the spark gap and decoherer in the years following the first world war until the advent of the transistor in 1948, the vacuum tube had no serious competition.

Both because the vacuum tube is still in wide use, and because many other areas of electronics and radio in particular are built upon a foundation derived from vacuum-tube experience, the General class amateur examination contains a number of questions dedicated to determining the applicant's knowledge of vacuum tubes and their operation.

Our first question, then, is how does a vacuum tube work? When that is out of the way, we can move on to learn the factors which limit a vacuum tube's usefulness—diodes, triodes, tetrodes, pentodes, hexodes, heptodes, and finally find out how are vacuum tubes used. By the time we get even a brief view of the realms opened up by our final question, we should have adequate knowledge of tubes and their operation to handle any questions such as those on the study list.

VACUUM TUBE OPERATION

Before we can determine how a vacuum tube works, we must first determine just what a vacuum tube amounts to. Even though we know that by "vacuum tube" we mean only those gadgets used in radio transmitters and receivers, that still leaves an almost unbelievable amount of territory—multifunction tubes, kylstrons, magnetrons . . . the list doesn't go on forever, but it might as well.

All of those special types of tubes share the fact that they consist of electrodes sealed into a tube full of vacuum. Unfortunately, as we have just illustrated, that's not precise enough to sort out the kind of "vacuum tubes" the exam is concerned with. Rather than try to define all the differences, let's start with the one common element and see what develops.

We already know that an electrical circuit, to be a circuit, must be complete. That is, it must have both an input and an output. Even the simplest vacuum tube, then must have at least two different electrodes sealed into its vacuum, one for input and one for output.

The first recorded vacuum tube was just about that simple. It was built by Thomas A. Edison in 1883, and its operation as a vacuum tube was purely accidental. Edison was trying to improve his newly invented incandescent lamp, and one of his experiments was to put a metal plate into the bulb near the filament (Fig. 5-1). He discovered that when the filament lit, a small electric current flowed between plate and filament. The discovery was duly noted, published, and became known to the world as the *Edison effect*. Since it did nothing either way for the operation of the light bulb, Edison apparently ignored it and went on to other inventions—thereby missing the chance to become known as the inventor of electronics atop all his other laurels!

Serious scientists did, however, sit up and take notice of the Edison effect. As a direct result of it, Sir J. J. Thomson spent several years in study and experiments, and in 1900 or so announced to the world his *electron theory*. That's the familiar structure of electrons, protons, and neutrons which we accept now as the *true* picture of the way things are made, and the reasons for electric current.

The electron theory explained the current of the Edison effect as an *evaporation* of electrons from the surface of the filament, caused by the heat energy present in the filament. As these electrons boiled off, some struck the metal plate—and any time we have electrons in motion, we have to have a current. That's the current Edison observed.

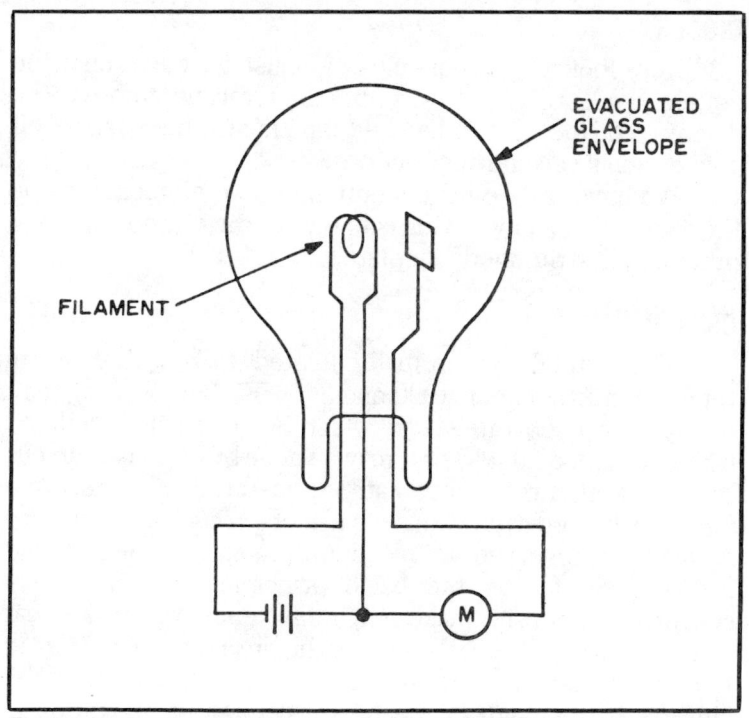

Fig. 5-1. Vaccum tube owes its existence to the Edison effect, discovered by the light bulb's inventor. Circuit which revealed the Edison effect is shown here. Metal plate was sealed into the bulb, near the filament but electrically insulated from it. When the filament lit, current flowed between plate and filament. More current flowed when the plate was connected to the positive pole of the filament battery, than when connected to negative. Edison didn't know why this happened; others followed it up.

Edison also noticed that when he connected the metal plate to the positive side of a battery, the current flow went up, and if the connection was to the negative side, the current went down. This showed that the moving particles had to be negative in polarity, since they were attracted to a positive pole and repelled by a negative one.

A few years later, in 1905, an Englishman named Fleming obtained a patent on a device making use of the Edison effect, which he called a *valve*. It was intended to detect radio signals by making use of the one-way property inherent in the current flow from a hot filament to a metal plate (anode). It worked in just the same way as today's diode detector circuits, and it established the vacuum tube as a part of the growing art of radio communciations.

DIODES

Let's look a little more closely at just what goes on in the simple diode, since it was the first vacuum tube to be discovered, the first to be used, and is still the basis of all conventional vacuum-tube action.

We have, as we have mentioned, two elements inside the vacuum for a diode. One is known as the cathode, and the other is called the anode, or plate.

Cathode

The cathode is normally heated to a rather warm temperature (between 1000 and 3000°F). The heating may be done by a separate *heater* which is electrically insulated from the cathode itself, or it may be done by forcing a current through a high-resistance cathode material. Normally, an *indirectly* heated cathode makes use of a *heater*, and the word *filament* is reserved to mean a combination heater and cathode. So far as the basic principle of operation is concerned it doesn't make much difference where the heat comes from so long as it's present, but in practice if ac is used to provide heating power it's better to keep the ac out of the signal circuits, and so the heater-cathode combination is most frequently encountered in ordinary tubes.

At the cathode temperature, some of the electrons of the cathode material literally boil off into the empty space surrounding the cathode. This forms a cloud of electrons known as the *space charge* around the cathode itself.

Plate or Anode

The plate or anode is separated from the cathode by distance, and it's far enough away that the space charge never quite reaches it.

Under these conditions, with no voltage applied between plate and cathode, the space charge is self-limiting. It always contains all the electrons emitted (boiled off) from the cathode, and so always has a negative charge which just balances the positive charge produced on the cathode by loss of the electrons.

If the temperature of the cathode is raised so that more electrons are emitted, the space charge will increase, but the current flow to the plate will remain essentially zero.

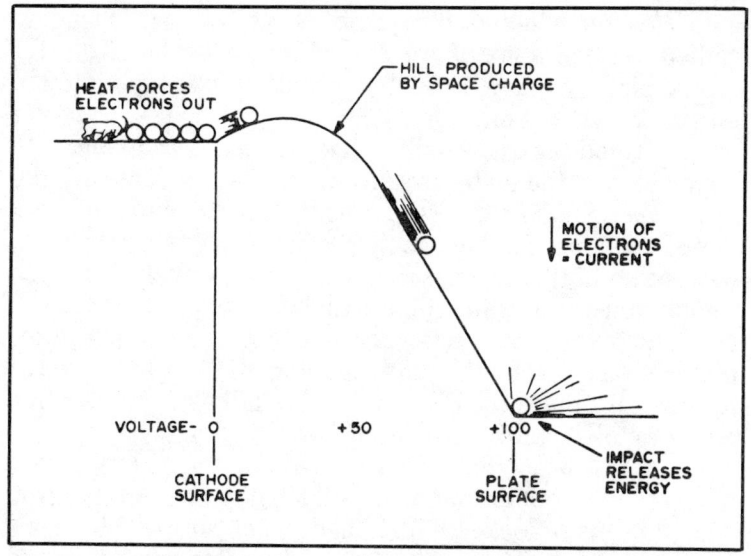

Fig. 5-2. Action of electrons inside a simple diode tube is shown here. Difference in voltage between cathode and plate surfaces forms a *voltage gradient* shown here as a downhill slope. Space charge puts a hump in this hill, near the cathode. Angry ram represents action of heat, boiling electrons off the cathode surface. Those electrons with enough energy to make it over the hump fall down the hill, and hit the plate with a crash, releasing more heat.

Similarly, if the cathode temperature is lowered, the plate current will remain unchanged at zero.

If we return the cathode to normal temperature and put a small positive voltage on the plate, the picture changes. The positive voltage on the plate attracts some of the electrons from the space charge, and a current flows between plate and cathode.

Not all the emitted electrons reach the plate, though. When we put voltage on the plate, this created a *voltage gradient* from cathode to plate something like a hill in the path of the emitted electrons (Fig. 5-2), and only those electrons which left the cathode with enough energy to "make it up the hill" go to the plate. Some are repelled by others in the space charge, and a few even return to the cathode as a result.

Making the cathode hotter doesn't change things much—but it we cool it down somewhat, the number of electrons in the space charge is reduced and we find that the "hill" is not so steep. As the temperature cools, we find one point at which all the emitted electrons go over to the plate.

We can achieve the same effect without changing cathode temperature if we simply increase plate voltage. When plate voltage is high enough, it will pull all the electrons over the hill.

In a diode, we have only these two factors available with which to vary the plate current. If either stays constant, the other exercises control (so far as it is able). With constant plate voltage, cathode temperature will determine the current up until the increased space charge with increased temperature makes the hill too high for the plate voltage to pull electrons over. With constant cathode temperature, plate voltage will be the ruling factor until it is great enough to pull all the electrons over the hill, at which time temperature again takes over.

This "hill" effect determines the maximum power dissipation rating for the tube, in a slightly indirect manner. The "hill" is *between* cathode and plate, so that electrons leaving the cathode must "climb up" the hill to get past it, and those which reach the plate are "falling down" the other side. (The analogy to a hill is not quite real, but if the voltages are plotted on a graph as in Fig. 5-2, it's close enough for all practical purposes.)

The more electrons that hit the plate, or the harder they fall, the more shaken will be the molecules of the plate material. This shaking up of molecules is what we generally call heat, and a large amount of it goes on in a typical vacuum tube. Plates often run at a dull red glow, and some tubes are designed to operate with their plates white hot—with the heat *all* coming from the impact of the electrons falling down the "voltage hill" between cathode and plate.

Any material eventually gets hot enough to melt, and the plates of vacuum tubes are no exceptions to this rule. For this reason, any tube is rated for a *maximum plate dissipation* which is the power in watts the plate can safely convert into heat. Power is the rating factor because it includes current (the number of electrons hitting the plate)—and voltage (the hardness with which they fall). It's also easy to calculate and to measure, in operation.

While we've explained, "power dissipation" ratings in terms of the simple diode, it's the same situation for any kind of tube. Not only the plate is involved, either. Any electrode which is more positive than the cathode (on the "downhill" side of a voltage hump) must dissipate the impact energy of the electrons that get to it, and such electrodes are individually rated for maximum power dissipation.

The diode's main usefulness in radio today is as a polarity-sensitive switch (Fig. 5-3). If the plate is positive to the cathode, it conducts a current, while if the plate is negative, the electrons are repelled and current flow stops. This makes it useful as a power rectifier to change ac to pulsating dc, and also as a nonlinear device for modulation, mixing, and detection of signals (which we'll explore later).

TRIODES

The first great advance over the diode came about when Lee DeForest surrounded the cathode with a coil of wire

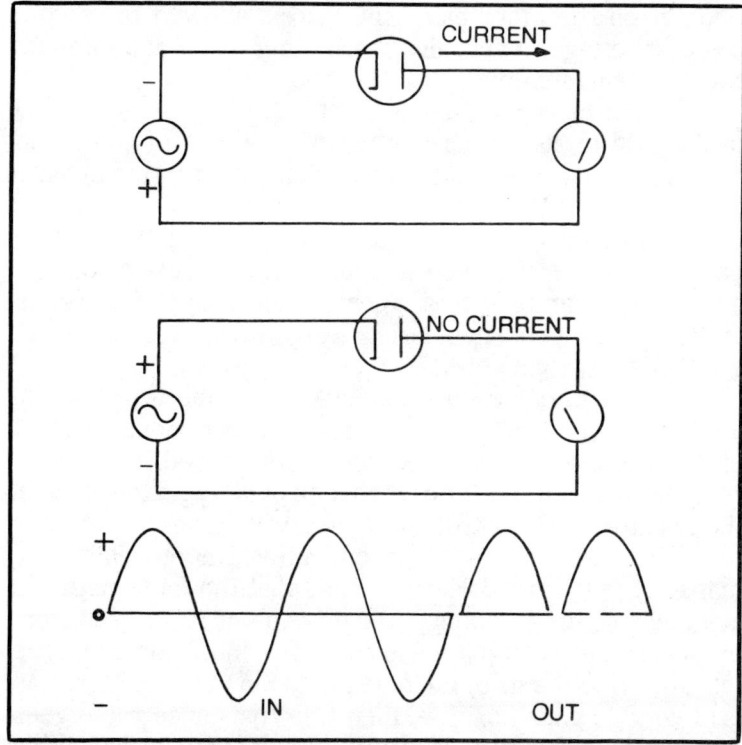

Fig. 5-3. Primary use of the diode in electronics today is as a switch. When the plate is positive current can flow. When the plate is negative (center) no current flows. This action can be used to change an ac input into pulsating dc output (bottom), or to detect the envelope of an amplitude-modulated radio signal, as well as for many more exotic uses not strictly a part of radio.

which he called the *grid* (Fig. 5-4). By applying a negative voltage between the grid and the cathode, it became possible to add a new dimension of control.

For proper operation of a triode, as the three-element tube is known, cathode temperature and plate voltage are kept in the range where plate voltage controls cathode current. The plate is kept positive to permit large currents, but the grid is kept negative so that it repels electrons from its neighborhood and so holds down the plate current.

The effective *size* of the grid, as it shows up in its interference with plate current, depends upon the voltage applied to it. The more negative the voltage, the more effectively the grid blocks current flow. It's possible with most tubes to bring plate current down to zero by putting an extremely negative voltage (known as *cutoff voltage*) on the grid. If, on the other hand, the grid is allowed to reach a positive voltage, it acts just like another plate, and loses its control of current flow.

For normal amplifier operation, the *bias* voltage applied to the grid is somewhere between the limits of cutoff and positive voltage, and is chosen so that the variation in plate current which results when small changes occur in the grid voltage is a reasonably true replica of the variations in the grid voltage. Picking the right combination of plate voltage, plate current, and grid bias to achieve this happy effect is one of the fine points of engineering design work which we won't go into deeply right here.

For our purposes, we can think of the tube as something a little different. As we mentioned in the previous chapter, either a tube or a transistor can be considered to be just a resistor in the circuit from cathode to plate, with the value of the resistor being controlled by the grid voltage.

This is a somewhat unconventional approach to the matter, but it works! Keep in mind that the more negative the grid-to-cathode voltage, the more ohms in the "resistor" presented between plate and cathode connections, and the less negative the grid, the fewer ohms in the "resistor." As grid voltage goes positive, then, the resistance value goes down and more current flows. If the tube is in series with a fixed resistor, this will reduce the voltage at the plate, and we have a change in plate voltage which is brought about by the change in grid voltage but which is 180 degrees out of phase.

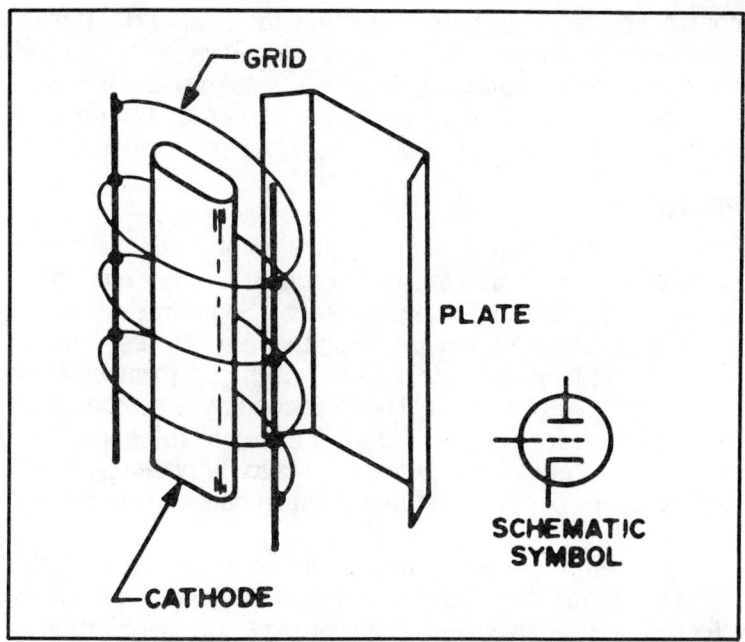

Fig. 5-4. Introduction of a grid between cathode and plate made it possible to control cathode current by means of a separate isolated electrode, and gave birth to electronic amplification as we know it today. Appearance of a simple triode is shown at left; schematic symbol is at right.

When used in this manner as an amplifier, the triode vacuum tube operates almost exactly like the transistor except that its resistance is controlled by the *voltage* on its grid rather than by the *current* injected into the base, and it uses somewhat higher voltages than do most transistors.

However, both the triode and the transistor share a common problem. The grid and the plate are both conductors, and they are separated by an insulating medium. This makes them form a capacitor—which permits some of the plate voltage to feed back to the grid. This plate-to-grid capacitance makes the triode tricky to handle as a radio-frequency amplifier (and the transistor too, for that matter, but we're talking about tubes at the moment).

TETRODE

To solve the problem, another grid structure was introduced between the original grid (now called the *control* grid) and the plate. The new grid, being added to screen the control grid from the plate, was called the *screen* grid. It was, like the plate, connected to a positive voltage so that it had little or no effect upon the electron stream passing through, but was bypassed through a capacitor to ground to prevent any signal from coupling back through grid-plate capacitance.

PENTODE

The screen grid worked nicely, and the resulting structure is known as a *tetrode* because it has four elements. However, it introduced a new problem. Sometimes the plate voltage is driven to a value *lower* than the screen voltage. Electrons still hit the plate hard—hard enough to knock a few *secondary* electrons free. These secondary electrons then went to the screen rather than returning to the plate. The result was an effective *negative* registance between plate and screen under certain conditions, which could cause oscillation.

The direct cause of this effect was the fact that the screen was the most positive thing around when secondary electrons were released. A direct cure was installation of still another grid, the third, between screen and plate. This *suppressor* grid is connected to the cathode, so that it will be more negative than either screen or plate. When secondary electrons leave the plate now, the negative charge on the suppressor grid drives them right back where they came from.

With five elements—a cathode, three grids, and a plate—the resulting tube is known as a pentode. Most present-day rf amplifier tubes are pentodes. Some power tubes are *beam power* tubes; they make use of special beam-forming plates attached to the cathode. These plates focus the electron beam in such a way as to create a *virtual suppressor* by space-charge effect between screen and plate.

Each of the additional electrodes introduces a small amount of noise into the signal. Normally this is of no consequence, but in critical application such as the first stage of a VHF receiver, triodes are still preferred despite their problems simply because of their low noise.

The many other kinds of tubes you may meet are, for the most part, combinations of the kinds we've examined here. They all start out with the elements of a diode, and all conventional tubes then add grids as necessary to do their job. Often two or more separate tubes are combined in the same envelope for convenience (the 6U8A, Fig. 5-5, is an example of a triode and a pentode sharing the same chunk of glass), but the essential structure remains unchanged.

FACTORS LIMITING A VACUUM TUBE'S USEFULNESS

The vacuum tube is a most useful gadget, but each individual one is limited in its use by a number of factors. One of these, power dissipation, we've already met. Some of the others include such things as the circuit efficiency, operating frequency limits, and power requirements.

The major limits on any specific tube are those set by its power requirements, power dissipation, and operating frequency limits. The power requirements include both the power necessary to heat the cathode and that required to operate the tube's plate circuit (and screen, if any). Power dissipation involves not only the ratings, but the method by which the tube is cooled. Frequency limits are usually one-sided. Almost all tubes will operate at frequencies down to and including zero (or dc), but every tube has an upper frequency limit beyond which it will no longer act as a tube should. Let's examine these in reverse order, looking at the frequency limits first.

Frequency Limits

The absolute frequency limit in any tube is set by the physical distance between its electrodes. The electrons in the tube do *not* travel at the speed of light; they take a

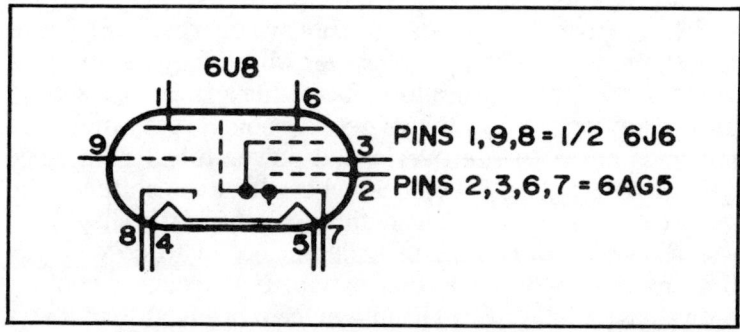

Fig. 5-5. Combination tube such as type 6U8 is actually two different tubes enclosed in same glass envelope and sharing the same heater for convenience. Triode half of the 6U8 is the same as half a 6J6, while the pentode portion is the same as a 6AG5. Portions can be used independently.

definite amount of time to make the trip from cathode to plate. If the signal frequency is so high that the distance from cathode to plate is an appreciable part of a wavelength, then everything goes sour. For instance, a positive-going signal at the grid should increase plate current—but if it's a half-wave from grid to plate, the plate current will be drecreasing at that instant rather than increasing, and cannot increase until a half-cycle later when the grid signal is going negative.

Almost no tubes actually make it up to this limit, though. Other factors impose even lower frequency limits on them. One is the combined effect of the cathode structure's own inductance and the capacitance from cathode to each other element in the tube. Taken all together, these effects add up to a low-pass filter circuit which prevents any signals above the filter's cutoff frequency from flowing through the cathode—and so makes the tube unusable at these higher frequencies.

Tube designers combat these limits by making tubes intended for VHF operation physically small, and providing multiple connections for all elements. An extreme example is the family of tiny *planar* triode tubes made for space use, which have no connector pins as such. Instead, each element is brought through the envelope as a ring (Fig. 5-6). Distance between elements is reduced to thousandths of an inch. The tubes operate far into the VHF region—but they still do have frequency limitations.

Power Requirements

Power dissipation comes about because the electrons which compose the cathode current have energy, and release it in the form of heat when they get where they're going, as we already saw. The limiting factor here is almost always purely physical; when things get just so hot, they melt. It's not uncommon to see overloaded power tubes with large dents in their glass envelopes, where heat has softened the glass to such an extent that anything can change its shape!

The type of cooling provided has a lot to do with the limiting effects of power dissipation. If a stream of cool air blows past a tube, it can handle much more heat than if it is sealed into an airtight box of polished aluminum. Very-high-power tubes are often liquid cooled, with water actually flowing through the interior of the plate structure just like an

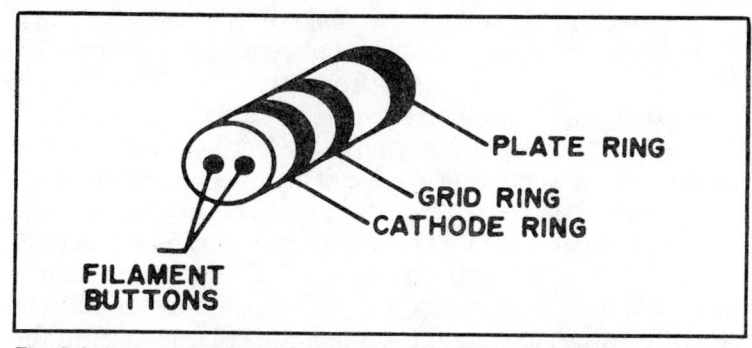

Fig. 5-6. In the quest for UHF operation, tube designers have come up with strange designs. One is this *planar* triode which is made of ceramic, with electrode connections coming out as rings around the tube body to eliminate lead inductance as much as possible. Tube operates well into the UHF region, but still has an upper frequency limit.

automobile engine's cooling jacket. Published ratings usually assume good ventilation is provided.

Power requirements limit the usefulness of tubes in a number of ways, most of them indirect. In comparison with a transistor, for instance, a tube requires much more power—so much so that the vacuum-tube portable radio is almost extinct now. More power is necessary for heating the filament of just one radio tube than is used by all the transistors of an average pocket receiver.

Similarly, in the case of mobile radio equipment, those tubes which require extremely high plate voltage supplies, or high current, are usually ruled out because operating power is limited.

In most instances, even when high voltage and high current are available, the comparative danger of high-voltage operation as compared with low-voltage circuits (transistors, etc.) tends to swing the choice away from the tube.

Circuit Efficiency

Circuit efficiency also limits the usefulness of any circuit, not just those using tubes. In general, the efficiency of any circuit is the ratio between power put *into* the circuit and power taken *out*. For vacuum-tube amplifiers, it's sliced a little thinner in the standard definition of plate circuit efficiency.

Plate circuit efficiency, according to this definition, is the ratio of *signal* power output to dc *supply* power input (Fig.

5-7). It has nothing to do with *signal* power input. Thus, an rf amplifier which operated on 1 kV plate supply and drew 500 mA current would take 500W from the supply. If it delivers 300W of rf output, its plate circuit efficiency is 60%.

While circuit efficiency up to 95% or better is possible on paper, almost no working circuit has ever been built which gets better than 70 to 80% efficiency. The trouble with the higher efficiencies is that they measure all rf power output, not just that at the signal frequency. In order to get figures above 75%, it's necessary to choose operating conditions which greatly increase the percentage of harmonics in the output—and all the "extra" output power consists of just those harmonics!

Most audio power amplifiers are far less efficient than this. Typical figures are around 20 to 25% for "moderate" distortion. We'll get into this a bit more a little later when we look at some of the ways in which tubes are used.

The definition of plate circuit efficiency contains a built-in loophole, and in the early days of SSB operation many operators took advantage of the same loophole (which was also in the FCC rules at that time). The loophole is this: signal power input is not taken into account.

Some types of amplifier circuits use only a small fraction of the input signal as input, and feed the rest right on through into the output circuit. Such an amplifier can, for instance, accept 100W of signal input power, amplify 10W of it by a factor of 30 to get 300W, and feed through the remaining 90W of input to the output. The output power will be 390W, yet the amplifier produced only 300 of them. If input power were to be reduced to 10W, output would drop only to 300. And in either case the dc supply power taken by the circuit would be the same. This means that the same circuit might produce 300W at an efficiency of 60%, or 390W at an efficiency of 78%, with no change in the circuit or its adjustments!

For the definition of efficiency, this loophole is closed by requiring that only the power output *produced by that circuit* be used in calculating efficiency.

The loophole no longer exists. Part 96.67 of the current edition of the Rules and Regulations states that power input shall now be measured to the stage or stages delivering power to the antenna, and the Commission has served notice that in the case of feedthrough amplifiers, this is interpreted to include all driver stages as well.

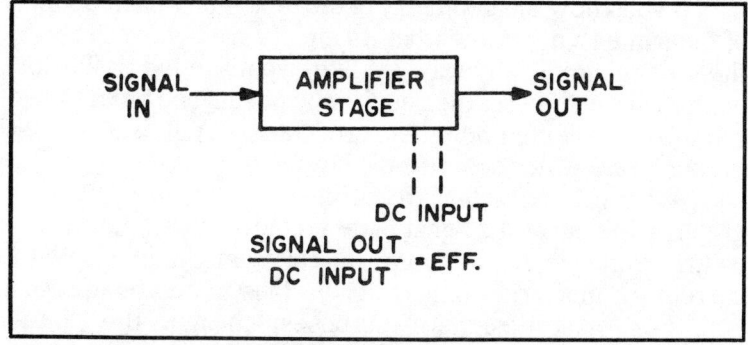

Fig. 5-7. Efficiency of an amplifier or a circuit is figured as shown here, by dividing signal output power by the dc power put into the stage. Input signal power is ignored in the calculation.

VACUUM TUBE APPLICATIONS

While at first glance it would appear that vacuum tubes are used for many purposes, we can compress all the uses for the ordinary garden variety of tube (that is, all except such special-purpose items as TV picture tubes, oscilloscope CRTs, tuning indicators, and the like) into one, with our *voltage-variable resistance* concept introduced in the previous section.

The diode, for instance, when it is being used as a rectifier, changes from being a very high resistance (when cut off) to a relatively low resistance (when turned on). The amplifier simply controls the flow of current between two terminals in response to the voltage between one of the two and a third. The oscillator is simply an amplifier connected in a special circuit. Even the digital logic circuits used in early electronic computers for timing and storage can be viewed as resistances controlled by voltages.

Tube Characteristic Curves

The exact characteristics of any individual type of tube such as, for instance, the 6C4 triode, depend upon the materials from which it is made, and primarily upon the shape and spacing of its electrodes.

These shape and space factors are generally called the *internal geometry* of the tube type, and they fix the amount of effect the grid voltage will have upon plate current, the maximum plate dissipation, the maximum cathode current, etc.

If you know the geometry of the tube—or what amounts of the same thing, know what its effects are—you can apply the tube in almost any way you like. To make life easier for equipment designers, the people who design and build tubes run measurements upon their products and publish *characteristic curves* which describe the key factors.

A typical characteristic curve for a triode (Fig. 5-8) graphs plate current against plate voltage for various values of grid voltage. A not-so-typical curve might graph plate current against grid voltage, for various plate-voltage values. For most designers' purposes, though, the plate-current/plate-voltage curve is best, so it's the one most often supplied.

However, if we know plate current and plate voltage at any instant, then by Ohm's law we can determine the effective plate-to-cathode resistance represented by the tube.

If the tube's plate is working into a reactance rather than a resistance, as for instance the primary of a transformer, or a choke-coupled output circuit, the plate voltage averages out to be constant and the manufacturer's curves can be used as they are. Changes of grid voltage then change only plate current.

If the plate works into a resistor, as in a resistance-coupled amplifier, it's a bit different and the picture gets messier. A *load line* must be drawn on the curve to determine the dc plate voltage present at any instant.

By using the curves, together with his accumulated training and experience, the equipment designer picks an *operating point* for the tube by proper choice of the plate and grid voltages, so that the change in grid voltage will cause a corresponding change in resistance between plate and cathode. That is, if a 0.5V increase in grid voltage causes a 10% drop in resistance, a 1V increase of grid voltage should cause a 20% resistance drop and a 1V decrease at the grid should result in a 20% increase of resistance. These figures are, of course, merely examples and do not correspond to any actual circuit.

When this has been done, the result is a circuit which provides for its output an amplified or stronger version of the input signal. The *linearity* of amplification is a measure of how accurately the output follows the input; another phrase used for the same effect is *distortion*.

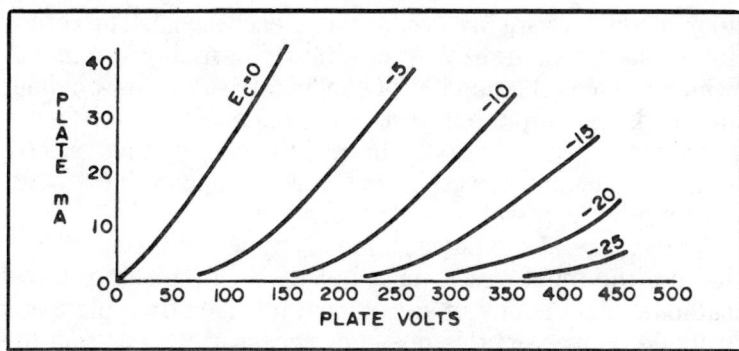

Fig. 5-8. Typical set of plate-voltage/plate-current curves for a triode tube shows the relation of plate milliamperes to plat volts for six selected values of grid voltage. Relationship for other grid voltages must be interpolated between these, if they are needed.

Notice particularly that this same process occurs in any kind of amplifier based on tubes or transistors, no matter what the circuit's name may be. Class A, B, or C amplifiers, as well as grounded-grid, grounded-cathode, and cathode-follower circuits, all act on this same basic process.

Amplifier Types

The terms *class A, class B*, and *class C* describe the operating point of the circuit. A class A circuit is intended to produce low distortion, and to give an output which is a faithful replica of the input. A class B circuit's operating point is much closer to the cutoff point (a true class B circuit operates exactly *at* cutoff); output is distorted but efficiency is higher. A class C circuit remains cut off most of the time and passes current only during the peaks of the input signal. Output is distorted beyond recognition but efficiency is highest under these conditions. Since the distortion can be removed from a continuous (unmodulated) rf signal by a resonant circuit, class C amplifiers are used for rf.

As we have already learned, any circuit requires two conductors to complete it; we can think of one as the *hot* lead and the other as a *return* path for the current. The usual return path in most radio equipment is the chassis, which we refer to as *ground* or *common* since it is often connected to ground, and provides a common return for all signal paths.

Our vacuum tubes, with their associated components, convert an input signal into an output signal. This means that four conductors are involved, two for each signal. The return for each signal, however, is almost invariably grounded, which reduces the number of conductors to three—an input hot lead, and output hot lead, and ground. This is fortunate, since a triode tube has only three elements; we have exactly enough conductors to assign one conductor to each element, with neither elements nor conductors left over.

Because of the physical means by which the tube does its job, the input *signal* must be applied between grid and cathode, and the output signal must be taken from plate and cathode. However, this does not necessarily mean that the cathode must be connected to ground, input to grid, and output to plate.

While it's true that the most conventional use of tubes follows just that assignment (called grounded-cathode operation), we can connect our common ground to any one of the three elements (Fig. 5-9). If, for instance, we ground the grid, then we must apply the input between cathode and ground. In order to take output with only one wire at the plate, we must pass all the output current through the input circuit. That is, since the input is connected to the cathode, the output signal's path must go through the input to get to the cathode in order to reach the plate.

Similarly, we could ground the plate. The input signal is now applied between grid and plate, while the output signal is taken from cathode and plate. For the input to get to the cathode, it must travel through the output circuit.

This means that in both the grounded-grid and the grounded-plate (usually called cathode-follower) circuits, the input and output circuits are directly connected. In the grounded-grid circuit, they are in series, so that all the current of one must pass through the other, while in the cathode follower, they are in parallel, so that the voltage of one becomes the voltage of the other.

In the conventional grounded-cathode circuit, the input and output circuits are isolated and do not interact so directly.

Gain

The apparent differences between these three different circuits are due primarily to this difference in relationships of input and output signals. In the grounded-grid circuit, a small

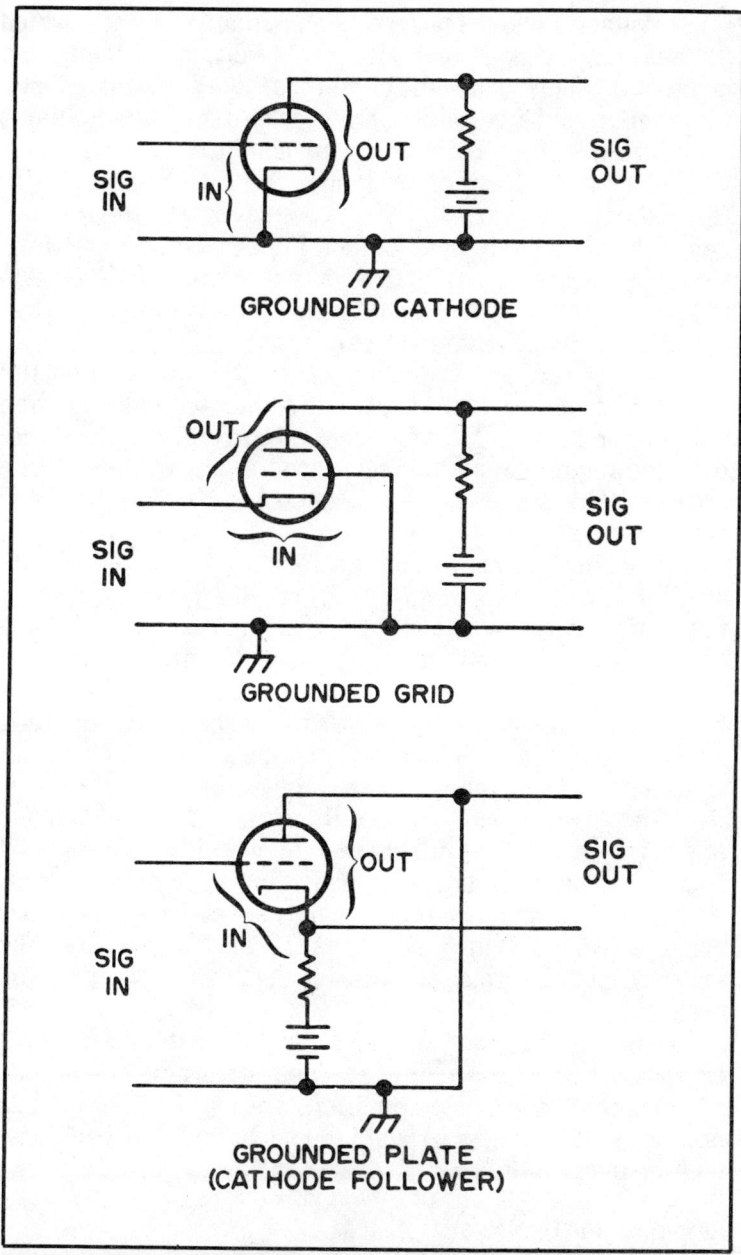

Fig. 5-9. Differences between the three ways of connecting tubes in amplifier circuits are shown here. These are simplified schematics and leave out all the necessities such as grid bias, coupling capacitors etc., to emphasize the similarities and differences of the three different circuit types.

input voltage causes a large change in plate current—which forces a large change in input current because of the series connection. This makes the circuit look like a low impedance to its input, and a very high impedance to the output. It also permits feed through of power from input to output.

Since it's voltage rather than current which interacts in the cathode follower, the effects are reversed. Input impedance is very high, and output impedance is very low. Coupling of signal from input to output is small; voltage gain is always less than 1, which means that a cathode follower actually introduces a voltage loss.

The gain of any individual tube is determined by the internal geometry of the tube, together with the applied voltages and currents. If voltages and currents are the same, the tube must provide the same gain in any of these three circuits. The stage, however, need not deliver the same gain.

For instance, a conventional grounded-cathode amplifier has its input and output circuits in series, so that the output current and input current must always be the same. Its gain can affect only the signal voltage. Thus the current gain of the grounded-grid amplifier cannot exceed 1, but high voltage gain is possible. While the gain of the tube itself remains high, the interaction between input and output signals outside the tube reduces stage gain.

The cathode follower has its input and output circuits effectively in parallel, so that the voltage in and out must be approximately the same. Its gain can affect only current. Voltage gain cannot be greater than 1, but high current gain is possible. Again, the tube gain is unchanged, but the input/output interaction outside the tube provides the restricting factor.

Since both the grounded-grid and the cathode-follower circuits produce less gain than does the grounded-cathode arrangement, and since amplifiers are usually intended to produce gain, this gives rise to the question "Why use these inefficient circuits?"

Amplification Quality

The grounded-grid amplifier reduces the stray coupling between input and output circuits inside the tube, because the grid acts as a shield between cathode and plate. This makes it possible to avoid having to *neutralize* a high-frequency amplifier, and in some cases makes it possible to operate a given tube at a higher frequency than would otherwise be possible (by modifying the effect of the built-in "low-pass filter" we examined earlier).

In addition, the low input impedance of the grounded-grid circuit is often convenient for a high-power rf amplifier, and the capability of feeding through power from input to output is also nice (although no longer offering a loophole in the FCC regulations).

Contrary to popular opinion, there is no appreciable difference in amplification quality between a grounded-grid amplifier and one using the conventional grounded-cathode circuit. Quality in each case depends upon proper adjustment of operating point and input signal level, not upon the choice of the common electrode.

The cathode follower's special properties depend upon its high input impedance and low output impedance. This makes it ideal for use in transforming impedances from high to low levels. The input capacitance is reduced by the same factor that input resistance is increased, making the cathode follower an excellent device for coupling energy out of critical circuits such as rf oscillators. The low output impedance makes it capable of driving a feedline without any intervening transformer, although other factors make this idea less attractive than it may sound at first.

Feedback

The cathode follower offers an ideal example of *feedback* in action; this is an important idea in all electronics, and seldom appears so clearly illustrated as in the cathode follower. Let's take a conventional grounded-cathode circuit and change it, step-by-step, into a cathode follower to see how feedback provides all the special characteristics of the cathode follower while the tube itself continues to operate just as it did before.

We'll start with an imaginary triode which operates normally with a plate-to-cathode potential of 100V, a cathode-to-grid potential of 5V, and a plate current of 10 mA. Under these conditions, with 10k resistor as its load, the tube provides a voltage gain of 21 times (Fig. 5-10). That is, a

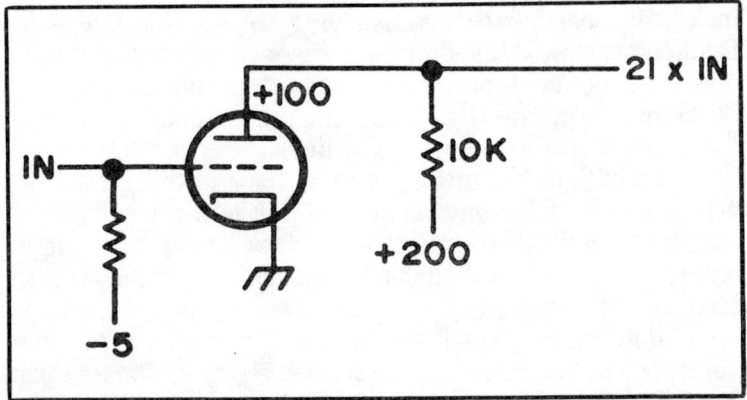

Fig. 5-10. This grounded-cathode schematic illustrates how feedback introduced by resistance in the cathode circuit reduces stage gain, although the tube itself continues to operate unchanged. Voltage applied to the grid maintains grid-to-cathode voltage at 5; as cathode voltage changes, grid voltage is changed to always be 5V less.

0.1V change in grid voltage will change the plate current by 210 µA, causing the plate voltage to change by 2.1V. A 1V change in grid voltage will change plate current by 2.1 mA, causing a plate voltage change of 21V.

Now let's move 1k of the plate load resistor around to the cathode circuit, leaving 9k in the plate lead (Fig. 5-11). This 1k in the cathode circuit is in both the cathode-grid circuit and in the plate-cathode circuit, so that the plate path still sees 10k.

If we change the grid voltage by 1V, the plate current will change by 2.1 mA. This will increase the cathode voltage by 2.1V (if the grid is going positive so that the current increases) and decrease the plate voltage by 18.9V. The plate-cathode voltage change is still 21V. The output voltage change would be less than that, however.

Unfortunately, the 2.1V change in cathode voltage is of such a polarity as to *reduce* the effect of the 1V input signal; were the entire 2.1V increase to occur, it would completely cancel the input signal which produced it. This, of course, cannot happen. What does happen is this:

The *effective* input signal is the one between grid and cathode, while the actual input signal is between grid and ground. When the grid voltage goes up 1V, it cannot do so instantly but must increase a few millivolts at a time. As it does so, the cathode voltage comes right along behind to buck it—and at some point they meet and level off so that 1V applied between grid and ground produces a grid-to-cathode voltage just right to permit the cathode-to-ground voltage at that instant.

In our example, this will happen when the grid-cathode voltage is a little higher than 0.322V. The cathode current increase caused by this voltage is a little more than 0.67 mA, and the rise in cathode voltage is about 0.676V. The two voltages add up to 0.998—which would have been 1.0 except that we rounded off our figures.

The feedback voltage appearing across the cathode resistor, then, reduces the effective input voltage to a smaller value. This in turn reduces the output signal voltage, since there is less input signal available to the grid. Stage gain is reduced—but the tube itself is still providing a 21-time voltage gain.

Input resistance of the stage increases, because resistance is defined as voltage divided by current (Ohm's law). Only the effective input voltage sees the original stage input resistance, but the entire input signal is affected. This means that the original input resistance must be multiplied by the same value that the voltage is divided by in order to keep current constant. In our example, the effective voltage is about a third of the actual voltage, so the effective input resistance is three times that of the original circuit.

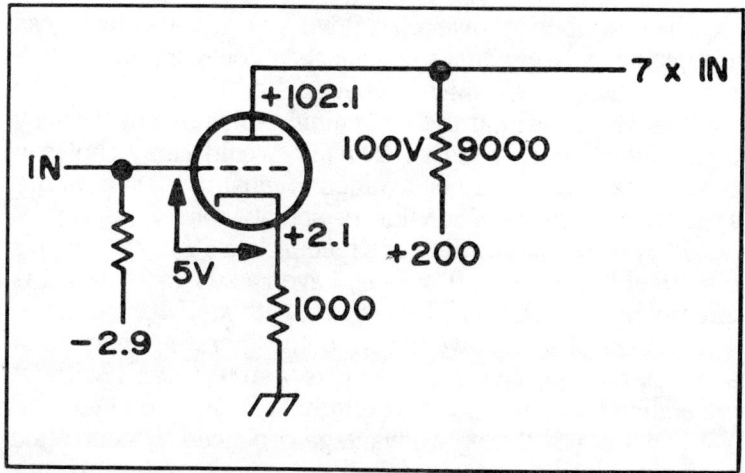

Fig. 5-11. Stage gain is reduced from that shown in Fig. 5-10 by the introduction of a resistor in the cathode circuit. But the grid-to-cathode potential is still 5V.

Now let's move some more of the resistance from the plate circuit to the cathode (Fig. 5-12). This time, let's make it 50/50, with 5k in each.

The feedback causes the effective signal input voltage to be smaller than the actual input, just as before, only more so. Where in our first example the cathode voltage rose 2.1 for every volt increase in grid-cathode voltage, it will now rise 10.5 because of the greater resistance.

Effective stage gain under these conditions is reduced from 21 times to about 1.83 times. That is, a 1V input signal from grid to ground will produce an apparent 1.83V change in plate voltage—which means a 183 μA change in plate current.

This means that the increased cathode resistance caused the leveling-off to occur with an effective input voltage of only about 0.087V. The remaining 0.913V of the input signal was bucked out by cathode voltage. Input resistance, similarly, is much greater now.

If we move all 10k over to the cathode circuit (Fig. 5-13), we find that effective stage gain is down to 21/22. That is, a 1V input signal will cause a 0.954V change in cathode voltage. The effective input between grid and cathode, then, is only about 0.05V. The tube is still producing its 21-time gain, because 0.05 times 21 is 1.05 (actually the input is a little less than 0.05, providing 0.95V out). Input resistance is now 21 times larger than originally.

The cathode follower circuit we have just developed can accept much larger input signals than could the grounded-cathode circuit with which we began. For instance, we had a 5V bias on the grid at the beginning, which meant that any input signal more positive than 5V would carry the grid positive (a condition to be avoided in most cases). With the approximate 20-time division of actual input voltage produced by the cathode follower's feedback, a 5V input signal is effectively reduced to 0.25V, and even a 20V input signal is effectively cut back to 1V from grid to cathode.

Where originally we ran out of the operating range with a 5V signal, we can now go up to a 100V signal without exceeding the same operating limits.

This is not the only advantage produced by controlled feedback. The changes in input and output impedance are also due to feedback's modification of effective voltage and

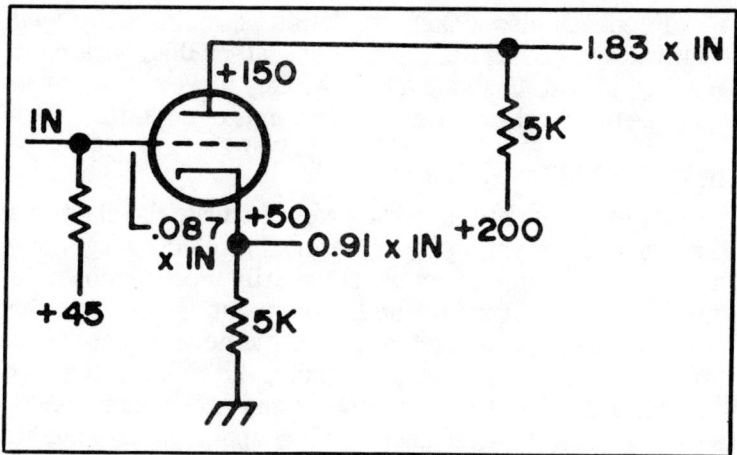

Fig. 5-12. As cathode voltage changes (because of the cathode resistor), grid voltage changes, but it is held to a differential of 5V.

current levels. Not so obvious is the fact that any distortion introduced by the tube is reduced, because it is *not* a part of the original input signal.

The grounded-grid amplifier's characteristics of low input impedance and high output impedance are also the result of feedback; in the grounded-grid circuit, it is current that feeds back rather than voltage, and this reverses the effect upon impedances.

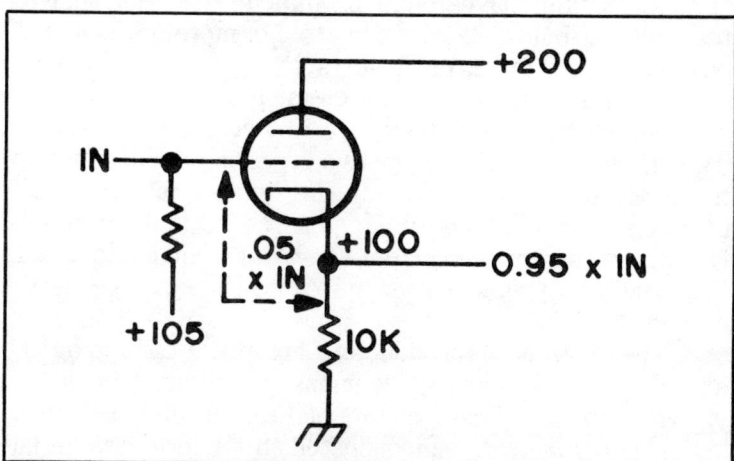

Fig. 5-13. With a high series resistance in the cathode circuit, the amplifier becomes a cathode follower, providing an output voltage gain of less than 1.0.

Feedback also makes oscillation possible, if the feedback voltage boosts the input signal rather than bucking it. We'll go into all this in another chapter, however, when we examine practical amplifiers and transmitter circuits.

TRANSISTORS

At this late date, if anyone were to make the claim that the transistor revolutionized the modern electronics industry, he would probably be summarily hooted down for offering such an insipid understatement. It is true that electronics really took off only after the development of the transistor, so such is properly a claim. But today, integrated circuits, made up from transistors and other components, have replaced the transistor in many applications—and are easier to use.

Note that some of the equations given in this chapter are merely approximations that work most of the time, so don't cry if the discussion seems incomplete, and you know more than is necessary.

ELEMENTARY SEMICONDUCTOR THEORY

First, let us examine some elementary semiconductor theory. Semiconductors are the tetravalent elements from the central section of the period table of elements. They are neither good conductors nor good insulators. Both silicon and germanium have found prominent use in transistor technology, but silicon seems to be more prevalent in devices made in more recent years.

To make a semiconductor element suitable for use in a solid-state electronic device it is necessary to add an *impurity* element. There are two types of semiconductors, made of either Ge or Si, differing mostly in the type of conduction created by the impurities. One type, called N-type, uses negative current carriers for conduction, while the other, P-type, uses a type of positive charge as the current carrier.

The N-type semiconductors are made by doping the material with an impurity that has the ability to add free electrons to the crystal structure of the pure semiconductor.

The *tetravalent* semiconductor atoms form crystal lattices by creating covalent bonds with each other, as shown in Fig. 5-14(A). To gain the extra electron needed for an N-type

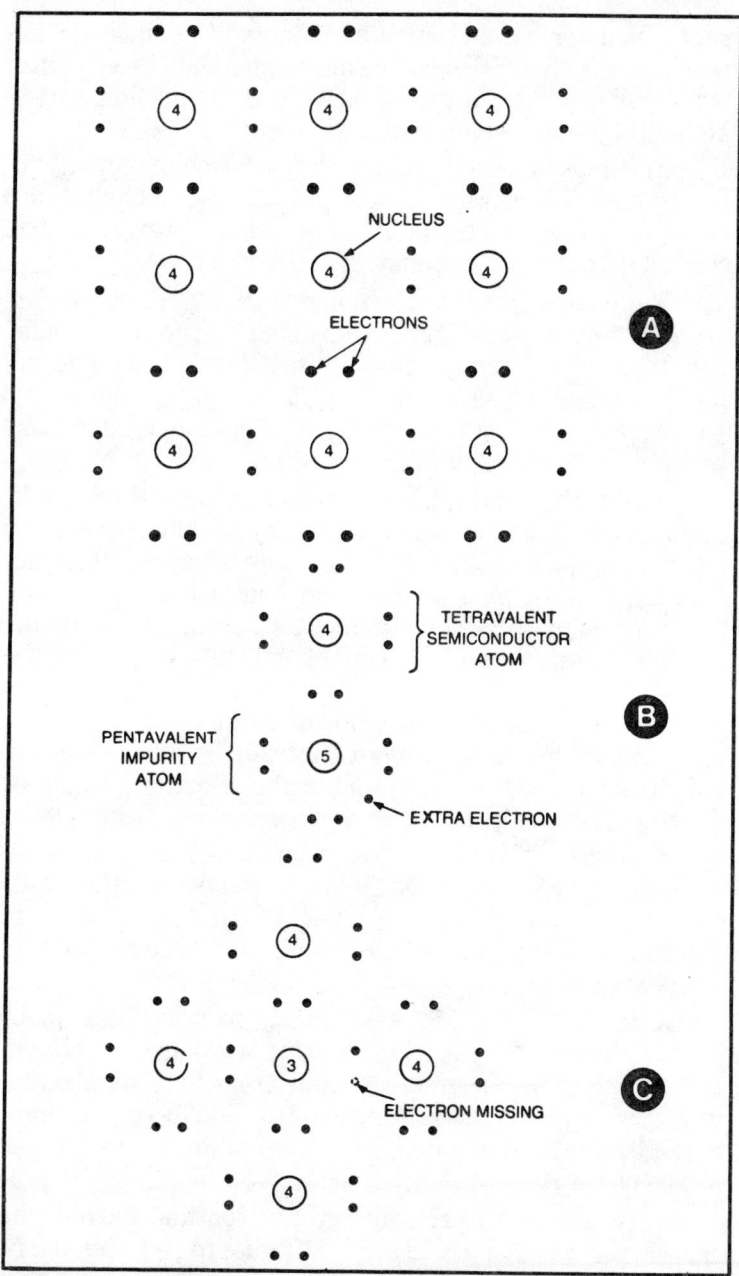

Fig. 5-14. At (A), covalent bonding in a semiconductor crystal; at (B), the effect of adding a pentavalent impurity; and at (C), the effect of adding a pentavalent impurity.

133

semiconductor we add minute amounts of pentavalent impurities, Fig. 5-14(B), such as antimony. Only a very small amount of material is needed, on the order of one impurity in 100,000,000 parts semiconductor.

The impurity also forms *covalent bonds* with the tetravalent atoms in the crystal lattice. In each case, though, there is one excess electron, and it is available to form an electric current if a potential is connected across the crystal.

Remember this fact: *Current flow in N-type semiconductors is by electron flow.* Conduction in the N-type semiconductor crystal is relatively easy to visualize in your mind, but many students have trouble with P-type conduction. I suspect the reason for this is that some authors tend to make the subject more difficult than necessary.

The P-type conduction is by something called a *hole*. Simply stated, a hole is a *place* in a crystal lattice where an electron *should be, but isn't*. This is shown graphically in Fig. 5-14(C). The *trivalent* atom of the impurity (i.e., gallium) forms covalent bonds with the tetravalent atoms of the semiconductor. When the trivalent atom takes it place in the orderly structure of the lattice, there will be one neighboring tetravalent atom which cannot form a bond, because all three electrons of the impurity atom are taken by other semiconductor atoms. This *place* where the electron would be expected, if the impurity were also tetravalent, is the *hole*.

How does a hole flow to form an electric current? After all, it is only a place! I thought you would never ask—holes don't actually flow, they merely appear to flow because an electron will occasionally fill a hole, only to leave another hole someplace else.

Consider Fig. 5-15. Suppose that initially there was a hole at atom A. An electric field (i.e., a voltage) is applied, and this tears loose an electron at atom B. This electron migrates under influence of the electric field to a point where it can be captured by atom A. This fills the hole at A, but leaves one at B.

Although it was actually an electron that moved, the *appearance* was that a hole moved, from A to B. Holes can be treated *as if* they were electron-size positive charges. This is, though, merely a convention that works most of the time—so accept it.

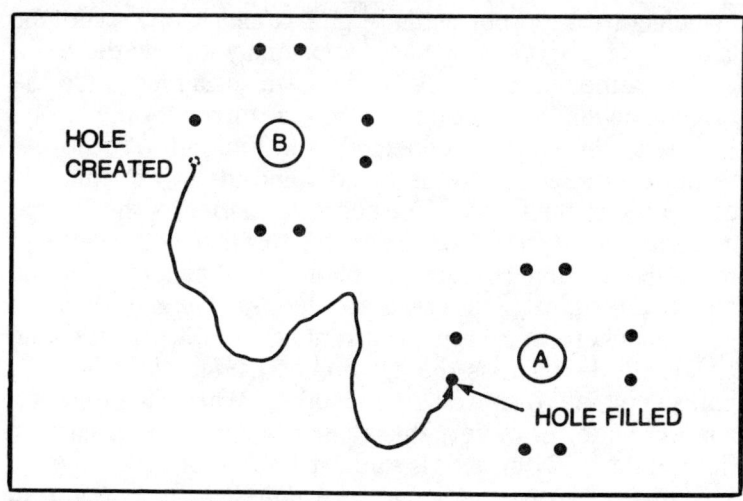

Fig. 5-15. How a hole "migrates."

PN Junctions

A *diode* is formed by joining together a section of P-type material and a section of N-type material. A modest example is given in Fig. 5-16. Please keep in mind that the

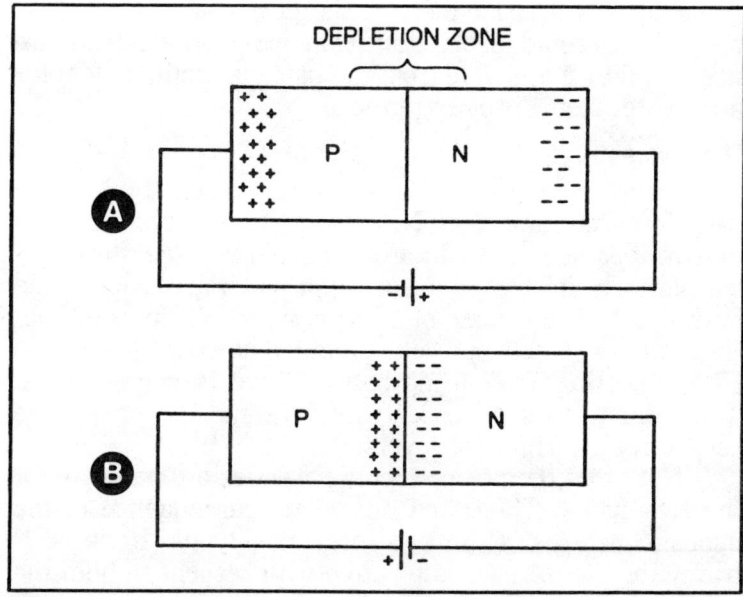

Fig. 5-16. At (A), a reverse-biased PN junction; and at (B), a forward-biased PN junction.

manufacturers do not actually grab a chunk of P-type and another of N-type and unceremoniously throw them together, although history tells us that this is almost what the original inventors of the transistor were forced to do!

With the battery connected as in Fig. 5-16(A), the (+) terminal is attached to the N-side, and the (−) terminal is connected to the P-side. The negative charges of the N-type material are attracted to the positive terminal of the battery, while the positive charges (i.e., holes) are "attracted" to the negative terminal. This creates a wide *depletion zone* near the junction where there are no current carriers. Current would ordinarily flow across the junction by having electrons and holes combine to neutralize each other. When the depletion zone is wide, however, this cannot happen, so no current flows. Such a PN junction is said to be *reverse biased*.

Figure 5-16(B) shows a *forward-biased* PN junction. In this case the negative terminal of the battery is connected to the N-side of the junction, and the positive terminal is to the P-side.

Here we have the respective charges repelled by the battery terminals, so they tend to pile up at the junction. Here we find that current flow across the junction increases tremendously because there are large numbers of electrons and holes to come together and combine. New electrons are injected into the N-side from the battery, while new holes are created from the other terminal.

The Elementary Transistor

Figure 5-17 shows a schematic representation of two basic forms of transistors: NPN and PNP. Both types consist of two PN junctions. In Fig. 5-17(A) we see the NPN transistor in which a section of P-type material is sandwiched between two sections of N-type material. In the PNP transistor just the opposite situation is found—a single N-type section sandwiched between two P-type sections. These transistors are essentially the same except that the respective polarities are reversed.

Note that the power supply polarities are opposite for the two types. This is an immediate consequence of the opposite natures of the two respective types. In an NPN transistor the *collector* is positive with respect to both the base and the emitter, and the base is slightly positive with respect to the emitter.

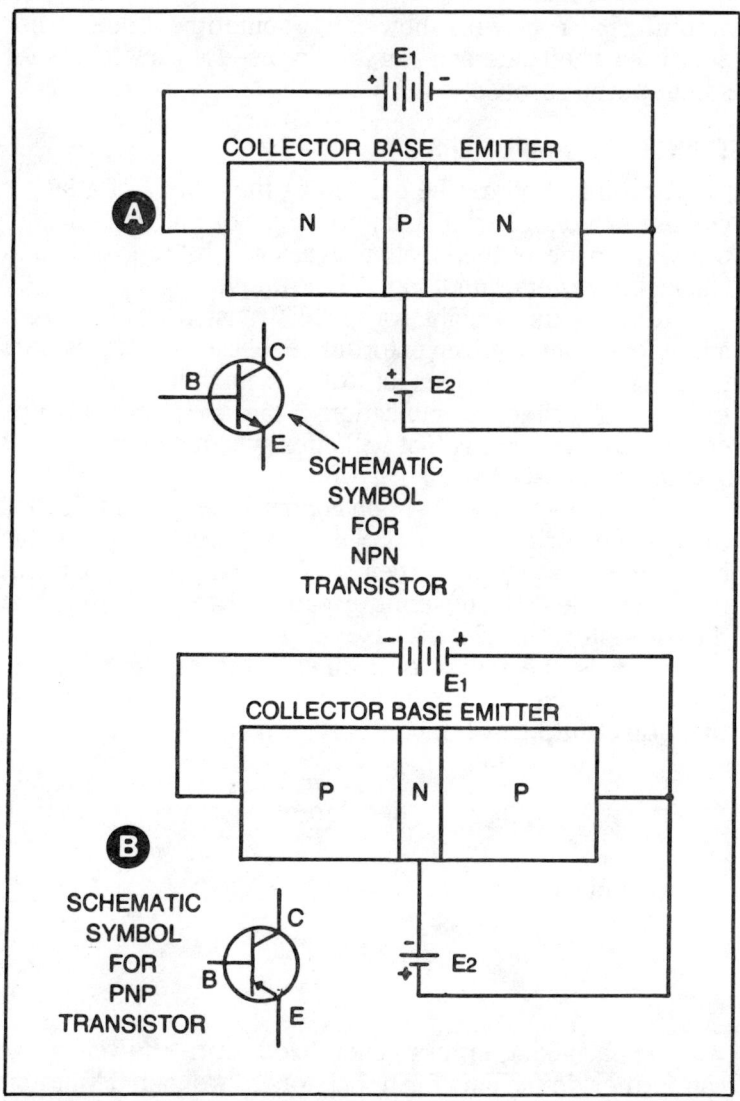

Fig. 5-17. Proper bias relationships for an NPN transistor (A) and for a PNP transistor (B).

In the PNP type of transistor of Fig. 5-17(B) we find the collector more negative than the base and emitter, while the base is slightly more negative than the emitter.

The polarity relationships shown are found when the transistors are used in their normal mode of operation as amplifiers. You will, however, sometimes see other polarities when the transistor is being used as a switch, or for some special purpose.

TRANSISTOR AMPLIFIERS

Amplification can be defined as the *control* of a larger current or voltage by a smaller current or voltage. Ideally, the waveshape of the smaller signal will be reproduced in shape, but larger in amplitude, at the output.

Most texts straightaway label transistors as current amplifiers, but this is unfortunate because it fixes into people's minds the idea of current amplification at the expense of voltage amplification. Transistors may also be connected into circuits that will offer substantial amounts of *voltage gain*, as well as *current gain*.

The base-emitter junction controls the current flowing in the collector-emitter path. Since the current flowing in the base circuit is only 2 percent to 5 percent of the current flowing in the collector-emitter path, it can be claimed that the transistor *amplified* the base current.

Gain is the measure of amplification in any type of amplifier device. In the simplest case we could define gain as the ratio of output over input. Voltage gain, then, is:

$$A_v = \frac{E_{out}}{E_{in}} \quad (5.1)$$

and current gain is:

$$A_I = \frac{I_{out}}{I_{in}} \quad (5.2)$$

When talking in less generalized terms, however, we can further define gain for transistors. Two gain definitions are usually given, alpha (α) and beta (β).

Consider Fig. 5-18. Here we see a simple transistor amplifier showing the respective base, emitter, and collector currents. Keep in mind that I_b is approximately $0.05 I_e$, while I_c is approximately $0.95 I_e$. By Kirchhoff's law it is also true that:

$$I_e = I_b + I_c \quad (5.3)$$

Alpha gain is defined as the ratio of collector current to emitter current, so by Eq. 5.3 will always be less than unity. Mathematically:

$$\alpha = \frac{I_c}{I_e} \qquad (5.4)$$

This will *always* be less than unity, and in our example above where $I_c = 0.95 I_e$ it would be exactly 0.95:

$$\alpha = \frac{I_c}{I_e} = \frac{0.95 I_e}{I_e} = 0.95 \qquad (5.5)$$

Example:
An NPN transistor has an emitter current of 25 mA, and a collector current of 23 mA. What is the alpha gain?

$$\alpha = I_c/I_e = 23/25 = 0.92 \qquad (5.6)$$

Beta gain, that usually quoted, is defined as:

$$\beta = \frac{I_c}{I_b} \qquad (5.7)$$

Beta gain will always have a value greater than unity—which may explain its popularity since it seems ridiculous to define *gain* with a number less than one, as in the case of alpha gain. Beta is also symbolized by the letters h_{fe}, so will be in the rest of this chapter.

Example:
An NPN transistor has 25 mA flowing in the collector, and 100 μA (microamperes) flowing in the base. Find h_{fe}.

$$h_{fe} = I_c/I_b = (0.025 \text{ amps})/(0.0001 \text{ amps}) = 250 \qquad (5.8)$$

Since Kirchhoff's law still holds, we can conclude that a relationship exists between alpha and beta, which is:

$$\alpha = \frac{\beta}{1+\beta} \qquad (5.9)$$

or

$$\beta = \frac{\alpha}{1-\alpha} \qquad (5.10)$$

Examples:
Find the h_{fe} if $I_c = 49$ mA, and $I_e = 50$ mA. First find the alpha gain, and then substitute it into Eq. 5.10.

$$\alpha = 49/50 = 0.98 \tag{5.11}$$

$$h_{fe} = 0.98/(1 - 0.98) = 49 \tag{5.12}$$

What is the alpha if a transistor has a beta of 250?

$$\alpha = \frac{250}{1 + 250} = 0.996 \tag{5.13}$$

Voltage Amplification in Transistors

A transistor is basically a current amplifier because a small base current can control a much larger collector current. The transistor can, though, also be used as a voltage amplifier. An example of a circuit that provides voltage amplification is shown in Fig. 5-19. Let us assume that E_o is equal to ($V_{cc}/2$) when E_1, the input signal, is zero. Let us further assume that E_1 is a sinewave—as shown.

When E_1 goes positive, the transistor collector current increases. The voltage across resistor R_c increases. But V_{cc} is a constant, so according to Kirchhoff's voltage law $V_{cc} = E_2 + E_o$. We must, therefore, expect an increase in E_2 to cause a decrease in E_o. The minimum value of E_o occurs when E_1 is maximum.

Similarly, when E_1 goes negative, the transistor collector current tends to reduce. This reduces the voltage drop across R_c, and increases E_o.

The value of E_o may swing above and below $V_{cc}/2$ between the limits 0 volts when the collector current is greatest, and V_{cc} when it is zero.

Biasing

Practical transistor circuits do not use batteries for both collector and base voltages. This is impractical, so most use a single power supply that has a value between 1.5 volts and 28 volts, with 12 volts and 28 volts being very common. These limits, incidentally, are merely common, and you may find transistor circuits with potentials greater or less than the extremes presented.

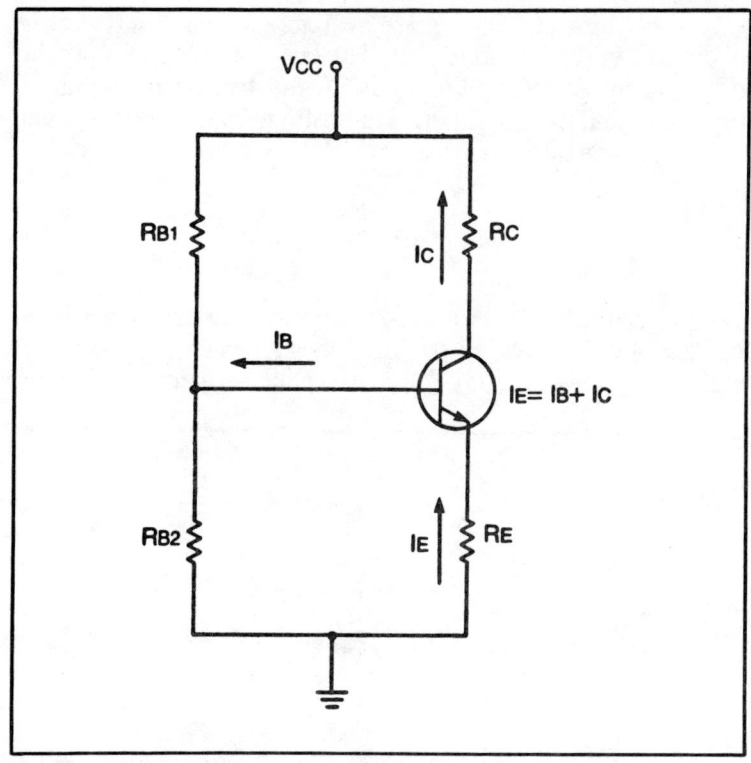

Fig. 5-18. Current distribution of a properly biased transistor.

Figure 5-20 shows three different methods for using resistors from a single V_{cc} supply to bias the base-emitter junction. In each case the collector load is designated R_c, the base resistor is R_b, and the emitter resistor is R_e. The formulas shown are approximations good only to 100 kHz, or somewhat less. At higher frequencies (rf) they tend to fall down a little bit, in fact a lot.

The circuit in Fig. 5-20(a) is a simple resistor bias method in which a base bias resistor is connected directly to V_{cc} (+) power supply. The output impedance is approximately equal to the collector resistor, R_c, provided that the impedance of the power supply is less than about one-tenth of the value of R_c at the lowest frequency of operation.

The input impedance can be quite high, on the order of the product of the beta gain and the emitter resistor, or:

$$Z_{in} = R_e \times h_{fe} \qquad (5.14)$$

141

Two different figures are available for gain in this type of circuit, voltage and current. The current gain is simply the beta rating, usually given as h_{fe} in the transistor manufacturer's specification sheet. The voltage gain, on the other hand, is given by:

$$A_v = \frac{R_c h_{fe}}{R_e} \tag{5.15}$$

The emitter resistor, R_e, is not actually necessary in all cases, but is intended to provide an increase in thermal stability. Unfortunately, it also reduces gain. Designers

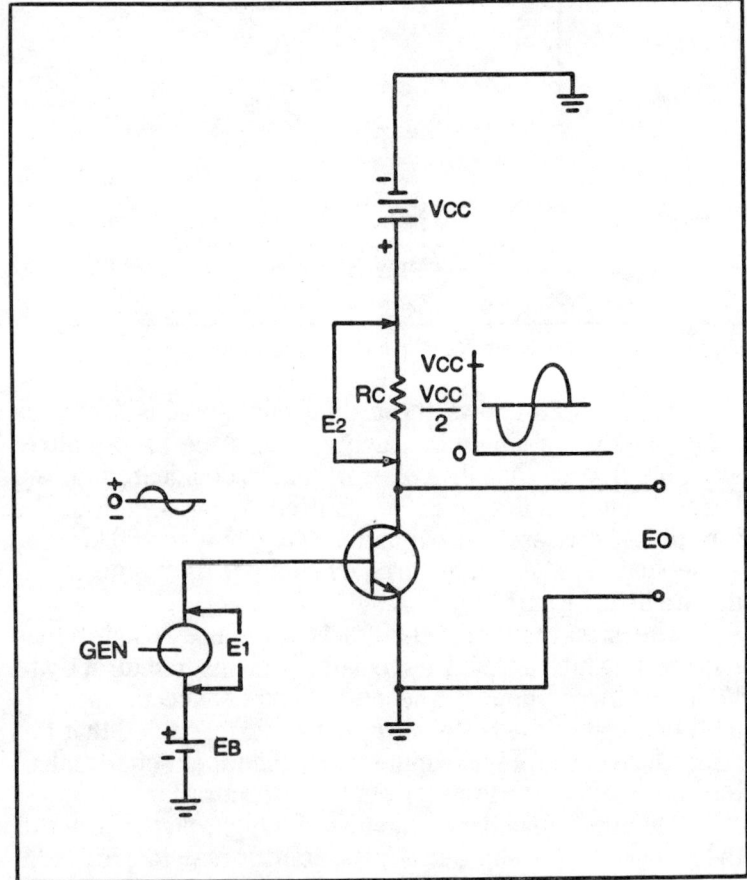

Fig. 5-19. Voltage amplification in a common-emitter circuit.

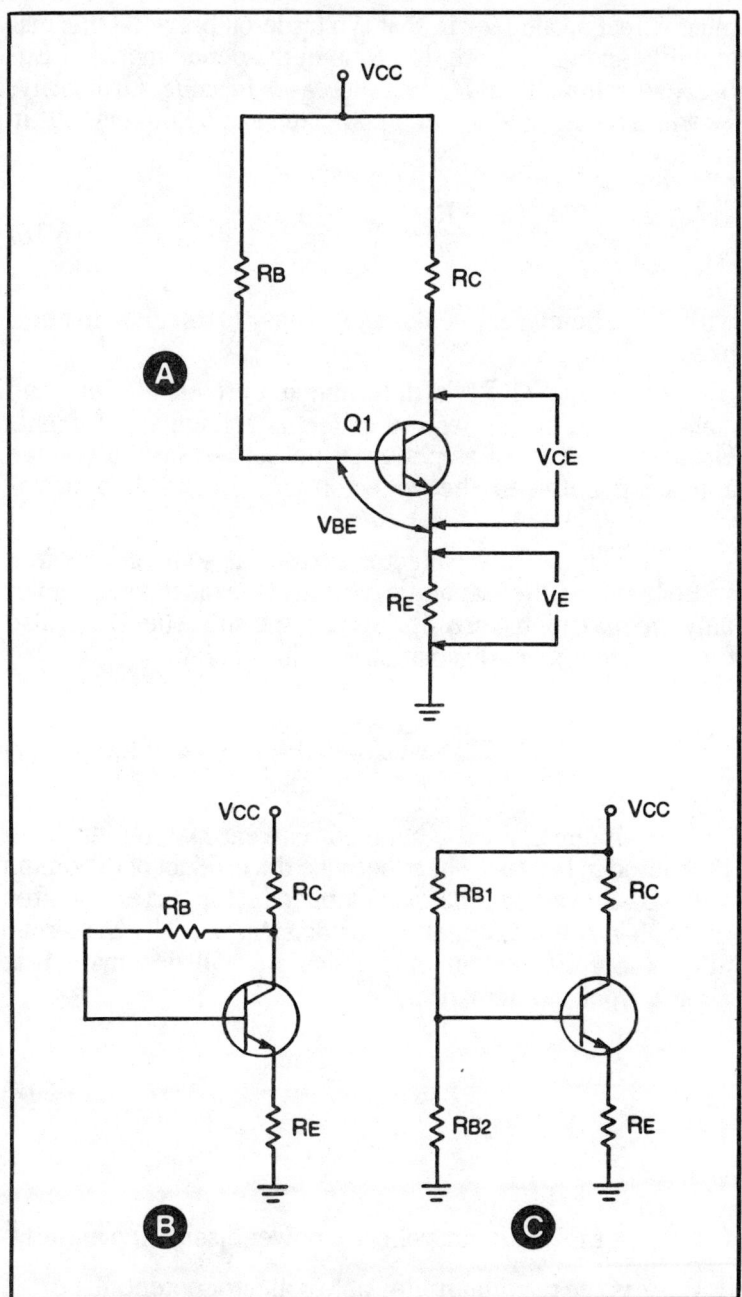

Fig. 5-20. At (A), simple biasing; (B), return to collector; and, (C) voltage divider biasing.

must select a value for R_e that is a trade-off between thermal stability and gain. Note that R_e is in the denominator of Eq. 5.15, so gain will reduce as it increases in value. Ordinarily, R_e will have a value between 50 ohms and 5 kilohms, but in any event

$$\frac{R_c}{20} \leq R_e \leq \frac{R_c}{5} \qquad (5.16)$$

with $R_c/10$ being very common, and easy to calculate in one's head.

The value of R_c is determined by computation, and consideration of the desired collector voltage and current. Ordinarily, for most amplifier purposes, we set the voltage appearing between the collector and ground at approximately $V_{cc}/2$.

We may set the collector current at some convenient value less than the maximum current. One must consider not only the maximum current shown in the spec sheet, but also the collector power dissipation (P_d). In general,

$$I_{c(max)} \times V_{ce/max} = P_{d(max)} \qquad (5.17)$$

The maximum *allowable* collector current may be less than that listed in the spec sheet because the product of maximum voltage allowed and the maximum collector current is often more than the maximum power dissipation. The maximum allowable collector current, which we will designate I_c to make us look mathematical, is:

$$I_c = \frac{P_{d(max)}}{V_{ce(max)}} \qquad (5.18)$$

Where:

P_d is the maximum collector power dissipation in watts

V_{ce} is the maximum allowable collector potential

I_c is the maximum allowable current in the collector

The base resistor value is determined by consideration of the required base current. The easiest way to approximate this current is:

$$I_b = \frac{I_c}{h_{fe}} \quad (5.19)$$

The beta (h_{fe}) is determined from the spec sheet for the transistor being used. We first determine collector current, and from that compute the voltage drop across the emitter resistor.

$$V_e = I_c R_e \quad (5.20)$$

The voltage drop across R_b is approximately

$$(V_{cc} - (V_{be} - V_e)) \quad (5.21)$$

The value V_{be} is approximately 0.6 volts for silicon transistors, and 0.2 volts for germanium transistors. The base resistor, by Ohm's law, is then:

$$R_b = \frac{(V_{cc} - (V_{be} + V_e))}{I_b} \quad (5.22)$$

Substituting Eq. 5.19 and 5.20 into Eq. 5.22:

$$R_b = \frac{(V_{cc} - (V_{be} + I_c R_e))}{\frac{I_c}{h_{fe}}} \quad (5.23)$$

$$R_b = \frac{h_{fe}(V_{cc} - (V_{be} + I_c R_e))}{I_c} \quad (5.24)$$

Let us point out that these are approximations. They are to put you into the correct ballpark, but not necessarily to home plate. You should build the circuit on a breadboard or proto typing chassis and make adjustments in the values that give the results nearer the desired results.

Figure 5-20(B) is a variation in which the base resistor is returned to the collector instead of directly to the V_{cc} supply. The parameters for this circuit are essentially the same as for Fig. 5-20(A), but the degenerative effect of the placement of the base resistor is said to improve stability. Keep in mind that the V_{cc} term in Eq. 5.24 must be replaced by a lower potential found at the collector end of the resistor R_c. This potential is $(V_{cc} - I_c R_c)$.

Figure 5-20(C) shows what is probably the best as regards thermal stability, but sometimes at the expense of some input impedance. For this circuit:

$$Z_{in} = R_{b2} (R_e h_{fe}) \qquad (5.25)$$

while the output impedance remains equal to R_c.

Also different are the respective current and voltage gains, which are:

$$A_I = \frac{R_b}{R_e} \qquad (5.26)$$

$$A_v = \frac{R_c}{R_e} \qquad (5.27)$$

Amplifier Configurations

There are three basic transistor amplifier configurations: common emitter, common collector, and common base. These are shown in Figs. 5-21(A), 5-21(B), and 5-21(C), respectively. The general properties shown in Table 5-1 hold true for these circuits.

These three circuits get their names from the element of the transistor that is common to both the input and the output circuit. In the common-emitter circuit the input signal is applied between the base and the *emitter*, while output is taken across the collector and the *emitter*.

Similarly, in the common-collector circuit, input signal is applied across the base-*collector* junction, while output is taken from the emitter-*collector* junction. In the common-base circuit, input signal is applied across the emitter-*base* junction, and output is taken from the collector-*base* junction.

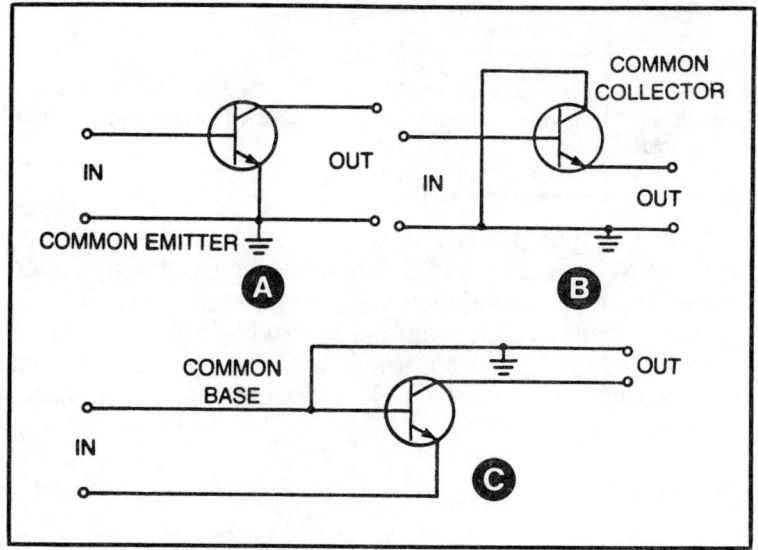

Fig. 5-21. Standard transistor configurations.

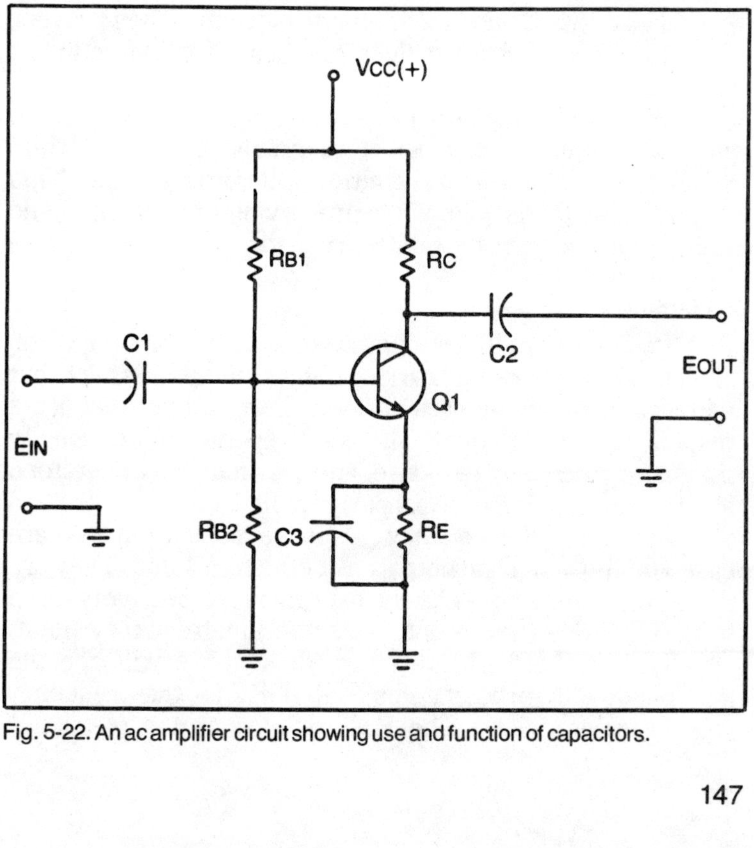

Fig. 5-22. An ac amplifier circuit showing use and function of capacitors.

Common Element	Relative Gain	
	Voltage	Current
Emitter	large	large
Collector	<1	large
Base	large	<1

Table 5-1. Relative Gain of Basic Transistor Configurations.

The output polarity in the common-emitter amplifier stage is *opposite* the input polarity. As the input signal goes more positive, the output signal goes more negative. This is *phase reversal*, and the output is said to be 180° out of phase with the input. This is *not* true in the common-collector and common-base circuits. In those the output signal is *in-phase* with the input.

AC Amplifiers

A transistor ac amplifier is shown in Fig. 5-22. It is essentially the same as the dc amplifiers discussed earlier, except that capacitors have been added.

Capacitors C1 and C2 are for dc blocking. They prevent the bias currents at the collector and base from affecting or being affected by the outside world.

Capacitor C3 is a bypass capacitor. Its purpose is to place the emitter terminal at ac ground potential, while keeping the dc bias on the emitter at its proper value. This will increase the ac gain while preserving the benefits of dc stability provided by the emitter resistor.

Frequency Response

Transistors will not operate out to any frequency you select. They are rather narrow minded in this respect, but then again, the device has not been invented yet that has a response from dc to light. There are several ways to specify transistor frequency response, and they must be understood before the devices can be properly applied.

Two of the frequency response measurements are based on the alpha and beta gains of the transistor. They are the points where the alpha and beta gains (respectively) drop to 0.707 times their values at some low-frequency point, usually 1000 hertz. More often, though, we will see the gain-bandwidth product, designated F_t, as the frequency response parameter. The F_t is defined as the frequency

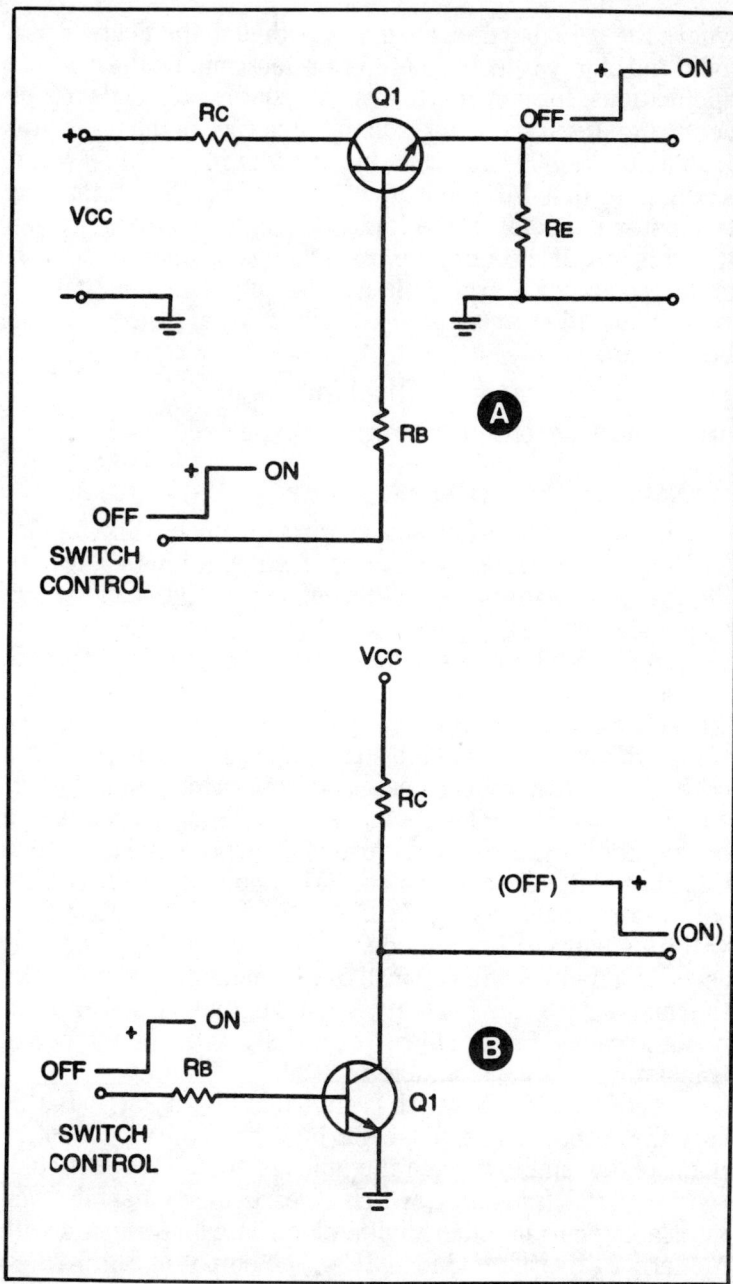

Fig. 5-23. At (A), a series transistor switch; and (B), a shunt-type NPN transistor switch.

where the gain (h_{fe}) drops to unity. Although the F_t gives us a valid and highly valuable method for determining the range of applications for any particular transistor, it tends to be confusing. Let us take, for example, the case of the transistor with a beta of 300 and a gain-bandwidth product of 50 MHz. The use of that MHz unit gives us the impression that the transistor is useful to the low VHF region—wrong. Recall that this 50-MHz spec is the *product* of gain and frequency. It is the frequency at which the gain is unity. Assume that you want to use the transistor at 1000 kHz. What happens to the beta? It drops to

$$h_{fe} = 50\,\text{MHz}/1\,\text{MHz} = 50,$$

hardly the value of 300 that might be expected.

TRANSISTORS AS SWITCHES

Under the correct set of circumstances the bipolar (NPN or PNP) transistor can be used as a kind of switch. Figure 5-23 shows two different ways this can be accomplished.

In Fig. 5-23(A) we see the series-pass type of electronic switch. When the base voltage is zero, the transistor is cut off, so no potential will appear across emitter resistor R_e. Although not a switch in the strictest sense, it will provide a potential at the output on command from an input signal. This could be useful if the load being driven is high power, and the driver signal must be lightly loaded. The potential applied to the base will cause transistor Q1 to be driven hard into saturation, so the voltage from the collector to emitter (V_{ce}) will be small in the on mode. Do not make the mistake of assuming that V_{ce} will be small in the on mode. Do not make the mistake of assuming that V_{ce} is negligible, however. It is, in some cases, but in others (especially with some types of transistors) it is considerable.

Another type of switch is shown in Fig. 5-23(B). In this case the output is inverted, being high when the input control point is low and low when the input is high. This particular type of NPN transistor switch is particularly useful when interfacing certain older digital electronic instruments with modern TTL or CMOS logic ICs. The output of Fig. 5-23(B) will be V_{cc} when the input is off, and close to zero (V_{ce}) when the input is on. If V_{cc} is set to a value that is compatible with

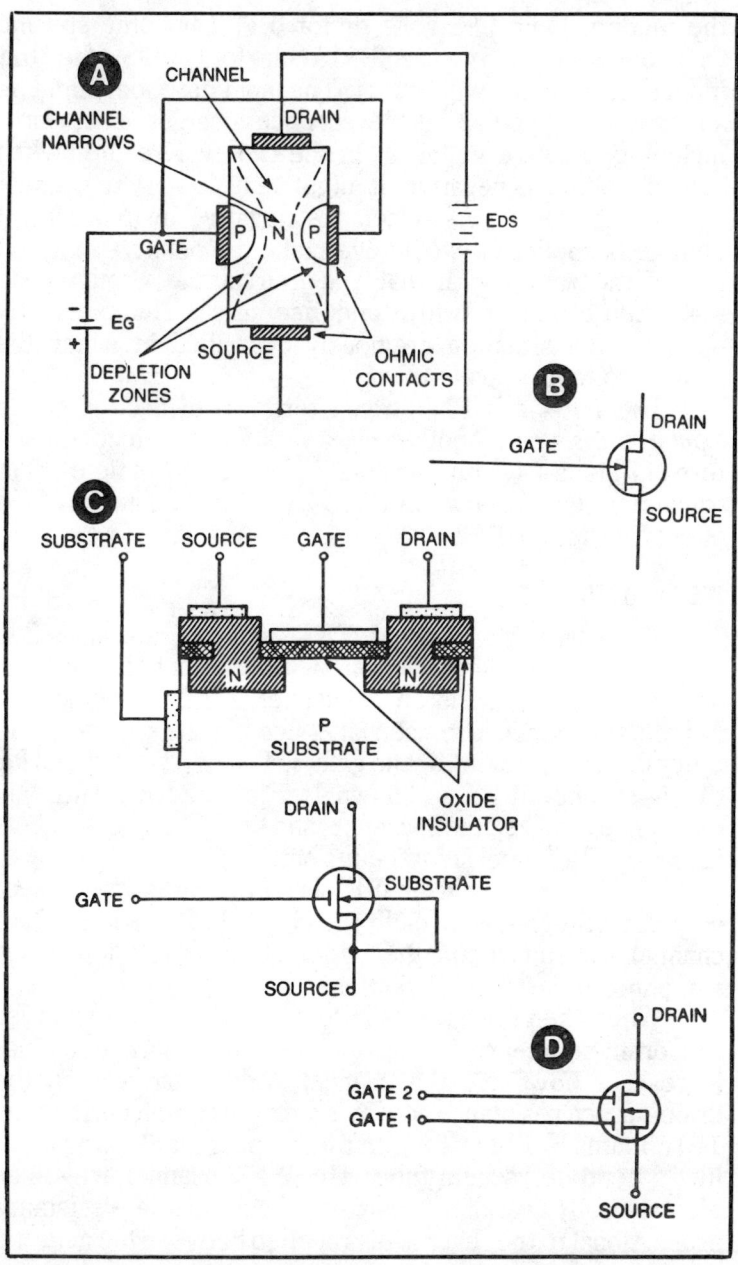

Fig. 5-24. At (A), the structure of a junction field effect transistor; (B) JFET circuit symbol; (C) structure of a MOSFET or IGFET and circuit symbol for a single gate MOSFET; and (D) the circuit symbol for a dual-gate MOSFET.

the modern logic (+5 volts dc for TTL, or some specific value between +4.5 volts and +15 volts for CMOS), then the input requirements will be met. The input potential might be any of several popular logic levels once considered standard, including negative voltages. In the case where the switch control voltage is negative, it might be necessary to connect a base resistor to V_{cc}. When the negative switch control voltage is applied, it would overcome the positive bias and turn off the transistor. In that type of circuit the output will be low when the input is low, and vice versa. This is exactly opposite the situation previously described in which the control voltage is positive.

The NPN and PNP transistors are of a class called *bipolar transistors*. Another class is the field effect transistors of Figs. 5-24(A) through 5-24(D). The type of field effect transistor (FET) shown in Fig. 5-24(A) is the junction field effect transistor (JFET).

JFET and MOSFET

There are two types of JFET, and they are classified according to the material making up the channel. The type shown in the figure is an N-channel JFET, meaning that N-type semiconductor material is used in the channel. The other material, making up the gate, is P-type. The symbol for the N-channel JFET is shown in Fig. 5-24(B). The only difference between the P-channel and the N-channel symbols is that the P-channel arrow points out.

The PN junction in the JFET is normally reverse biased, so a depletion zone forms in the channel. The width of the channel depends upon the width of the depletion zones surrounding the two gate sections.

When the reverse bias is low, the channel is wide, so the drain-source resistance is low. As the reverse bias increases, however, the channel width narrows, so the drain-source resistance becomes extremely high (as much as 100 kilohms to 1 megohm). In this respect the JFET is much like a pentode vacuum tube. The JFET channel acts as an electronically variable resistor which has a resistance proportional to the reverse bias applied between the gate and the source.

If the gate-source terminals are forward biased, a high current will flow that may well destroy the device. This is exactly the behavior expected of an ordinary PN junction, so be careful.

The other type of field effect transistor is shown in Fig. 5-24(C). This is sometimes called the insulated gate field effect transistor (IGFET), but most frequently it is called the metal oxide semiconductor field effect transistor (MOSFET).

The MOSFET operates in either of two modes, enhancement or depletion. Although some MOSFET devices will operate in only one of these, there are some which will operate in either, depending upon whether the voltage applied to the gate is negative or positive.

The MOSFET does not have any actual gate region. The gate is merely a metal ohmic contact attached to the insulated oxide layer. It creates an electrostatic field in the substrate region between the drain and source regions. The gate is, and acts like, a capacitor. When a positive voltage is applied to the gate, conduction of electrons in the channel is enhanced. Turning the gate on harder causes it to become even more enhanced. In the depletion mode there will be a thin layer of conducting semiconductor material between the drain and source. A negative potential applied to the gate will create a depletion zone in this material.

Those MOSFET devices which operate as depletion-enhancement field effect transistors have an intermediate value of current flow between full on and full off when the gate voltage is zero. Increasing the voltage in the positive direction turns the device on and allows operation in the enhancement mode. Increasing the voltage in the negative direction increases the depletion zone, so increases channel resistance. This will decrease channel current.

The JFET transistor can be handled just as you would handle any other semiconductor device, but the MOSFET is a bit sensitive. The thin metal oxide insulator between the gate electrode and the substrate/channel is very thin. The breakdown voltage is usually less than 100 volts, but static charges that can build up on your body are usually a lot more than that. As a result, the MOSFET can be damaged by merely touching it!

Some MOSFET devices use zener diodes inside the package to shunt excess voltages around the delicate gate insulator. An example of such a transistor is the RCA 40673.

It is a dual gate MOSFET (see Fig. 5-24(D)), but each gate has a pair of back to back zener diodes to prevent damage from stray static charges.

Figure 5-25 shows a manner in which the JFET (and by implication the MOSFET) can be used as an amplifier. Bias to the PN junction forming the gate is indirectly created by the source resistor. The channel current flows through this resistor, so will cause a voltage drop that places the source at a slight potential that is positive with respect to ground. The current through the gate resistor is minimal because the junction is reverse biased. The voltage drop across R_g, therefore, is almost zero, certainly a lot less than the voltage across R_s. This places the gate at dc ground potential. The bias situation is, then, the gate at ground potential and the source slightly positive. We want the gate slightly more negative than the source. Placing the source more positive than the gate is exactly the same thing.

The principal advantage of both types of FETs is an extremely high input impedance. The reverse-biased junction of the JFET creates a depletion zone that has a channel-gate impedance well into the megohm region, only a small leakage current flowing across the barrier.

The MOSFET has an even higher input resistance because it has a capacitor for an input, the capacitance formed by the metallic contact of the gate and the substrate. This capacitor has a metal oxide insulator for its dielectric. Some of these devices have input impedances on the order of 10^{12} ohms (that's *tera*ohms!), and even cheap MOSFET devices have extremely high values.

The MOSFET and JFET are used wherever the extremely high input impedance would confer some special advantage. Many scientific electronic instruments are driven from electrodes or special transducers which have extremely high source resistances. This makes it mandatory that the amplifier input that receives the signal must be even higher, on the order of 10 times greater if possible. Biopotential electrodes, for example, typically have source impedances on the order of 10 kilohms to 100 kilohms. Some chemical electrodes (which are actually transducers, not electrodes) have source impedances of almost 100 megohms.

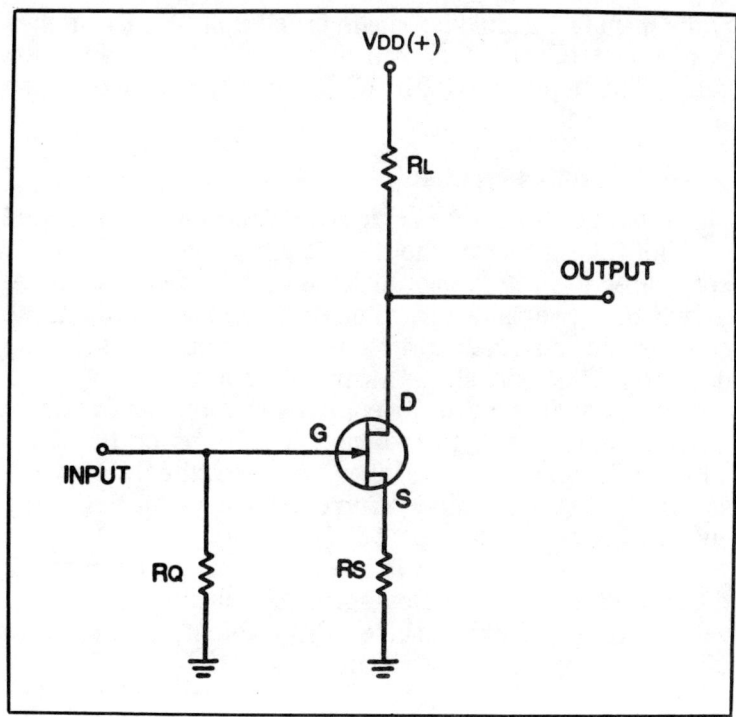

Fig. 5-25. Typical amplifier circuit using a JFET.

There are some circuits where the input impedance of an ordinary operational amplifier can be improved several orders of magnitude by connecting a pair of JFETs or MOSFETs between the input terminals and the inputs of the operational amplifier.

Another application of the JFET is as a low loss electronic switch, such as shown in Fig. 5-26. When the switch control voltage is zero (off), the negative bias is able to cut off the JFET. This effectively pinches off the JFET, and presents an extremely high channel resistance. This allows no transmission of signal through the channel. When the control voltage is +12 volts, the effective bias on the transistor is forward bias, and the negative potential is overcome. In this case, the channel resistance is extremely low, the switch is on, and the signal is transmitted through the device.

These types of solid-state switch are bidirectional, so the terminals are marked in/out and out/in, respectively.

Some manufacturers offer single or multiple JFET switches in a single IC package, known as multiple transmission gates. The popular CD4016 IC is a quad electronic switch such as described here.

Silicon Controlled Rectifiers

Figure 5-27 shows a device called the silicon controlled rectifier (SCR). It is another type of electronic switch, but is not bidirectional. It is an ordinary solid-state rectifier diode, except that it remains turned off until a current flows in the gate terminal. It then turns on, and current flows in one direction. The figure shows a sample circuit.

When switch S1 is closed, even briefly, current flows through resistor R1, to the gate. This turns on the SCR, allowing current I_2 to flow. When turned on, the SCR will act as any other rectifier diode. Current will pass only when the anode is positive with respect to the cathode.

The current will flow in the anode-cathode circuit until the current level drops below some critical minimum current value. If the current drops below this level, the SCR will turn off, and the circuit resumes its initial condition.

The SCR can be turned off in any of several ways. For one thing, switch S2 can be opened. This will reduce the current to zero, so it turns off. Another method is to apply a negative-going pulse to the anode. This will algebraically add with the potential across the diode, and if the pulse forces the current below the minimum level, the device will turn off.

A related device is the triac, which is a bidirectional SCR. It is basically a *pair* of SCRs with the gates tied together. It will pass current on *both* halves of an applied ac waveform.

SCRs and triacs are both part of a family of devices collectively called *thyristors*. They are used in *motor controllers, proportional heating controllers, light dimmers*, and several other assorted applications. They operate by allowing only a portion of each ac sinewave to reach the load.

There are a large number of other electronic devices that are considered discrete semiconductor components, but they are beyond the scope of this book, and besides, they are not all that useful in our *context*, i.e., passing the general class license examination.

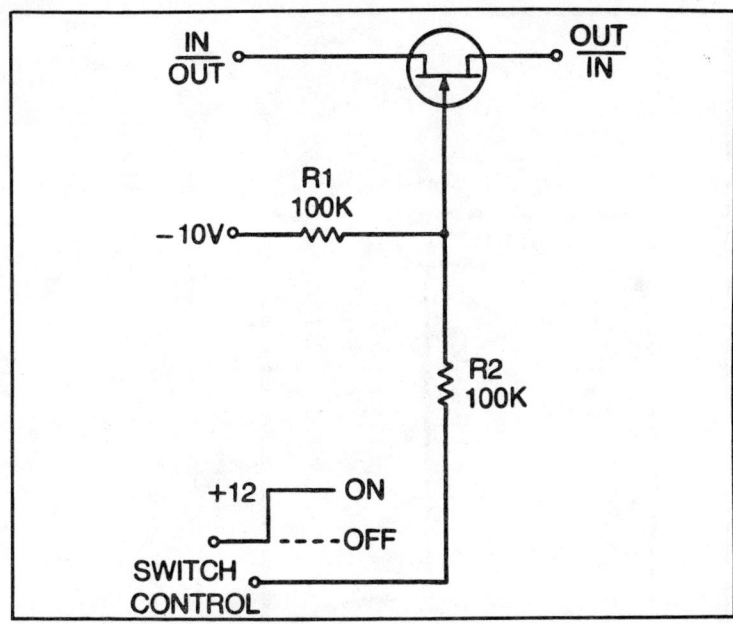

Fig. 5-26. Series switch using a JFET.

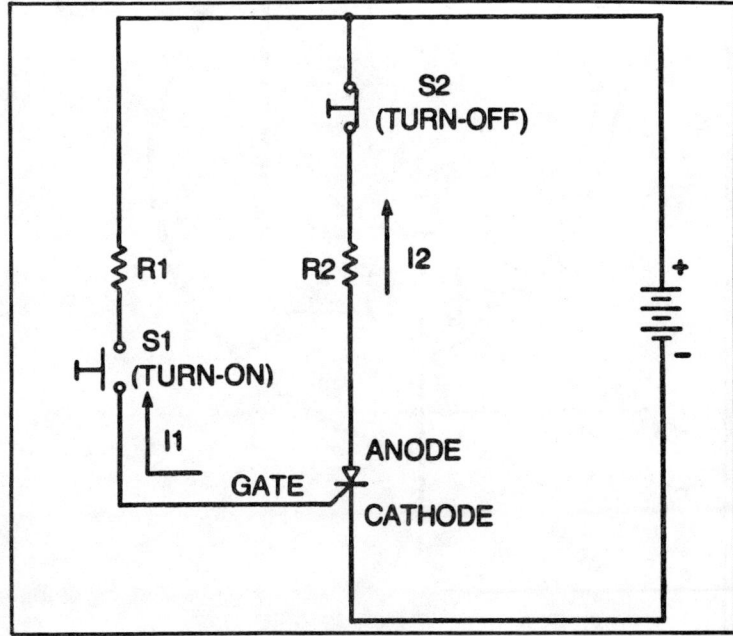

Fig. 5-27. Silicon controlled rectifier (SCR) circuit.

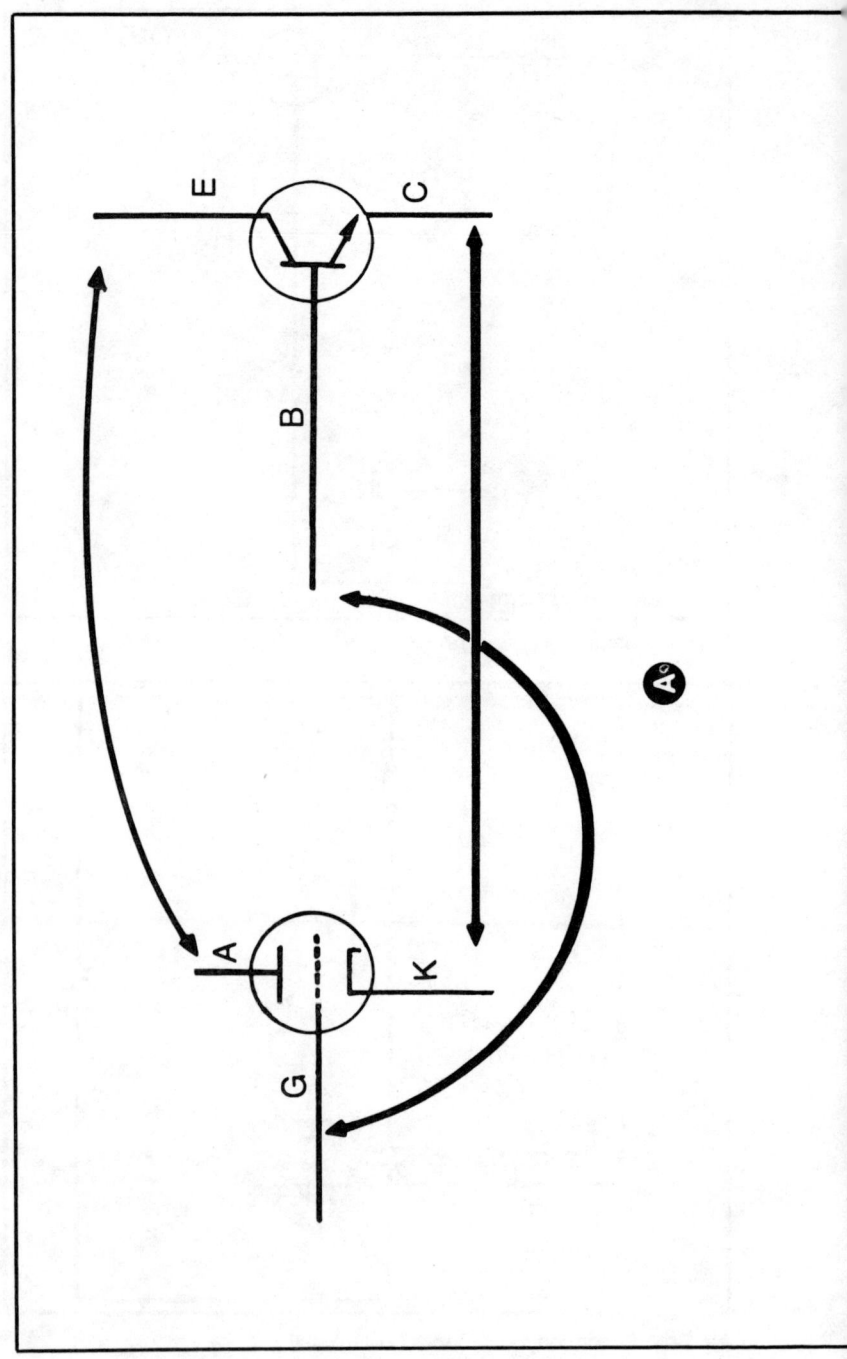

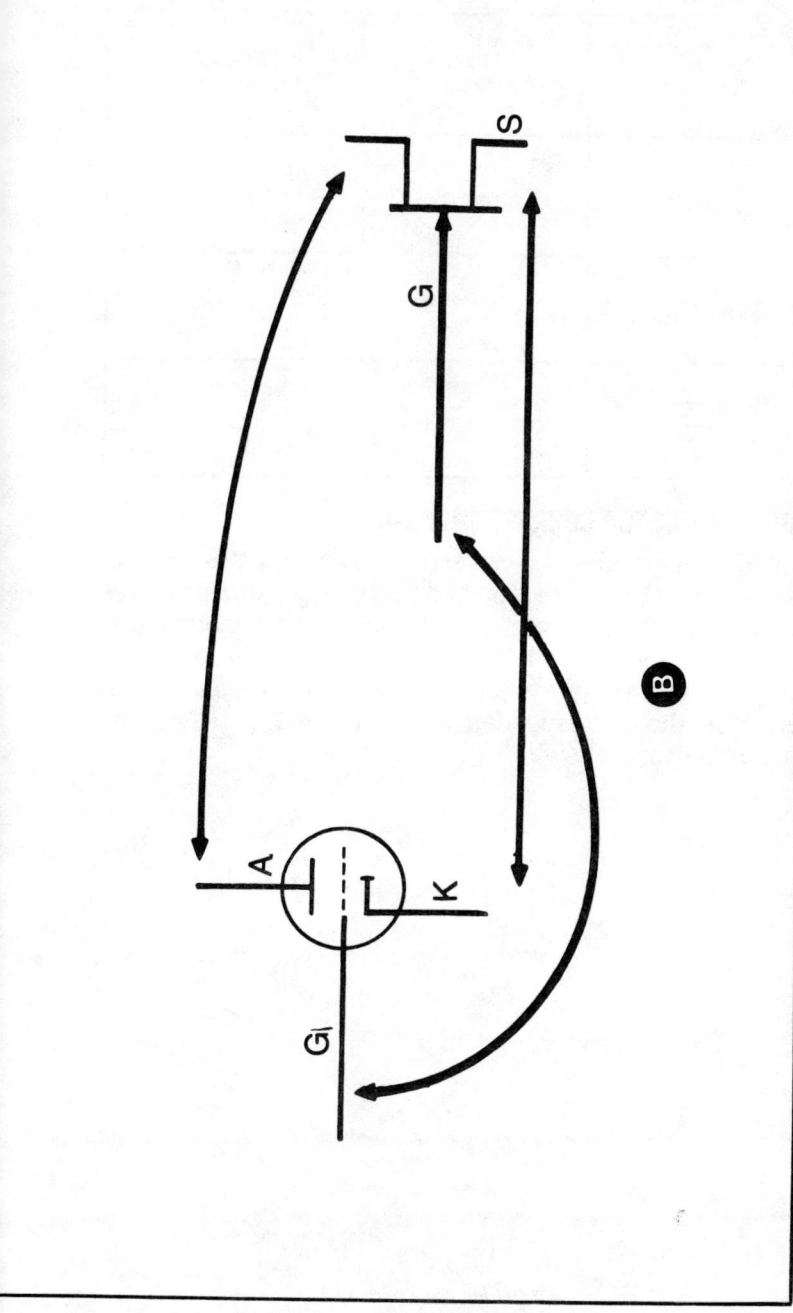

Fig. 5-28. "Corresponding" elements in tubes and transistors.

Table 5-2. "Corresponding" Elements in Tubes and Transistors.

VACUUM TUBE	BIPOLAR TRANSISTOR	FIELD EFFECT TRANSISTOR
GRID	BASE	GATE
CATHODE	EMITTER	SOURCE
ANODE (PLATE)	COLLECTOR	DRAIN

TUBE-TRANSISTOR CORRESPONDENCE

Several people who have recently taken the General Class examination have reported that one question involved the correspondence of vacuum tube and transistor elements. Although notions of such "correspondence" tend to torture reality a bit, they are none the less popular—or useful! Memorize the correspondences shown in Fig. 5-28, and summarized in Table 5-2.

6

Power Supplies

To operate any radio equipment, whether transmitting or receiving, you've got to have power—and it has to be just the right kind. While receivers can get by with a mere thimbleful of energy, transmitters have healthy appetites. The more potent the signal output, the heftier the rig's gulps of power.

So every different kind of equipment has its own power requirements, and the result is that the art of providing the proper power structure to keep a station simmering isn't as simple as might be expected.

Like all other aspects of radio theory, the FCC expects its licensees to be familiar with the theory of dc power supplies, and devotes a number of questions in the General Class examination to that subject.

For openers, we'll try to learn how ac converted to dc. We can then address ourselves to the techniques of making our new dc usable, and find out how filters operate. Next, we will delve into how power supply performance is rated to gain some definitions of necessary terms. Finally, we'll see how power supply performance can be improved and look into methods of regulating voltage and current.

CONVERTING AC TO DC

Most electronic equipment operates from direct current. FCC rules require *pure dc* for the plate supply of any

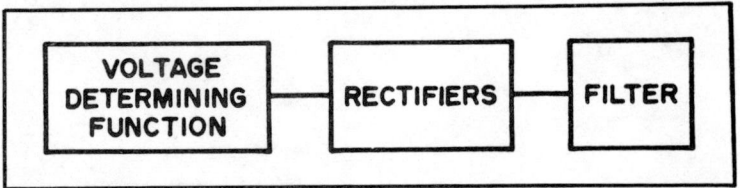

Fig. 6-1. Any power supply must include these three functions. Voltage is usually determined by a transformer, which may step input ac either up or down. Vacuum tubes require a step up, usually, while transistors require a step down. Rectifiers change the ac output of the voltage-determining function into pulsating dc, and the filter smooths out the ripple in the rectifier output to produce *pure dc*.

transmitter operating below the VHF region, and good operating practice requires it up into the microwaves. Yet dc of the proper voltage and power level doesn't come out of the wall outlet on demand. We must have circuits which take the ordinary ac available in any home and convert it to the required dc. Such circuits are known as *power supplies*, and they're the subject of our discussions in this chapter.

In general, any power supply is composed of three distinct functions as shown in the block diagram of Fig. 6-1. These are (1) the voltage-determining portion, (2) the rectification portion, and (3) the filter. Operation of each depends upon the characteristics of the other two, yet many choices are possible for each block of the circuit.

Voltage-Determining Function

The voltage-determining portion usually consists of one or more transformers which step the household ac from its 115 or 230V value up or down as required.

The amount of change required depends both upon the desired output voltage and the rectifier arrangement, as well as upon the type of filter employed. The transformer itself, however, is essentially the same as those employed for impedance matching and examined in the previous chapter, except that power transformers need operate at only one frequency and so are somewhat simpler to design.

Rectifiers

The stepped-up or stepped-down ac goes from the voltage-determining block to the rectifiers, and there's where wide choice comes into play. At least three different rectifier arrangements are in common use with single-phase power supplies; all three are shown in Fig. 6-2.

The half-wave rectifier is the simplest of the three, but is also the least efficient since it throws away half the ac cycle. This circuit is seldom used for circuits which are intended to deliver any appreciable amount of power, although it finds frequent application in bias supplies and other low-current uses.

The centertapped full-wave rectifier is probably the most commonly used circuit. While it requires a center-

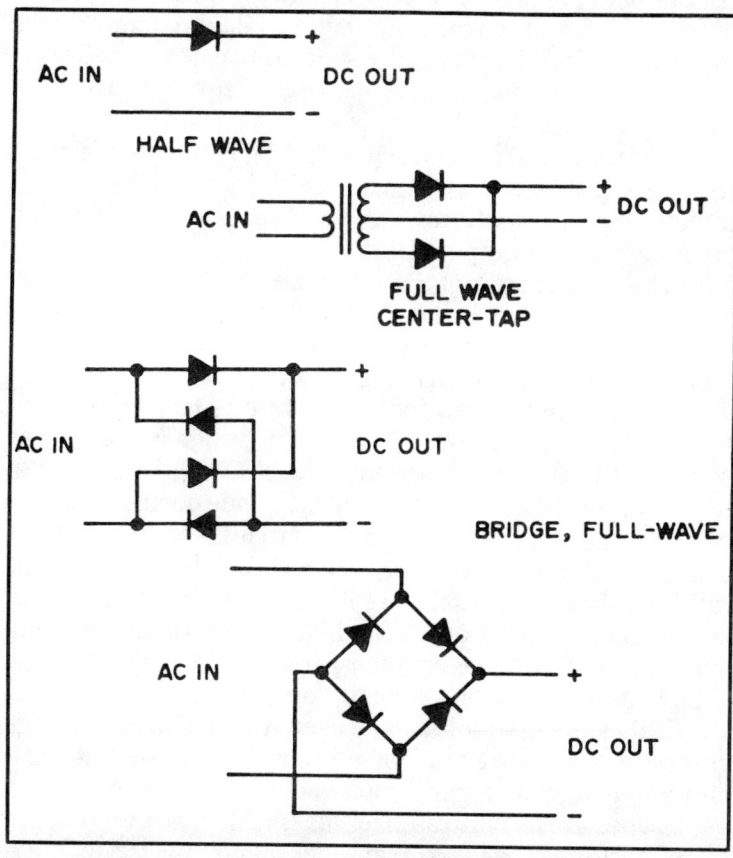

Fig. 6-2. These three circuits represent the only rectifier hookups in common use. Half-wave circuit is used only in ac/dc receivers and extremely lightweight gear. Full-wave centertap circuit is functionally just about two half-wave circuits back to back, with a transformer providing two-phase ac. Bridge circuit uses four diodes, but steers the ac input through to the output without the need of a transformer while retaining full-wave advantages.

tapped transformer capable of supplying twice the voltage desired at the output, it makes full use of the input ac cycle and places no severe demands upon any circuit components. When virtually all rectification was accomplished by vacuum tubes or mercury-vapor bottles, this circuit was almost universal.

The full-wave bridge rectifier, like the centertapped circuit, makes full use of the input ac cycle, but with some rather significant differences. The centertapped circuit is, essentially, just a pair of half-wave circuits of opposite phase operating in parallel. The bridge, however, is a completely different kind of circuit, in that its diodes "steer" the incoming ac in the proper direction so that it always comes out at the same output terminal regardless of which input terminal it entered.

Because of this steering effect, the bridge does not require a double-voltage transformer. If used with the same transformer as a centertap full-wave circuit, the bridge will produce twice the output voltage.

The bridge circuit does, however, place additional requirements upon some of its diodes, which limit its attractiveness if tube-type diodes are to be used. Full output voltage appears between cathodes and filaments of the "off" diodes; most tube-type diodes are not rated for this stress. With solid-state rectifiers, though, the bridge is not limited by this difficulty—and since the advent of solid-state silcon rectifiers, the bridge circuit has gained wide popularity.

All three of these rectifier circuits involve diodes, which are electronic one-way valves. Diodes come in three major flavors, with subflavors in some cases. They may be high-vacuum tubes, of either high, medium, or low impedance; mercury-vapor tubes; or solid state, such as silicon, germanium, or selenium stacks.

Tubes were traditionally used when extremely high reserve voltages were involved, but in most new designs only solid-state diodes are employed.

Tubes are usually rated for maximum direct current per plate, maximum peak current per plate, maximum peak inverse plate voltage, and maximum rms supply voltage per plate.

Solid-state diodes may be rated for peak inverse voltage (PIV), rms supply or input voltage, average forward current, peak one-cycle surge current, peak forward current, forward voltage drop, and thermal resistance. Of all these, the most important are those which correspond to the tube ratings: PIV, average forward current, and peak one-cycle surge current.

PIV is the maximum voltage which can be applied *in reverse* to the diode before it breaks down and permits current flow "against the stream"; when exceeded, instant destruction of the diode usually results. If the diode is connected to a capacitor, diode PIV should be at least twice the peak value of applied ac voltage; otherwise, PIV should be at least equal the peak of the applied ac.

Two or more diodes can be connected in series to increase their PIV ratings, provided that voltage equalizing resistors are connected in parallel with each as shown in Fig. 6-3. These resistors assure that each diode gets only its share of the applied voltage; otherwise, most of the voltage would appear across the diode with highest back resistance.

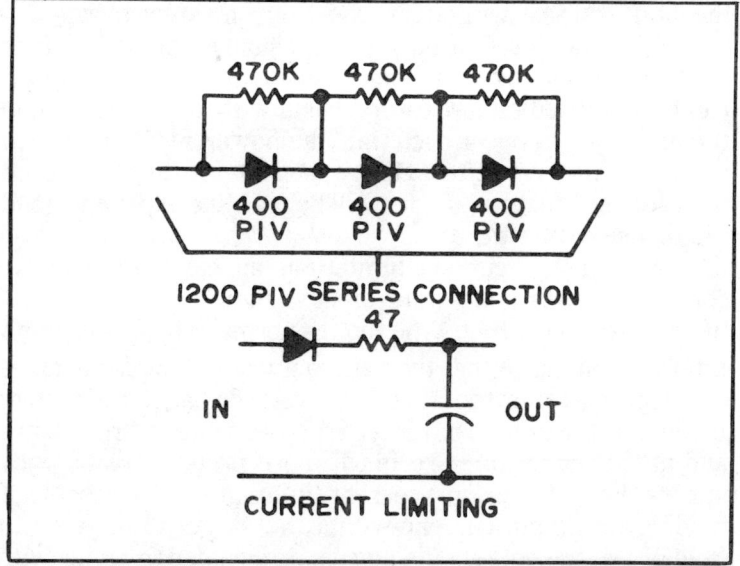

Fig. 6-3. Semiconductor diodes require their own special tricks. To extend the PIV of diodes, two or more can be connected in series, but resistors must be paralleled with each to assure that back voltage divides equally among all units rather than piling up on the unit with the highest reverse resistance. When capacitor-input filters are used, current through the diode must be limited by a low-value resistor, to stay within the one-cycle surge ratings of 10 to 50A. A value of 47 ohms at 1W usually suffices.

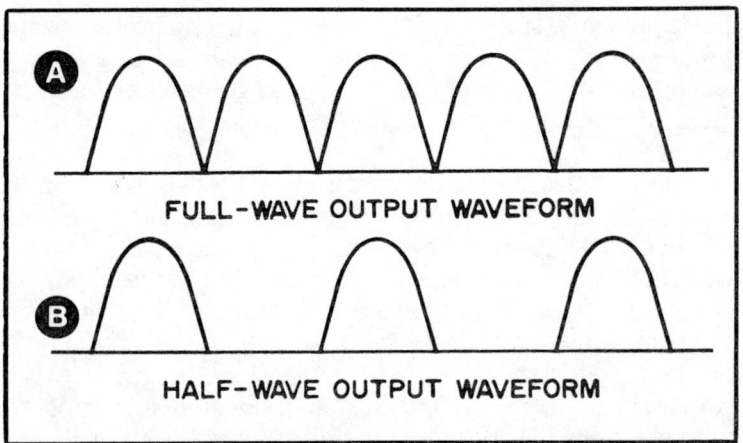

Fig. 6-4. These waveforms are produced by the rectifier circuits shown in Fig. 6-2. Half-wave circuits "wastes" half the input ac cycle; full-wave circuit makes use of both halves to produce a higher output voltage, and also produces double the ripple frequency which makes filtering easier.

Average forward current is the maximum current which the diode can pass without overheating, on a steady basis.

Peak one-cycle surge current, usually at least 10 times greater than average current, is the maximum current which can be tolerated on a one-time basis without destruction of the unit. Surges occur each time the power supply is turned on, as the filter capacitors charge, and if this rating is ignored with semiconductors, diodes will behave like expensive (and rapid) fuses every time.

While surge current limitations appear more often in connection with solid-state diodes, they apply to all rectifiers. However, tube-type diodes normally have such high internal resistance that they automatically limit themselves to surge currents too small to cause damage. Solid-state diodes, on the other hand, have much less internal resistance and at the same time are much more prone to damage by surges. For this reason a current-limiting resistor capable of holding maximum current within the surge rating even in case of a dead short should always be included in series with solid-state diodes as shown in Fig. 6-3.

Regardless of the type of rectifier used, the output of the rectification part of the power supply is dc rather than ac. This dc is, however, not yet usable because it is *pulsating* rather than pure. The waveform of the dc at this stage, were it to be fed into a resistive load, would look like Fig. 6-4 which shows both half-wave and full-wave rectifier-output waveforms. The continual change in level of this power makes it unusable for our purposes; that's why our power supply contains the final block, the filter.

Filters

When our dc emerges from the rectifier circuit, it's pulsating as shown in Fig. 6-4, and cannot be used. The filter circuit evens out the voltage and current waveforms, turning it into *pure* dc required by FCC regulations.

Filters are composed of capacitors and/or resistors and inductors, with the capacitors being connected in parallel with the output of the power supply, and the inductors or resistors in series. This arrangement makes possible two different layouts for the filter's input circuit, as shown in Fig. 6-5. Either the series inductor can be the first component encountered at the input or the parallel capacitor can be first. If the inductor appears first, the circuit is known as a *choke-input* filter, while if the capacitor appears first, the filter is called a *capacitor-input* type.

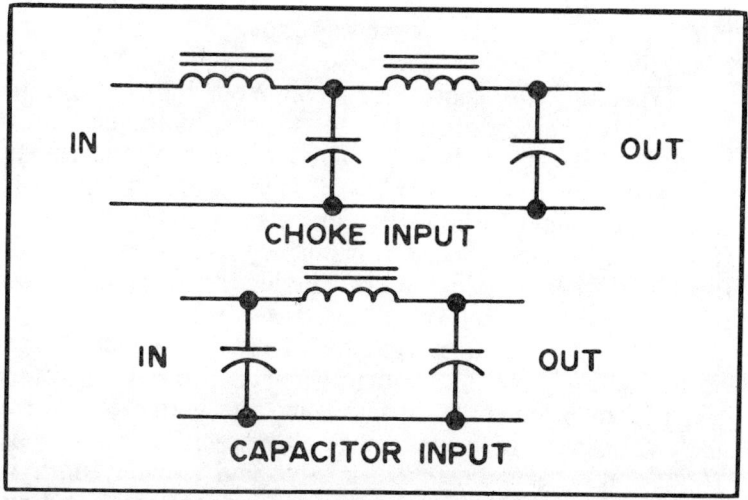

Fig. 6-5. These schematics show the two types of filter circuits most often encountered in ham equipment. Choke-input filter produces better regulation but lower output voltage; capacitor-input filter is harder on diodes and the transformer but produces a higher output voltage. Capacitor input is widely used at low power levels but choke input is almost exclusively employed at high power.

Choke-input and capacitor-input filters have very different characteristics, caused by the differences between inductors and capacitors. Compared to a choke-input filter, a capacitor-input filter will produce a higher output voltage at light current drains, but with poorer voltage regulation since at heavy current loads, output voltage for both types is similar. The capacitor-input filter requires higher voltage ratings for the rectifier diodes, and imposes a heavier current load on the transformer.

Because of these differences, capacitor-input filters are most often used for receivers, test equipment, and solid state transmitters where current loads are not likely to vary widely. Vacuum tube transmitters, on the other hand, use choke-input filter circuits more frequently.

The differences between the choke-input filter actions and the actions of the capacitor-input filter are best understood by looking at the waveforms which show what happens inside the power supply during a single cycle of ac input. The waveforms for a capacitor-input filter are shown in Fig. 6-6. Full-wave rectification is assumed; half-wave rectification merely emphasizes the differences.

Waveform A in Fig. 6-6 merely shows what the full-wave ac would look like across a resistor, in the absence of any filter.

When the filter capacitor is connected, with a moderate load current being drawn, the voltage across the capacitor follows waveform B; from point B1 to B2 the capacitor is charging, and from B2 to B3 it is discharging. Current through the diodes, however, flows only during the time from B1 to B2 because the transformer voltage is less than the capacitor voltage for the remainder of the cycle, so the rectifier current waveform follows that shown as C.

With very little current being drawn, the discharge portion of the capacitor-voltage waveform (that from B2 to B3) becomes almost horizontal, and the entire waveform shown at B rises toward the peak of the rectified waveform. Waveform C than becomes smaller and smaller, until it disappears at the limit of zero load current, with output voltage becoming equal to ac peak input voltage.

Under heavy current load, the discharge curve (B2 to B3) steepens, pulling the entire waveform toward the zero line and increasing the amplitude and duration of the pulses

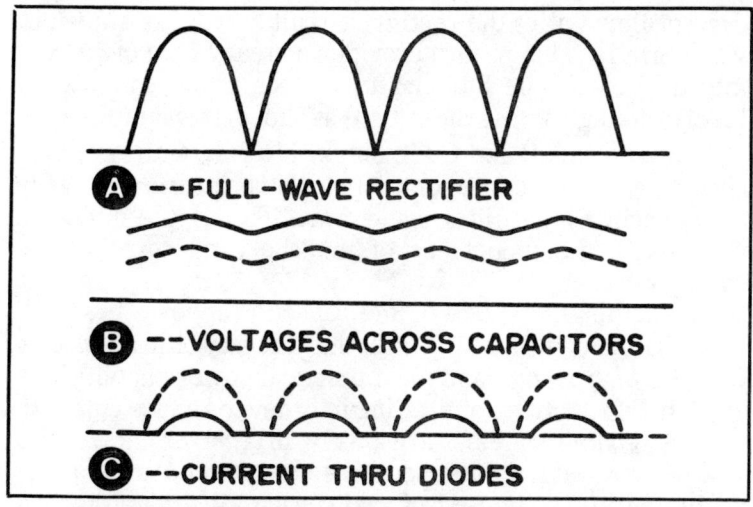

Fig. 6-6. Waveforms encountered in a capacitor-input filter with both light (solid) and heavy (dotted) load currents. Note how both voltage and current waveforms vary as the load current is changed. With zero load current and perfect capacitors, the output voltage would be pure dc equal to the peak voltage of the rectified signal; with the actual capacitors and a light load current, the output is slightly less than peak.

in waveform C. These current pulses eventually become so large as to limit the performance of the power supply.

The ac component remaining in waveform B is known as the *ripple* frequency of the power supply, and is determined by the timing between voltage peaks in waveform A. In a half-wave circuit, the ripple frequency is the same as the frequency of the input ac, while in a full-wave circuit, the ripple frequency is twice that of the ac input. Amplitude of the ripple is determined by the peak-to-peak excursions of waveform B, and so depends upon the current being drawn from a capacitor-input filter—the less the current drain, the lower the ripple amplitude!

In the capacitor-input filter, then, we have seen that current is drawn through the rectifier circuit only when the transformer voltage exceeds the capacitor voltage, and as a result flows in pulses.

The choke-input filter's waveforms are shown in Fig. 6-7. Again, waveform A merely repeats the full-wave ac waveform in the absence of any filter.

Since the inductance of the choke acts to oppose any change in current flow through the choke, it will tend to keep current flow out of the rectifier circuit steady, as shown in waveform B. This in turn provides a steady flow of current into the rest of the filter circuit, which produces a steady level of voltage across the output, as shown at waveform C.

Waveforms B and C are obtainable only with "perfect" chokes, and in practice, some ripple will be present just as in the capacitor input filter waveforms. This ripple is due to changes in the inductance of the choke as current flow through it changes.

Whether the choke or the capacitor appears first, the combination of a choke in series and a capacitor in parallel is called a *filter section* and most practical power supplies use at least two sections of filtering in order to reduce ripple to the desired low values. Occasionally, in receivers, the choke will be omitted from the second section, to produce a "pi-section" filter composed of two capacitors separated by a choke.

In addition to the chokes and capacitors, every power supply should include a bleeder resistor across its output. One of the most important purposes so far as the individual user of the equipment is concerned is safety; the bleeder provides a path for eventual discharge of the filter capacitors, so that they cannot retain their possibly lethal charge for indefinite times. This, though important for safety, is not the only reason for including the bleeder resistor.

The *electrical* purpose of the bleeder resistor is to establish a minimum load upon the power supply, which will maintain current flow through the filter circuit at or above a certain critical level. This is necessary because the inductance of the chokes in the filter varies with the current through them; by maintaining a minimum current at all times, smaller values of inductance may be used.

Whether the filter uses choke or capacitor input, both the inductance and capacitance values required are very large in comparison to those required for radio-frequency circuits. Inductance values are usually measured in henrys rather than millihenrys or microhenrys, and capacitors are in the range from 2 to 200 microfarads. As high as 100,000 microfarads are used in some solid state transmitter power supplies. These large reactive elements are necessary both because the ripple frequency to be filtered out is low (120 Hz is the highest ripple frequency normally encountered in

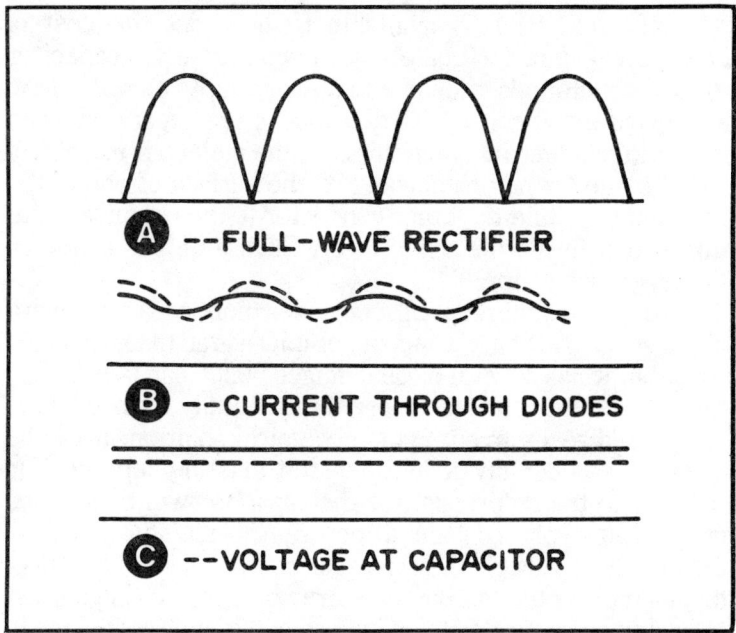

Fig. 6-7. Waveforms found in a choke-input filter are very different from those of a capacitor-input circuit. Input choke maintains current through the diodes essentially constant, so the capacitor voltage does not vary widely between light (solid) and heavy (dashed) load conditions. With a half-wave rectifier, output voltage would be only half as great as with a full-wave; this difference does not show up nearly so much with capacitor-input filters.

power supplies operating from commercial ac power), and the energy drain is high, which forces the circuit to operate at a low impedance level.

To achieve the high inductance values, *swinging* chokes are often used, particularly in choke-input filters. When a swinging choke is used, the value of the bleeder resistor becomes a critical item in filter design.

A swinging choke differs from an ordinary inductor in that it has a smaller-than-normal air gap in its core. This smaller air gap permits the choke's inductance to fluctuate as current through its winding varies; with low current, inductance is high, and as current increases, inductance drops. Typical swinging chokes vary over a 10-to-1 inductance range through their rated current range. Only enough copper and iron are necessary to provide the minimum inductance, in this design, which gets by because high inductance is necessary only at low current levels.

No such trick is available to help cut the cost of capacitors—but fortunately a special type of capacitor exists, which finds wide use in power supply circuits. Like all capacitors, it consists of two conductors separated by an insulating dielectric, but in this case the dielectric is a film of metal oxide formed chemically on the surface of one of the conductors, while the other conductor is the chemical solution which forms the oxide layer. Such a unit is called an *electrolytic* capacitor.

An electrolytic capacitor provides more capacitance in a given space and at a lower cost per microfarad than any other type, but it has several disadvantages which partially cancel this advantage. Since the capacitor is literally formed by the action of direct current upon a chemical solution, it can be used only on dc; any ac in the circuit must be kept small in relation to the dc present, or the capacitor will be shorted out. All electrolytics depend upon internal moisture for their action, even though they may be called "dry." The "dry" means merely that all the moisture stays inside, in contrast to early designs called "wet" which had grave tendencies to drip electrolyte all over everything.

In addition, the highest voltage rating seldom exceeds 500V peak, and any voltage in excess of this can cause instant punch-through. The resulting arc does two things immediately; it dries out the capacitor and vaporizes the chemical. A few milliseconds later, the former capacitor is a mess of sticky goo all over the interior of a once-clean chassis.

Unlike other capacitors, electrolytics have rather high leakage current ratings. This may be as much as 1 mA for every 4 microfarads of capacitance, but varies widely with temperature, age of the individual capacitor, operating voltage, and many other factors. The need to maintain polarization and the high leakage are the two factors which restrict electrolytics' main applications to power supply filters.

Electrolytics are manufactured in an assortment of voltage ratings from 3V up to 525V (working ratings, with surge voltages of up to 10% higher permissible), and in capacitances from 1,000,000 microfarads (yes, that's one full farad) down to a small as one microfarad. In general the high capacitance values are obtainable only at low voltage, and vice versa. Typical units found in high-voltage power supplies range from 8 to 50 microfarads, at 350 to 450V. For low-voltage supplies, typical values might be 1000 microfarads at 15V.

When voltage is first applied to an electrolytic capacitor, leakage current is very high. The current *forms* the dielectric film, however, and leakage drops rapidly. This ability to form a new dielectric gives this capacitor a self-healing characteristic in case of momentary overvoltage, provided that the overvoltage doesn't start an arc which prevents the healing action.

Electrolytics are manufactured by wrapping an aluminum foil sheet and an electrolyte-soaked cathode material together, and applying dc with carefully limited current until the dielectric film is formed. The film tends to dissolve when the capacitor is idle, but usually re-forms when power is applied unless the power surge is too great and causes overheating. For this reason, it's advisable to keep power supplies in operation intermittently when equipment is to be shut down for extended periods of time, or alternatively to carefully re-form the filter capacitors before applying full voltage again.

POWER SUPPLY PERFORMANCE RATING

Performance of a power supply, like that of any other item of electronic equipment, can be measured only by comparison to known standards. The process of making such comparisons amounts to *rating* the performance of the supply.

Some of the obvious factors involved in such a rating are the output voltage and output current available from the supply, as well as the input voltage and current required. Not so obvious, however, are some nevertheless important factors, such as *regulation* and *ripple content* of the supply's output.

Let's start the discussion by defining our terms. Then we'll be in a position to find out how to fit values into the phrases, or interpret them should we encounter a statement such as, "This circuit offers a high degree of regulation but cannot tolerate input voltage variations."

Regulation

One of the major terms—and unfortunately, one with many contradictory meanings—is *regulation*. All of its meanings deal with *changes* of output voltage from a given power supply, but that's about all they have in common. A highly regulated supply maintains fairly constant output voltage, and a supply with poor regulation has an output voltage which can be expected to vary widely—but just what makes the output vary isn't too clearly defined.

One of the meanings often attached to regulation is a measure of the change in output voltage as the input ac voltage is varied. For instance, if a supply is designed to operate with 115V ac input and produces 250V dc output under its design conditions, and still produces approximately 250V dc even when input voltage goes down to 100 or climbs to 130, that supply has good regulation in this sense.

A more commonly intended meaning for regulation deals with changes in output voltage as output current is varied, with input voltage held constant. Our well-regulated (for input variations) supply of the previous paragraph might produce its 250V dc output at a load current of 100 mA, yet drop to 200V at a current of 200 mA and climb to 300V if current drain is reduced to 50 mA. In this case, its regulation (in this sense) would be rather poor.

That's what we meant by "contradictory" meanings, since we've given an example of a well-regulated (for input variations) which is poorly regulated (for load changes).

Compounding the situation is the addition of adjectives to the word regulation, to create such phrases as *static* regulation and *dynamic* regulation.

While the phrase *dynamic* regulation appears in the FCC study list as something you are expected to be able to define, we have not been able to find any mention of this phrase in the standard engineering texts and references, such as Terman's *Electronic and Radio Engineering*, Eastman's *Fundamentals of Vacuum Tubes*, the *Radiotron Designer's Handbook*, *Electronic Designer's Handbook* by Landee, Davis, and Albrecht, or *Reference Data for Radio Engineers*, all of which have extensive sections on power supply design and measurements. Neither could we find the phrase in three separate editions of the ARRL *Radio Amateur's Handbook*, or two editions of the Editors and Engineers' *Radio Handbook*. This makes it a bit difficult to provide a guaranteed definition of the phrase!

Some authors have attempted to separate the sometimes-contradictory meanings of regulation by attaching *static* to indicate changes caused by changes in input voltage, and *dynamic* to indicate changes due to variations in current drawn. The idea is that input-voltage variations are likely to occur more slowly, and to persist longer, than are current variations caused by circuit operation, thus justifying their being called static and using dynamic for the more rapidly varying current changes.

This does not, however, remove all confusion, because the idea of a static change is in itself a contradiction. And at least a few stubborn souls have reversed these conventions as well, using dynamic to mean input-voltage changes and static for changes of load.

Another meaning possible for dynamic regulation involves the reaction of the power supply to a rapidly changing load such as that produced by a class B modulator or an SSB final. Many supplies which exhibit good regulation under conventional testing go wild under such rapidly changing loads, because they are unable to keep up with the changes. In this context, *static* regulation would be that measured by imposing various loads for relatively long times while making measurements, and *dynamic* regulation would be that shown in action under loads which were continually changing—and this is *probably* the context meant by the FCC.

While we're compounding the confusion, it must be brought out that regulation as a general property of power supply is one thing, confusing though it may be—but regulation as a factor to be measured is something else. In the U.S., engineers measure power supply regulation in percent, and the regulation percentage is defined as the ratio between the difference of unloaded and loaded voltage, and the loaded voltage, all times 100. That is, a supply which delivers 250V without load, and 225V when loaded, has a regulation percentage of (250−225)/225, times 100, or 25/225 times 100, or 11.1%.

The only thing you can be certain of when the word *regulation* is used is that changes in the output voltage of a power supply are being discussed.

Output Impedance

Hand-in-hand with regulation, but fortunately with much more precise meaning, is the term *output impedance* as applied to a power supply. The output impedance of a power supply is defined as the no-load voltage minus full-load voltage, divided by full-load current, and is expressed in ohms. If the power supply we used as an example to show the confusion possible with regulation percentage achieved that performance with a 100 mA current drain at full load, its output impedance would be 250−225 or 25V divided by 0.1A, 250 ohms.

The importance of the *output impedance* is that the power supply acts to any external circuit just like a short circuit in series with a resistor of the corresponding value. If 10 mA is drawn through a 250-ohm resistor, the resulting voltage drop is 2.5V. Similarly, if 10 mA is taken from our example supply, the voltage should drop 2.5V from its no-load value, or to 247.5V.

Output impedance is especially important with regard to both static and dynamic regulation in the final context we examined. In both cases, good regulation demands low output impedance—well below 100 ohms in most cases.

Ripple Content

Ripple frequency is a term we've already met, as is *ripple amplitude*. Ripple amplitude is usually specified as a percentage of ripple, which is the ratio of the peak-to-peak value of the ripple component only compared to the average value of the dc output voltage. A more meaningful way to rate ripple is directly in terms of peak-to-peak ripple voltage and frequency. In some cases, any ripple voltage over a microvolt or so is too much; in others as much as 25 to 30V peak-to-peak of ripple may be acceptable. It all depends on what the resulting dc is to be used for.

Usual values of ripple percentage range from 0.1 to 5%; most charts for filter design appearing in the handbooks are calculated for 5% ripple; but, if two sections of filtering are used the result will be 5% of 5%, or ¼ of 1%, ripple.

IMPROVING PERFORMANCE

Now that we have our terms defined, we can turn our attention to the performance factors to which they refer. The major factor is, as one might expect since it is surrounded by the most confusion, regulation. Ideally, a power supply should produce an output voltage which is constant, regardless of changes in either input voltage or load. This would be perfect regulation, or zero output impedance, and in general it cannot be achieved. But it *can* be attained over a surprisingly wide range of load currents, by use of some special circuitry we'll examine shortly.

Unless special regulator circuits or components are used, however, the regulation of the supply must necessarily be less than perfect. Just how much much less depends upon the entire design of the supply. A choke-input filter provides better regulation over its range of operating current than does a capacitor-input one but requires a higher-voltage transformer to achieve the same output voltage. Large filter capacitors produce better dynamic regulation than do small ones (it's difficult to get too large an output filter capacitor; 500 microfarad still leaves room for improvement on a 500V supply!) but may produce more loading of the transformer and rectifiers, and at any rate are more costly. A comparison of the output regulation for choke-input and capacitor-input filters appears as Fig. 6-8, and dynamic regulation of small and large output filter capacitors is illustrated by Fig. 6-9.

If better performance than that shown in Figs. 6-8 and 6-9 is necessary, then a fourth block must be added to Fig. 6-1—a regulator circuit.

When the utmost in performance is required from a power supply, some form of regulator circuit is usually included between the filter and the output terminals.

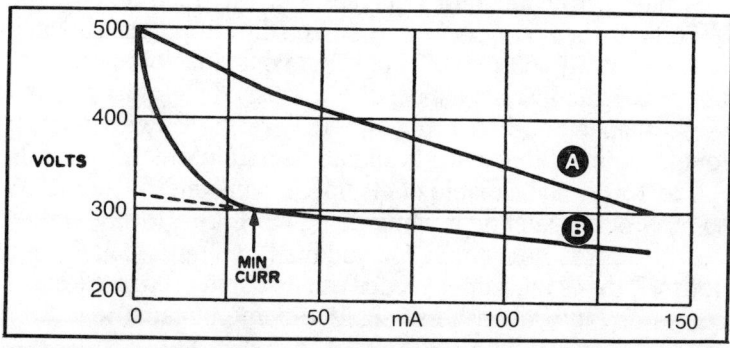

Fig. 6-8. This graph compares the output voltage at various load currents for a typical power supply using (A) capacitor input filtering and (B) choke input filtering, with all other factors held constant. Note that minimum current drain is necessary in order to pull the voltage of the choke-input filter circuit down into the regulation region, and that the capacitor-input voltage is always higher than that from the choke-input circuit.

REGULATORS

Regulator circuits may be as simple as a zener diode, or more complex than many communications receivers. They may regulate the output voltage, the output current, or both. They may guard against changes in output with changes in load, with changes in input voltage, or both. In general, a wide choice of regulators is available.

The simplest voltage regulator for many purposes is a simple neon bulb. The neon gas which provides the bulb's glow has an unusual characteristic of maintaining constant voltage, regardless of current (within limits, of course) through it. In most cases, this is about 55V. A resistor must be placed in series with the bulb, to limit the current through it to the maximum for which the bulb is rated, as shown in Fig. 6-10. Output voltage remains constant at the bulb's *maintaining voltage* from zero current drain up to the point at which the current drain through the resistor reduces voltage below the maintaining level and the bulb goes out. So long as the bulb glows, output voltage is regulated.

The familiar VR tube is simply a variation of the same principle. A mixture of argon, neon, and xenon gas is used, and the electrodes in the tube are shaped to permit higher current operation, but the circuit remains the same. VR tubes come in ratings of 75, 90, 105, and 150V, and may be series-connected to produce additional values.

The semiconductor equivalent of the neon bulb or the VR tube is the *zener diode*. Any silicon diode operated beyond its PIV rating will exhibit the same constant-voltage effect, but zener diodes are especially processed to improve their performance, and to bring the PIV rating down to lower levels. Zener diodes are available to regulate voltages from 3.9 to 150V, and capable of dissipating anywhere from 0.1W to 50W or more of heat while doing so. Like VR tubes, they can be series-connected for "oddball" voltage values, and both VR tubes and zener diodes can be connected in the same series string if need be to achieve some unusual level.

Regulation of these devices is much better than that attained from a bare power supply, but is far from perfect. Output voltage across a VR tube may vary as much as 2V from the full-load the full-load to the no-load condition, which is nearly 3% variation for a 75V tube. While this is better

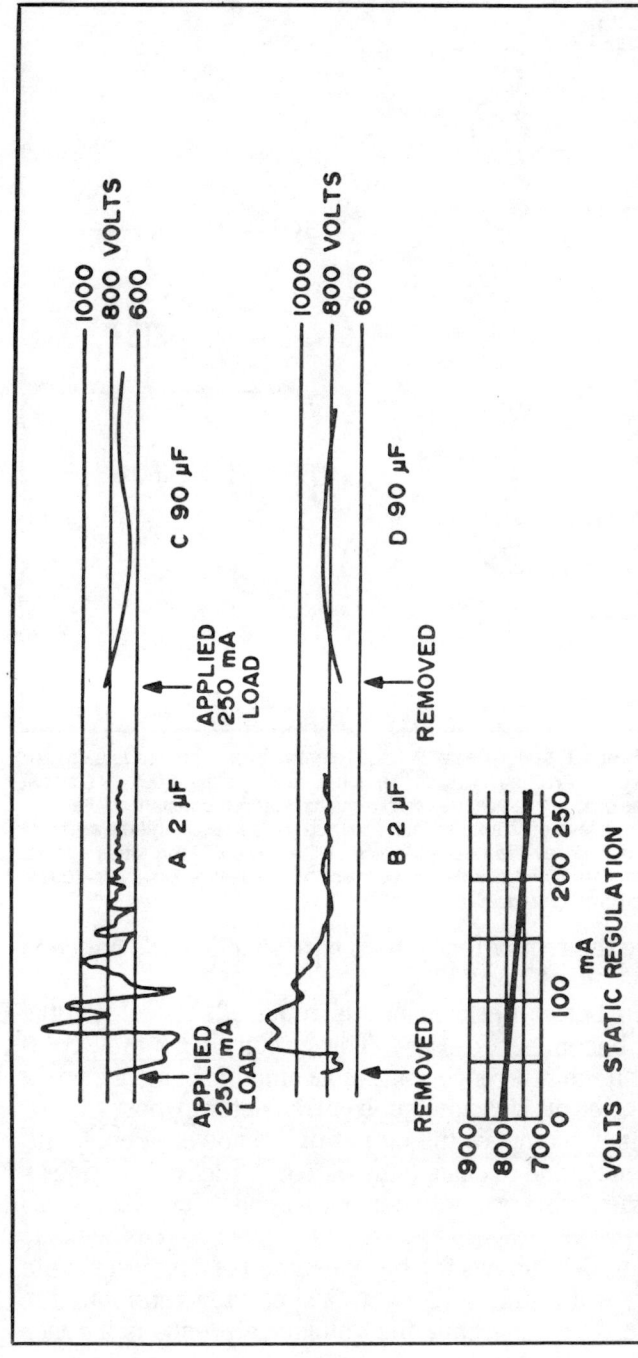

Fig. 6-9. These waveforms illustrate the meaning of *dynamic regulation* more effectively than could many, many words. The graph shows static regulation of the test supply, which remained the same for any value of output filter capacitor. Waveform A shows the transients which appeared on the power lead when full load was suddenly applied with 2 microfarad output capacitance, and B shows removal of the load. C and D are the same, application and removal of full load, but with 90 microfarad output capacitance.

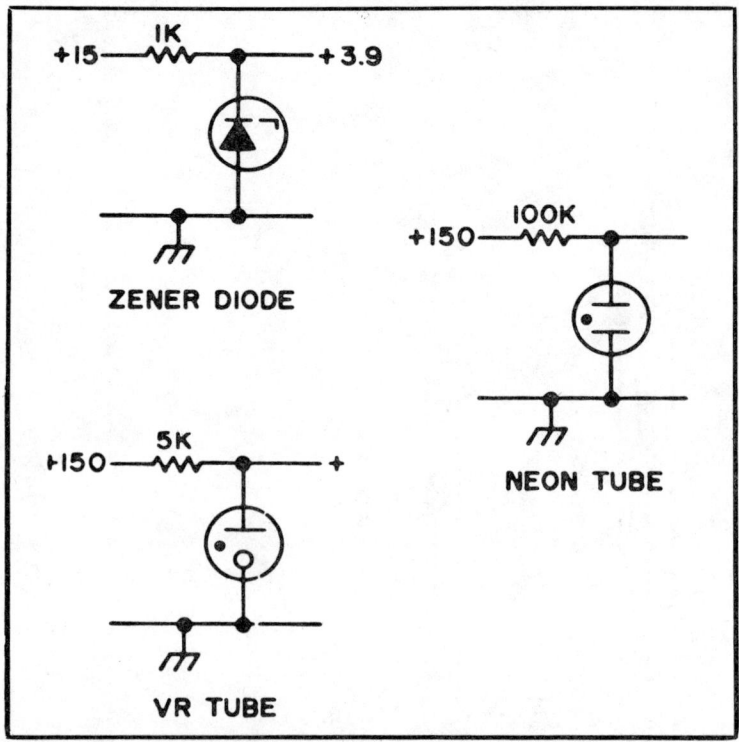

Fig. 6-10. Three different types of voltage references or simple regulators are shown here, drawn in such a way as to emphasize the similarity of their actions. Each of these produces a voltage which remains relatively constant over a wide range of current levels; so long as their minimun current requirements are met, the voltage is fixed by the characteristics of the device. Neon tubes vary in voltage, from 50 to 70V. VR tubes come in specified sizes, as do zener diodes, but the choice is much wider with zeners.

than triple the regulation of many bare supplies, it's not good enough for many purposes.

When better performance is required, the electronic regulator becomes necessary. This circuit, in general, combines a voltage reference source, an amplifier, and a control device (series pass vacuum tube or transistor in most cases), to continually compare the output of the power supply with the reference, and to change the output in such a direction as to drive the difference between output and reference toward zero.

Figure 6-11 shows the block diagram of a typical simple regulator, while Fig. 6-12 shows a simplified version of the schematic. In this circuit, the voltage reference is a zener

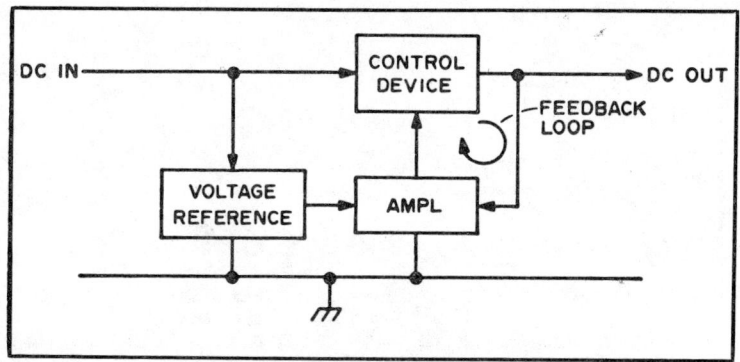

Fig. 6-11. This block diagram identifies the major functions within a typical electronically regulated power supply's regulator circuit. Input to this circuit is obtained from a conventional power supply. Output is controlled by an amplifier stage driven by the difference between the output voltage and an internal reference voltage. The result is that output voltage remains constant over a wide operating range.

diode. Its regulation is much better when used as a reference, however, than when used directly as a regulator, because the load current on the reference remains constant. When near-perfect performance is required, special voltage-reference diodes guaranteed to maintain voltage within a fraction of a percent of rated levels are available.

The control device is an ordinary power transistor. Figure 6-12 shows an NPN unit, but PNP are used in some cases. You can view the control transistor as being an emitter follower, in which the emitter voltage "follows" the voltage applied to the cases, or you can look at it as being a variable resistance which forms the upper leg of a voltage divider, with the lower leg being composed of the load connected to the power supply.

The cathode of the zener is connected directly to the base of the control transistor, so that whatever voltage appears at the base is reproduced (except for a small offset voltage established by the V_{be} voltage at the emitter, which is the output terminal of the circuit.

A regulator circuit such as this normally acts to reduce ripple in the output, as well as to clamp the voltage level constant regardless of variations in load current, because any ripple which gets through is also detected as a variation in output voltage, and is canceled just as in any other output-level change.

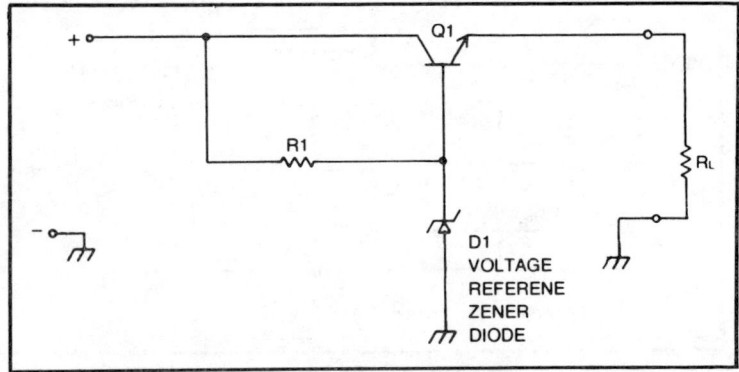

Fig. 6-12. Simplified schematic of the regulator shown in Fig. 6-11. This circuit omits such items as parasitic suppressors, bypass capacitors, etc., to emphasize those parts which do the job of regulation.

A regulator circuit of this type acts to *improve both the static and the dynamic regulation of the supply, and reduces output impedance of the supply to nearly zero.*

Figure 6-13 uses a *voltage-sampling network*, which is simply a voltage divider connected across the output of the supply, with the amplifier input connected to its midpoint. This network permits output voltage of the supply to be higher than the reference voltage, because it establishes the voltage fed to the amplifier as a fixed fraction of actual output voltage, and the regulator circuit actually clamps the voltage fed to the amplifier at the reference level.

In practice, most regulated supplies use such sampling networks, and include a potentiometer at the junction point (see inset in Fig. 6-13) to permit variations of the actual output level by permitting small changes in sampling fraction. Some supplies have a wide adjustment range, while others permit only a narrow margin.

Note that a regulated supply of this sort requires an unregulated input voltage higher than the desired output voltage. This additional voltage is necessary in order to provide operating voltage for the control device, and represents "wasted" power so far as the external load circuit is concerned. Current capability of the regulated supply is also limited by the control device.

CURRENT LIMITING

While the term *regulated power supply* usually means a supply in which the output *voltage* is regulated, it's also

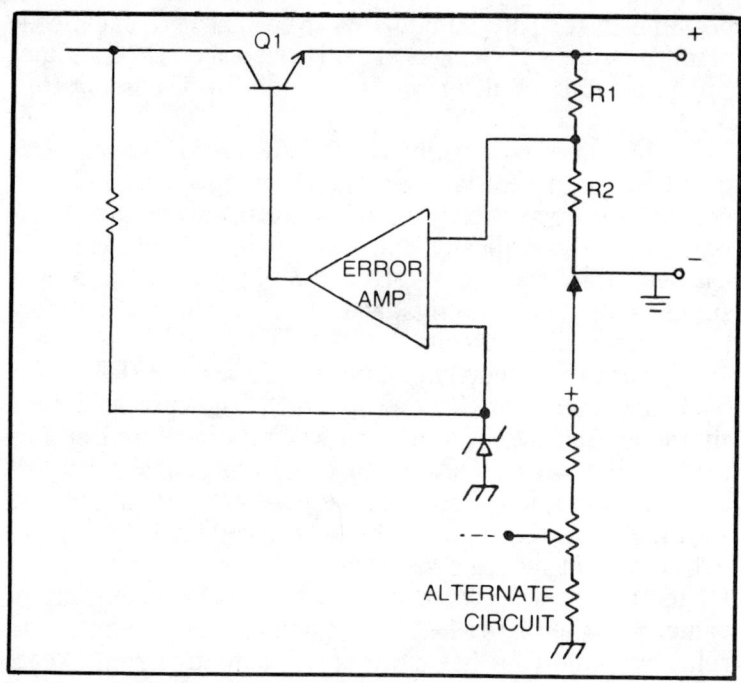

Fig. 6-13. Modifications to the regulated power supply include a voltage-sensing network which permits the output level to be higher than the reference voltage, and a series current-sensing resistor which allows the output impedance to be reduced to (or even below) zero ohms.

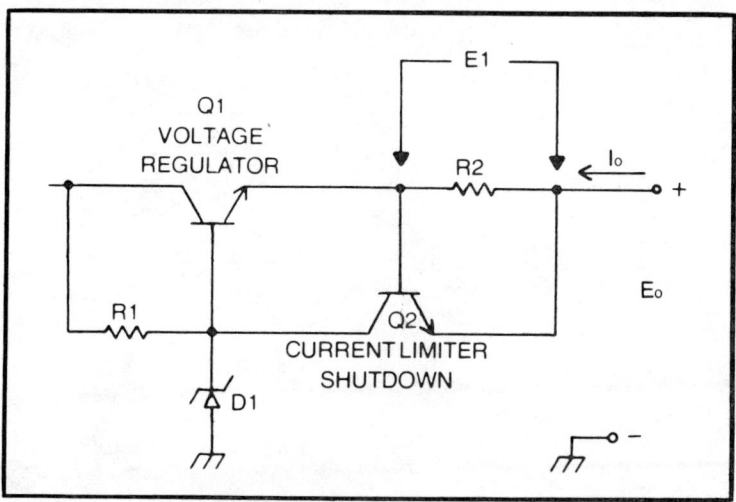

Fig. 6-14. Current regulator circuit uses two transistors, one of which is a current limiter. See text for operational details.

183

possible to regulate output *current* so that it remains constant regardless of changes in load resistance. This is sometimes called *current limiting* as we did in discussing Fig. 6-10.

The need for current regulation occurs far less frequently than that for voltage regulation, however, and so in general use, regulation has come to mean only voltage regulation unless specifically called out as applying to current. Current limiters, however, are widely used in power supplies for solid state rigs, and serve to protect the supply from output short circuits.

The simplest current limiter uses a resistor (R2 in Fig. 6-14) and transistor (Q2 in Fig. 6-14). The current limiter shutdown transistor will turn on when its base-emitter voltage exceeds 0.7 VDC. When this occurs, the collector-emitter path of Q2 shorts out the base-emitter path of Q1: shutting down the supply! The Q2 *b-e* voltage is set by the output current and the resistance of R1. The designer selects R1 to produce 0.7 volts drop at the current output limiting value. So, when I_o reaches maximium, the supply limits, and refuses to allow further increases in output current. Neat, huh?

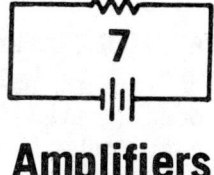

Amplifiers

By now it should be obvious that in this wacky world of electronics and ham radio, everything depends upon everything else—which makes it most difficult to name a single component as "the cornerstone" on which the whole setup rests.

But were it necessary to do so, the amplifier would be a prominent candidate for the position. Without amplifiers, neither radio nor any other application of electronics as we know them today could exist. Amplifiers are an essential ingredient of every oscillator, of every transmitter, and of every receiver, and in their absence we could expect little more performance than Marconi achieved eight decades back.

Because amplifiers are so essential, a number of questions on the General class examination are intended to test the applicant's knowledge of them, and many others which do not deal on the surface with amplifiers as such nevertheless require a knowledge of amplifier principles to answer.

In this chapter, we're going to discuss amplifiers, and cover the study list questions which deal directly with this subject. We will start with vacuum tube amplifiers, and then finish the chapter with a discussion of solid-state types.

For a start, we'll discuss what makes an amplifier an amplifier. This basic knowledge will be necessary in order to classify amplifiers into the three main classes.

Once we know what we're talking about and how they are classified, we can wind up the discussion on a more practical note with how and why amplifiers misbehave. In getting answers to this, we'll cover most of the more common types of problems encountered—but we hasten to point out that it's impossible to cover all the problems, because many are highly inprobable and so occur at such extended intervals that no one person could hope to list *all* the possible troubles with amplifiers. Fortunately, a solid knowledge of basic principles goes a long way toward helping cure these "rare diseases" of the circuits.

AMPLIFIER DEFINITION

Several acres of paper could be (and probably have been) covered with words in unsuccessful attempts to provide a detailed description of an amplifier. It gets complicated because an amplifier need not contain either tubes or transistors (although most of those you're likely to meet in amateur radio and particularly in the FCC exams do), and for that matter need not even be an electronic device!

For instance, one of the simplest imaginable "amplifiers" of mechanical force is a lever, or pry-bar. Our amplifiers are, in some ways, merely the electronic equivalent of levers.

In electronics, though, the range is a bit more limited, and we can define an amplifier in general as a circuit which increases the voltage, current, or, power level of an electrical signal. Most such amplifiers have at least two *ports* or sets of terminals, one for *input* and the other for *output*.

As we've defined it here, then, an amplifier is a device with an input and an output port (Fig. 7-1), and any signal fed into the input will appear at a higher power level at the output. This definition isn't really tight enough to get through engineering courses with, but it's good enough for everyday use.

It leads, naturally enough, to another question, though. If an amplifier boosts the level of a signal, then what, pray tell, is a *signal*? As we use the term throughout this study course, a signal is a sequence of electrical voltage, current, or power levels which, by their variation, carry some sort of information. This information may be the mere fact that the signal is present (as in a power signal), or it may be as complex as a composite video/audio TV broadcast signal.

Fig. 7-1. Any amplifier can be considered to be a black box with two ports, one for input and the other for output. Amplitude of the signal applied to the input is changed by the amplifier as the signal passes through to the output.

We sometimes define signals only in terms of *sequences of power levels* rather than in terms of voltage or current variations, because voltage and current have little significance of their own when we speak of signals because only the combination of voltage and current (or power) is meaningful. For instance, a simple transformer can double the voltage present in a circuit, but it doesn't change the power level much, and the change it does introduce is a loss rather than a gain. Therefore, a transformer is *not* an amplifier. Some genuine amplifiers, however, are current amplifiers or voltage amplifiers, but they also increase the power level.

The only types of amplifiers we'll be going into much detail about here are those which employ vacuum tubes or transistors as their *active devices* or amplifying elements, although other kinds are possible. Some of these other kinds include magnetic amplifiers, which change the coupling in special kinds of transformers, and diode amplifiers, which make use of special properties of certain special kinds of diodes. The common types, though, use tubes or transistors, and since those are the ones covered on the FCC exams, they're the ones we'll concentrate upon in this course.

In these amplifiers, the *power* characteristic of a signal which we mentioned a few lines back is of critical importance, since almost all common amplifier circuits make use of their active devices as variable resistances which valve steady new power from a power supply into the output circuit, under control of the input signal. We've already examined this action in vacuum tubes and in transistors. The only new point we're making now is that this variable-resistance property of the device is the normal situation in most practical amplifier circuits you're likely to run across.

As we saw much earlier, the relationship between voltage and current at any one point in a circuit may be described in terms of resistance or, more generally, impedance. Since

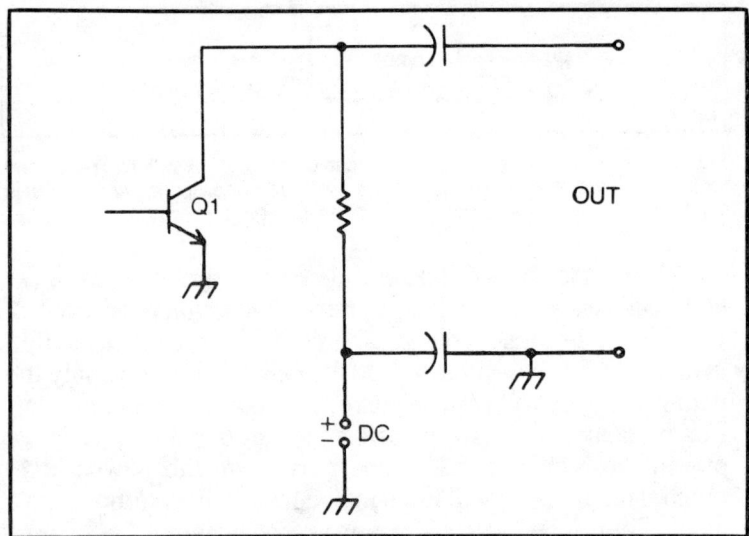

Fig. 7-2. The most common amplifier output coupling circuit at audio frequencies is the resistance-coupled version shown here. Lower capacitor is not obvious on most schematics since it is called a bypass capacitor, but it serves the purpose of taking output as a voltage drop across the load resistor.

most practical amplifiers operate with ac signals, impedance is the way it's specified. The characteristics of an amplifier's ports are usually described in terms of their impedance level and either the voltage, current, or power to be applied there. With impedance specified, voltage, current, and power are virtually interchangeable units, since the current or the power may be calculated if voltage and impedance are known, and similarly voltage may be calculated if current or power and impedance are known.

This specification is usually given in terms of the maximum signal level, and it's not at all uncommon to find an amplifier's input rated in voltage/impedance while the output is rated in watts. Many hi-fi rigs, for instance, are rated for a maximum 50W output at 16 ohms, and maximum 0.5V input at 500k. These port ratings are important, but they do not normally tell us much about the amplifier's performance. Performance of an amplifier is usually described in terms of gain and distortion, but this may vary depending upon the use to which the amplifier is to be put.

The gain may be given either in decibels, or as a ratio, so that it becomes difficult to compare amplifier ratings. A typical rf linear amplifier might be rated for 10 dB gain, input

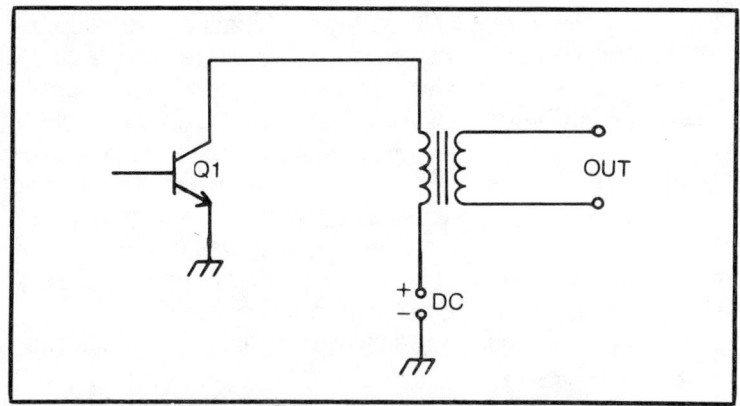

Fig. 7-3. Transformer-coupled circuit shown here is sometimes used in audio amplifiers, and often employed at rf where tuned tank circuits form the transformer. A transformer can produce the highest efficiency of all coupling circuits, but is more costly than a resistor and two capacitors.

and output impedances of 52 ohms, and require 100W input for maximum output. Such an amplifier would produce 1 kW output (10 dB gain with input of 100W). A hi-fi rig, on the other hand, might completely neglect to mention gain and merely give output and input levels together with distortion percentages.

Regardless of ratings, any normal amplifier operates by regulating the flow of current in its output circuit. If the amplifier is intended for use as a voltage amplifier (one in which the power gain is used to step up the signal voltage rather than signal current), the resulting variations in output current flow are converted back into voltage variations by means of a coupling device which may be a transformer, a resistance, or an impedance. Most of today's amplifiers use resistance coupling (Fig. 7-2); as the current through a resistor in series with the active device varies, so does the voltage drop across that resistor. This varying voltage drop is coupled out through a capacitor (to isolate the dc component of the signal) to the output port.

With transformer coupling (Fig. 7-3) the active device's operating current is supplied through the primary of a transformer. As this current varies, the changes induce a corresponding current in the transformer secondary, which is connected to the output port.

Resistance coupling is normally an inherently high-impedance, although certain special circuits can bring the impedance level down to low levels. Transformer coupling, on the other hand, seldom produces extremely high impedance levels. It is used primarily to produce medium-to-low-impedance ports, as for instance in the output circuit of an audio amplifier where the active device's current must drive a 16-ohm load. Rf amplifiers often use transformer coupling because an rf transformer can be composed of tuned circuits for selectivity; sometimes they use impedance coupling (Fig. 7-4) which is similar to resistance coupling except that an rf choke (inductor) replaces the resistor.

All of these coupling schemes are used with both tubes and transistors.

As you can see, amplifiers operate in many different ways, even though their basic principles are all similar. Because of this, it's been found necessary to classify amplifiers into various types for study and discussion. Many classification schemes are in use, and some of them are so widely accepted that they are part of the FCC examinations.

AMPLIFIER CLASSIFICATIONS

Just as we have many kinds of amplifiers, we have many kinds of amplifier classifications, because amplifiers are classified into groups, each of which has some character or property in common.

We finally restricted our definition of amplifier, for this chapter at least, to "a vacuum-tube or transistor device which boosts the power of its input signal"—and that in itself was a classification.

Within this classification, we could describe amplifiers according to their output coupling circuit. This would produce classifications such as *resistance-coupled*, transformer-coupled, and impedance-coupled, which we were using only a few paragraphs back.

Another way would be on the basis of the frequency range handled by the amplifier: audio, video, rf, i-f, dc, etc.

We could divide them into voltage amplifiers, in which the power boost shows up as increased voltage with no decrease in current, and power amplifiers, in which the power boost is used directly.

All of these classifications are in wide use, but none of them addresses the problems of amplifier operating conditions directly. Thus, an rf power amplifier may be linear, or it

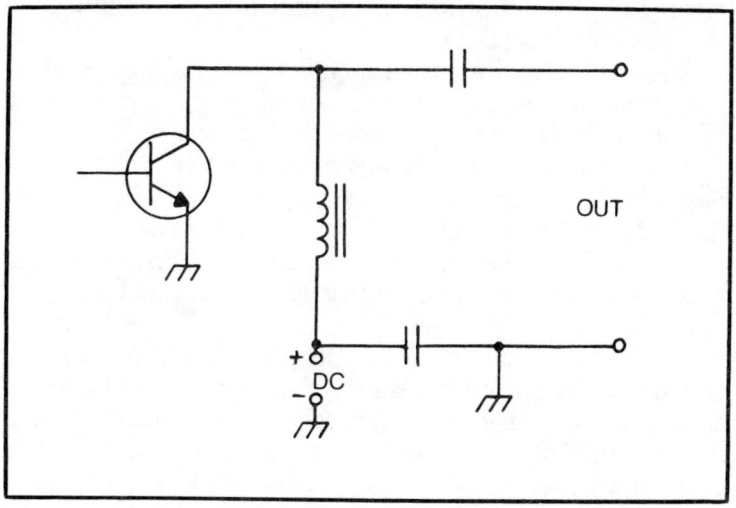

Fig. 7-4. Impedance-coupled circuit shown here is a cross between transformer and resistance coupling. The ac output signal sees a high impedance at the choke, and capacitors couple out the resulting voltage drop.

may produce distortion, and these classifications will not help us determine why it acts as it does. To handle such needs, we must have a classification system which is based upon operating conditions—and one exists in an almost universally standard form.

Unfortunately from our point of view, one authority defines these classes in a manner which is significantly different from the definitions agreed upon by most other authorities—and most hams have learned the classes from the one unique authority. This leads to an inordinate amount of confusion, disagreement, and downright unpleasantness at times. The best way to avoid this trap is to become familiar with both sets of definitions, and to know that two different sets exist; then you can pick your set according to the other fellow's rules, and discuss theory freely.

Incidentally, the key differences between the two sets of definitions do not appear in the FCC study-list questions, so they should pose no problems during the actual examination.

The classification system of which we speak divides all amplifiers into three broad classes, called *class A, class B* and *class C* (Fig. 7-5), and goes on to define an overlap group called *class AB*, which is a twilight zone between class A and class B. The definitions are published by engineering societies the world over.

A class A amplifier is an amplifier in which the operating conditions are such that plate or collector current in a specific tube or transistor flows at all times.

A class B amplifier is an amplifier in which the grid or base bias is approximately equal to the cutoff value so that the plate or collector current is approximately zero when no input signal is applied, and plate or collector current flows for approximately one-half of each cycle when input signal is applied.

A class C amplifier is an amplifier in which operating conditions are such that plate or collector current flows for appreciably less than one-half of each cycle when input signal is applied and plate or collector current is zero in the absence of input signal.

A class AB amplifier is an amplifier in which operating conditions are such that plate or collector current flows for appreciably more than half but less than the entire electrical cycle of the input signal.

We'll look at these definitions in much more detail shortly. First, however, let's look at the conflicting set of definitions for classes A, AB, B, and C, as published in the widely circulated handbook for radio amateurs published by the ARRL:

"A class A amplifier is one operated so that the wave shape of the output voltage is the same as that of the signal voltage applied to the grid . . .

"A class AB amplifier is a push-pull amplifier with higher bias than would be normal for pure class A operation, but less than the cutoff bias required for class B . . ."

(Class B operation is defined by means of a schematic diagram and four waveforms.)

"A radio-frequency power amplifier . . . can be used with an operating angle of less than 180 degrees. This is called class C operation."

It may appear that we're overemphasizing the differences in these sets of definitions, but in nearly 20 years of listening to on-the-air bull sessions get downright acrimonious simply because two people attempting to discuss amplifier operation didn't mean the same thing at all by class A or class AB or class C, we feel that the existence of multiple definitions is an essential fact to know.

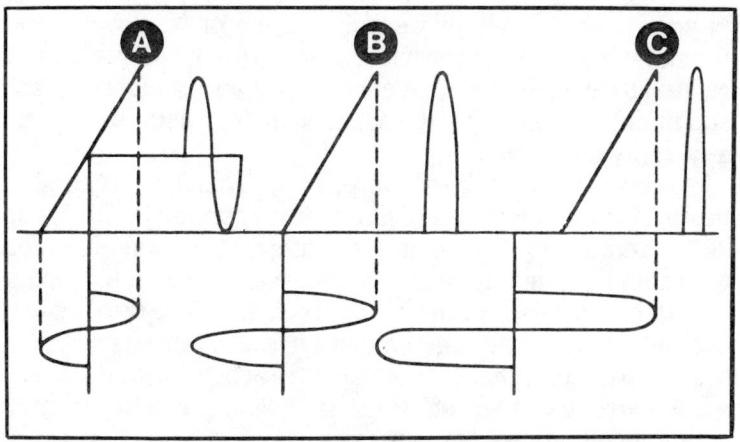

Fig. 7-5. Comparison of class A, class B, and class C amplifier operating characteristics is shown here for "ideal" amplifiers having perfectly linear characteristics across a limited range. The diagonal line represents the amplifier's action. Vertical lines indicate the grid bias setting of the amplifier; in class A, bias splits the operating range in half, while in class B only the positive-going half of the input signal is amplified, and in call C only a part of the positive-going half makes it through. No actual amplifier has such a straight characteristic, but this illustrates the differences in the classifications.

The major significance of the differences is that the engineers' version is based *entirely* upon an operating characteristic which can be precisely measured and specified; the flow of plate or collector current in the amplifier throughout one cycle of an input signal determines that amplifier's operating class. If plate or collector current always flows, it's class A. If current is cut off for part of the cycle, but less than half, it's AB. If cutoff is for approximately half a cycle, it's B. And if cutoff is appreciably more than half a cycle, it's class C.

The *amateur* definition, on the other hand, defines each different class by means of a different characteristic. It says that a class A amplifier is one which is free of distortion (that may not be what was intended, but it's precisely the meaning of the words used in the definition, and it's the way most people read them if you listen to the resulting arguments), while a class AB amplifier must be push-pull, and class C applied to rf power circuits.

Obviously, we're somewhat biased in favor of the engineers' definitions. They appear to be much more precise, and should you happen to be interested in radio engineering from a career viewpoint, they'll do professionally as well as in your hobby. Now let's examine them more closely, and then see how the somewhat looser amateur definitions are related to those of the engineers—because in a most general and highly oversimplified way, the amateur definitions usually apply also.

The class A amplifier concept was originally intended to be that of an "ideal" amplifier which reproduced its input signal at the output, without distorting it in any way. In order to accomplish this, it would be necessary for gain to remain constant regardless of the signal level, and plate or collector current could never cut off during the input-signal cycle (because if it did, that part of the cycle appearing at the input while current was cut off would fail to appear in the output, which would constitute distortion).

The engineers' definition of class A stems directly from one of the requirements of the "amateur definition"—but the reverse is not true. An amplifier can meet the engineers' definition even if plate current varies widely throughout the signal cycle, just so long as it never cuts off. With a wide variation of plate current, gain will also vary between wide limits, and the output signal will be distorted. In this case, the output signal will *not* have the exact same wave shape as the input, so the amateur definition is not fulfilled.

The class B amplifier represents an attempt to improve the efficiency of amplifier circuits, and dates from the days when radios all operated from batteries (before ac power supplies were widely available). A class A power amplifier cannot develop much output power in comparison with the power it takes from the power supply, since plate or collector current must flow even when no signal is present in order to remain class A, and this plate current is wasted so far as developing output is concerned. As a result, class A amplifiers normally produce a maximum of about 1W out for every 4W drawn from the power supply.

By drawing power only half of the time and leaving the tube transistor cut off for the other half of the input cycle, less power is wasted. On the other hand, with only half of the input signal appearing in the output, distortion of the amplified signal is extreme. Use of two tubes or transistors in a push-pull circuit reduces the distortion to acceptable limits, because while one is cut off the other is providing output and vice versa—and the push-pull connection is necessary in order to use class B operating conditions in audio amplifiers.

In an rf power amplifier, however, distortion of the individual cycles of rf does not matter because the tuned circuits have a "flywheel" effect which irons out the distortion. Distortion in rf power amplifiers usually means a distortion of the signal *envelope* or modulation, and this is not affected by the class B circuit. Therefore, it's possible to use single-tube class B amplifiers in transmitters, and some *linear* amplifiers operate this way.

The idea class B circuit is about twice as efficient as the same tubes would be if operated class A; that is, it provides up to 2W out for every 4W taken from the power supply. This means that batteries for such a circuit would last twice as long.

The class C operating conditions are just like class B, only more so. One textbook describes them, very accurately, as *switching* operation. The tube or transistor is operated with bias so far into cutoff that plate or collector current cannot flow except at the very peak of the input signal cycle. Thus, this class of amplifier draws very little current most of the time. On signal peaks, the tube or transistor switches from off to on and permits a brief pulse of current to flow. Since it is essentially a switch, this current pulse can have extremely high values—a full ampere isn't uncommon. The high-energy pulse causes the tuned rf output circuit to *ring* at its resonant frequency, and with another pulse coming along at the peak of every cycle, most respectable power levels can be developed.

The class C amplifier can theoretically produce as much as 3.8W of output for every 4W taken from the power supply, but practical circuits seldom deliver more than 3 for every 4. This is still three times as efficient as the same tube would be in class A, and since amateur power limits are set on the basis of input power rather than output power, it means you can get three times as much signal with class C.

Because of the extreme distortion and resulting need for tuned circuits, class C amplifiers are normally used only for rf power amplifiers.

In addition to the three major operating classifications of A, B, and C (class AB is considered to be a cross between A and B), another operational classification is used to indicate whether grid or base current flows in the amplifier. This classification consists of a numeric suffix applied to the classification letter; a 1 indicates that grid current does not flow, while a 2 means that it does.

Normally, class A amplifiers are A_1, and class C amplifiers are C_2. Most class B amplifiers are B_2. The only place the number has great meaning is with class AB, where about as many circuits operate AB_1 as do AB_2. However, the presence or absence of grid current does not necessarily go with the operating class in all cases. It's quite possible to design a class A_2 amplifier, which permits grid current to flow and yet operates class A, or a C_1, which operates class C without any grid current.

At this point it might be well to point out another basis of classification often used, that into *linear* and *nonlinear* amplifiers (Fig. 7-6). A linear amplifier is one which amplifies its input signal without introducing any distortion, while a nonlinear one does introduce distortion of some type. Since nothing is perfect, all actual amplifiers are nonlinear to some degree, and therefore the term *linear* is always used in a comparative sense. A good hi-fi amplifier is linear, for instance, even though it does have some small percent distortion. In speaking of rf amplifiers for modulated signals, the distortion is measured with respect to the modulation rather than the individual rf cycles, and so a linear rf amplifier is one which does not distort the modulation rather than one which does not distort the individual cycles of the input signal.

The linear/nonlinear classification helps us to list the ways in which class A, B, and C amplifiers are used. For instance, almost all linear audio amplifiers operate in class A, which squares with the amateur definition of class A as being a linear amplifier. Many rf linears at low power levels also operate class A, and in general the class A operation is used almost exclusively when very low distortion is required. Most receiver circuits are Class A.

Class AB and B amplifiers are widely used to produce moderate power levels with limited distortion. The output stages of many public-address amplifiers and almost all hi-fi power stages operate in class AB, and almost all SSB final linear amplifiers run as either class AB or B. In addition, class B amplifiers are sometimes used in battery-operated

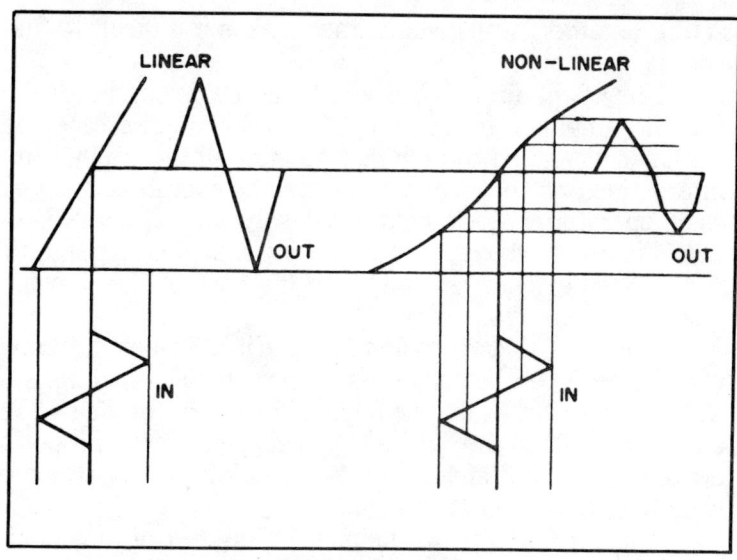

Fig. 7-6. Here we have linear and nonlinear amplifiers compared, with triangular signal waveforms instead of sine waves to bring out the differences which are not obvious with sine waves. A linear amplifier, left, faithfully reproduces the input signal at the output. Nonlinear amplifier, right, puts bends into the straight sides of the waveform. These bends represent second-harmonic distortion. In general, any distortion is described as nonlinear operation.

equipment to conserve battery power; many transistorized portable radios use class B output stages for this reason.

Class C operation is confined almost exclusively to rf power amplifiers which can be nonlinear; this means, in practice, all rf power stages in a CW or FM transmitter, as well as in an AM transmitter which uses high-level modulation (if modulation is applied before the final output stage, all the following stages must be linear, requiring A, AB, or B operation). Class C is hardly ever used in SSB transmitters, since the nonlinearity would destroy the modulation of the signal.

AMPLIFIER MALFUNCTIONS

Amplifiers, like all other physical objects, faithfully follow Murphy's First Law that says, "If anything can possibly go wrong it will." In consequence, they continually misbehave.

However, any specific case of "wrong-going" may be due to not just one but several of the possible causes all acting together, which sometimes makes it difficult to find and fix the trouble.

One of the first things which can go wrong is for the specific amplifier to be badly designed; in this case the designer has failed to take into account all the tricky little interactions which can affect the circuit's operation, and you as an operator are just about hopelessly lost. The cure is to forget it, and start over with a properly designed amplifier (if you're the designer as well, then it's back to the drawing board with our sympathy).

Since this kind of problem, though it's the most common in the case of new or one-of-a-kind circuits, is not within the range of the FCC exam, we'll ignore it the rest of the way in and assume that the misbehaving amplifier was properly designed and built and did, in fact, work perfectly at some time before the troubles began.

Another frequent wrong-goer is the user of the equipment; in this case nothing at all is wrong with the amplifier itself, but it's being used under conditions which never were intended by the designer. This is the most frequent cause of amplifier misbehavior—and we'll get back to it in much more detail a little later.

The final frequent cause of trouble is failure of one or more components in the circuit. A blown tube or transistor is relatively easy to find and fix—but a bypass capacitor which has changed in value (because of age or overheating) just enough so that it no longer does its job properly may provide more than its share of hair-pulling before it's detected.

From the standpoint of the operator, the only one of these three probable causes of problems which he can do anything about is the second—misuse of the equipment. The cure for the first is to redesign the circuit, and for the third is to locate and repair the defective part. While both of these are legitimate activities for a radio amateur, neither of them is involved directly with the *operating* of the equipment.

Because of this, and also because misuse accounts for most of the amplifier problems and bad signals in existence at any one instant, we'll concentrate on just the one problem.

Before we can talk about misuse of an amplifier, we must know what the proper use for that amplifier is. For instance, the purpose of a linear power amplifier used to bring a single-sideband signal from 10W up to 500W is defined, and "proper use" of this particular rig would be that use which accomplishes the power amplification without distortion and unwanted spurious output signals.

In general, the amplifier's name (such as linear, power, etc.) together with the designer's specifications for input and output levels will tell you all you need to know about the circuit's proper use amounts to "that use which accomplishes the purpose of the circuit."

Misuse, then can be defined as any use which defeats the circuit's intended purpose. In most cases, this boils down to one or more of three major mistakes:

1. The amplifier is adjusted to operating conditions such as bias or voltage levels which are outside the range anticipated by the designer.

2. Input signals are applied at levels either lower or higher than the range intended by the designer.

3. More output is attempted than the designer intended the circuit to produce.

Let's see how these three mistakes affect several typical amplifier circuits. For starters, let's see what happens when a class A amplifier intended for low-distortion amplification of an audio signal is misused. Such an amplifier might be in the speech-amplifier chain of any phone rig, or in a receiver.

If the amplifier's bias can be adjusted, and is set to a value higher than the designer intended, the stage will draw less current than it was designed for, and so cannot develop its rated power output (or voltage output, if it's a voltage amplifier).

Similarly, if the bias is set too low, a normal input signal may overcome the bias and change the operation condition from class A to class AB. The effect of bias adjustment, then, is to reduce the allowable input and output signals—and if the value is very far from the intended level, excessive distortion may be introduced into all signals (Fig. 7-7).

However, in most class A amplifiers, the bias level is not adjustable by the operator, so this particular example won't be met often in practice.

If the input signal is out of the range intended by the designer, other bad things happen. If the input signal is too weak, the probable effect would be simply low output, possibly contaminated with noise, but most class A amplifiers are

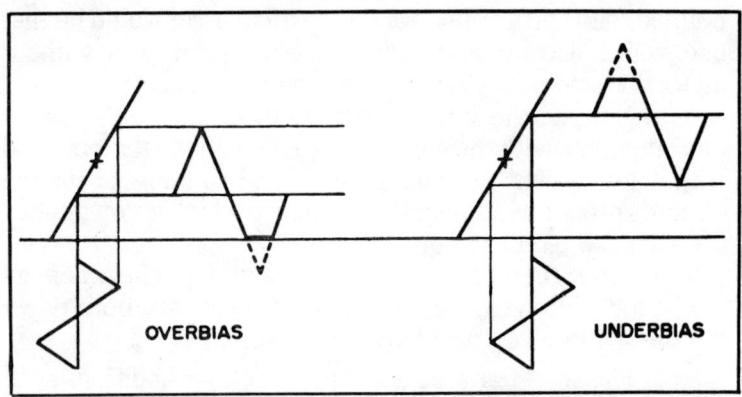

Fig. 7-7. Shown here are the effects of overbias and underbias on a class A amplifier, with a triangular input signal of maximum rated level which could be reproduced linearly were the bias proper as indicated by the "X" in the center of each amplifier transfer line. Overbias causes the negative-going peaks of the signal to "bottom" against the cutoff point, which is actually class AB operation, while underbias causes the amplifier to saturate and flat-top the signal.

intended to work with signals which range right down to the noise level anyway, so the designer has probably intended his circuit to work with input signals approaching zero volts.

Too much input signal, on the other hand, produces all sorts of ill effects, and you can find horrible examples all over any of the bands.

When input signal level in a class A amplifier is too high, either or both of two problems arise (Fig. 7-8). The positive-going peaks of the input signal completely overcome the amplifier's operating bias, which permits grid current to flow. If the designer has intended grid current, and allowed for it, this is not in itself bad—but most class A amplifiers are not meant to be used this way, and so grid current in 99 out of 100 of them indicates excessive input signal. When grid current flows, the input signal is distorted before it ever reaches the active device, and the distortion can't be removed later. Even if grid current does not flow, it's possible to saturate the active device so that it no longer changes in resistance as the input signal changes, and this too leads to distortion.

Meanwhile, the negative-going peaks of the input signal add to the operating bias, and may add to it enough to change the operating conditions to class AB instead of class A. This also produces distortion in the output signal.

An oscilloscope will show you which of these two problems is present in an amplifier. If the positive peaks are causing the problem, the voltage waveform of the output signal will tend to be flattened off at its *negative* peak; if it's the negative peak of the input signal doing the dirty work, the output signal will be flattened at its positive peak.

In either case, the cure is simple: reduce the level of the input signal. The normal control for doing so is the audio gain control on the transmitter panel, or the volume control on the receiver. If your transmitter has no audio gain control (and many do not) you can either speak a little more softly, or connect an attenuator pad (a sort of resistive voltage divider) between the microphone and the amplifier.

What about the output conditions? In our class A amplifier, which is normally used to provide voltage gain, this isn't usually adjustable. If it were, taking too little power

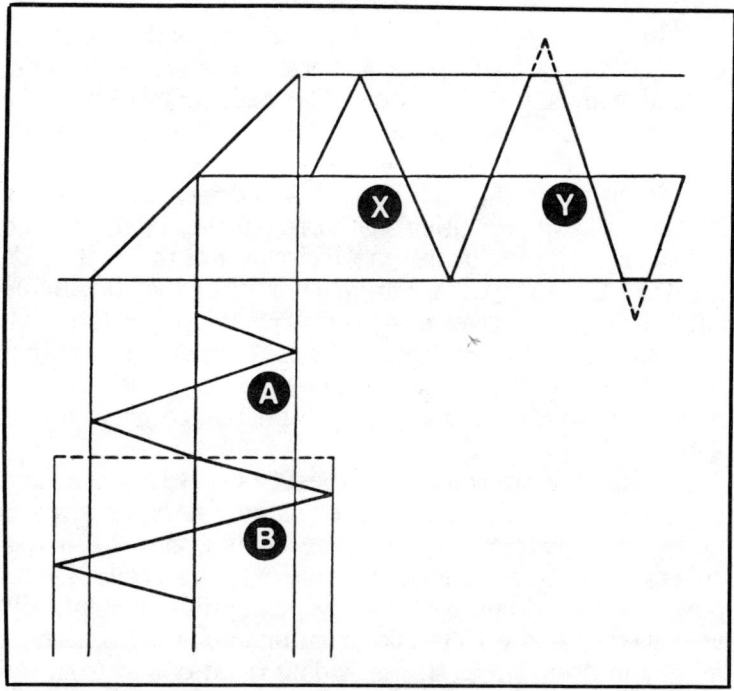

Fig. 7-8. In this example the bias is properly set but the input signal is excessive, Signal A is at maximum level, and produces linear output X. Signal B, at too high a level, runs outside the linear range and produces distorted output Y with both bottoming and flat-topping evident.

out would tend to produce the same symptoms as putting in too much input signal, while attempting to get out too much power would show up as heating of all the amplifier components in addition to distortion.

Now let's see what happens when we misuse a class C rf power amplifier. For openers this time, we'll assume that it's the final output stage of a CW transmitter.

In many class C amplifiers using fixed bias, the bias level can be adjusted by the operator. Too low a bias level will result in low power output, and too high a bias level will generate excessive harmonic output. Most vacuum tube class C circuits, however, use a self-bias arrangement in which the operating grid bias is developed by the flow of grid current through a *grid leak* resistor; with these, both the bias level and the input signal are set by the same control, which may be called the *drive* adjustment, or the driver tuning. Transistor class C amplifiers use a series resistor to self-bias.

Input signal level is not as critical with a class C amplifier as it is with class A, because we are not so concerned with signal distortion. The main thing is to have enough, without overdriving the circuit, and the designer usually specifies a broad range of operating conditions.

In circuits using self-bias, loss of drive causes loss of bias as well, with resulting high current in the amplifier stage and damage to components in the circuit. For this reason it's good practice to adjust the drive level (input signal) with the amplifier input voltages turned off. If input signal is too small (too little grid current), bias will be too low and power output will be low. If it's too large (excessive grid current) harmonics will be excessive and the final-stage grid may be damaged.

With a CW transmitter, it's best to start with the amount of drive signal the designer recommends. After drive is developed, plate voltage may be applied and the output adjustments made, and then drive may be reduced until power output drops by a barely perceptible amount, and advanced by some 15 to 20% from that point. This assures minimum drive while still providing rated output from the amplifier, and will probably result in an input signal level within the designer's intentions.

For an AM transmitter, this adjustment of drive and loading may not be adequate. A modulated amplifier requires enough drive to supply the modulation peaks as well as the carrier level, but we'll go into that later when we examine the modulation process.

Output signal level adjustment in a class C amplifier is the most critical adjustment of them all. It should not be attempted until the input signal level has been brought into the correct range, because the tube or transistor may be destroyed in the absence of input signal.

Most transmitters have two adjustments for output tuning, one marked "tuning" and the other marked "loading." The normal practice is to set the loading control for minimum output, then rapidly adjust the tuning control until plate current dips sharply. This dip indicates that the output tank circuit is tuned to resonance, and is acting as a high impedance.

While this is all right for tuneup, which normally is done rapidly, it can damage the equipment if extended operation is attempted with too-little power being taken out. Whether you use it or not, a 500W amplifier is developing its 500W worth of current and voltage swing, and if you take only 50 of them out, the rest are going to be looking for mischief inside the amplifier. Tuning coils may overheat and melt their plastic supports, or capacitors may arc over.

To take more power out, the loading control is adjusted to increase plate current, meanwhile readjusting the tuning control to keep the dip at its minimum value. The readjustment is necessary because the two controls interact; increasing the loading reduces the Q of the tank circuit, and changes the reactance at the same time. This process should be continued until the amplifier is producing its rated output, and then halted.

It's possible to load most class C amplifiers until the dip disappears, but if you check with an indicating wattmeter you'll find that pretty shortly after rated power output is reached, the output power begins to fall off even though the indicated input power keeps climbing. The extra power simply goes into the amplifier to be dissipated as heat, and can damage the components.

Notice that while distortion was the key result when a class A amplifier was misused, equipment damage is the key result in class C operation. The tipoff to problems in class C is excessive plate current. If excessive plate current can be controlled by the tuning and loading controls, it probably indicates an attempt to get too much power out of the rig. If not, it probably indicates loss of operating bias, which can be due to loss of input signal.

Of course, a badly designed rig can draw too much plate current for other reasons. If the amplifier is subject to oscillation, either self-oscillation near the operating frequency or parasitics at far removed spots in the spectrum, this can cause excessive current. Similarly, component failure which removes the input signal can cause it. But the most common cause is simply trying to get something for nothing.

How about the "linear" used with an SSB transmitter? These rigs have a reputation for being tricky, but actually they're little if any more complex than a class C rig. The major difference, in fact, is that their designers set them up to operate in class AB, and to produce as little distortion as possible between input and output.

Since the bias level is the major factor controlling which class a specific amplifier is operating in, the bias of a linear amplifier must be set to the level specified by the designer. Most such amplifiers have a bias control accessible to the operator, which is intended to be adjusted for some specified value of plate current with no input signal applied. If bias is too low (too much plate current), the power capabilities of the amplifier will be reduced, and if it is too high, distortion will increase.

The input signal, also, is critical—but only at the upper end of the range. Linears are designed to accept input signals down to zero, but the maximum input signal level must not be exceeded if operation is to remain linear. Too much input signal (overdrive) causes distortion just as it does in a class A amplifier, with flattened peaks. In an rf amplifier, this distortion shows up as illegal harmonics outside the ham bands, as well as *splatter* and *buckshot* which may be within the bands but is not within the channel you are using. Such spurious signals are prohibited by FCC rules. The distortion produced by overdriving a linear amplifier is sometimes called *flat-topping* the signal, but overdrive is not the only cause of flat-topped signals.

Misadjustment of the output level can also cause flat-topping and other distortion problems. A linear amplifier is more critical in its output adjustments than in any other. In order to develop rated power output, the active device (tube or transistor) must work into its rated load impedance, and this impedance is set by the output tuning and loading controls.

If the adjustments are set for too little loading (less than rated output power), the load impedance will be too high and distortion will be produced. Similarly, if the amplifier is too heavily loaded, distortion will also result. Too heavy loading produces less distortion than too light loading, however, so in case of doubt it's best to err on the side of too much loading. This should not be confused with too much output—overdriving in an attempt to get more than rated power out is a sure way to generate a bad signal.

The distortion produced by underloading shows up as *peak clipping* which is a form of flat-topping. That resulting from overloading is *intermodulation* which can put the unwanted sideband back into the signal. To guard against such problems, many operators tune up linear amplifiers with the aid of an oscilloscope, which can show them the proper combination of all input and output level adjustments.

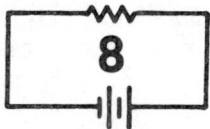

8
Transmitters and Their Operation

The usual reason anyone wants a ham ticket is so that he or she can legally operate a radio transmitter. After all, no license is required to run a receiver, as thousands of shortwave listeners can testify—but they're missing much of the fun!

Since this *is* the normal reason, it's only natural to expect that the FCC examinations for all grades of amateur licenses would include several questions dealing with the theory and operation of transmitters. The expectation is correct, and the General exam is no exception.

For starters, we should define precisely what a transmitter is. This will get us off the ground, putting some of the bits and pieces of knowledge we've picked up in the preceding chapters of this study course into something resembling an actual operating gadget. We'll discover that a transmitter includes, among other things, an oscillator and an amplifier. Logical follow-on subjects, then, are oscillator operation, and amplifier characteristics. By the time we've explored these, we'll be fairly solid on the theory end of transmitters, but there's still something to learn about rules, so our final subject for consideration will be power measurement.

While the entire emphasis in this chapter is on the applications of the circuits we are examining to transmitters, many of the circuits are also used in receivers, and when we get around to receiver theory we'll be referring back to this section.

TRANSMITTER OPERATION

In the most general sense of the word, a radio transmitter is any device which transmits radio energy. The first transmitters used spark gaps to generate this energy, and Marconi spanned the Atlantic with the output of a spark gap transmitter. We've still got some of them around, too, although they're not legal for deliberate use as transmitters (and haven't been for nearly 50 years)—we call them *automobiles*, and their rf output we call *ignition noise*. In fact, many early amateurs used salvaged Ford Model-T spark coils for their amateur transmitter!

However, when we speak of a radio transmitter these days we mean a collection of circuit and apparatus which is used to generate and transmit radio energy—hopefully in accordance with federal and international rules and regulations. Almost all transmitters in action today have five major functional portions, although they may differ in most of their details.

These five major parts of any modern radio transmitter are an oscillator, an amplifier, a power supply, control circuits, and a modulator.

The oscillator controls the radio frequency upon which the transmitter operates, by generating rf energy at some specific frequency. The amplifier boosts the power level of the oscillator's output up to the desired level to be fed to the antenna, and in many cases changes the frequency in some specified manner while doing so. The power supply furnishes operating power to the rest of the components. The control circuits permit the transmitter to be turned on and off, and otherwise to be controlled during operation. The modulator permits the transmitter to carry information; it's a subject unto itself, and we'll look at it next time around.

Figure 8-1 shows these major components of the typical transmitter in block-diagram form. A small portable transmitter such as those found in hand-held portable (Part 15) transceivers may have only one transistor in the oscillator, another in the amplifier, use a 9V battery as the power supply, have only an on-off switch for control, and use the receiver audio section as a modulator, but all five of the major functions are there. Similarly, a 50 kW broadcast transmitter may use three or four tubes in its oscillator, a dozen or more in the amplifier, have a power supply which takes up the best part of a good-sized room, make use of elaborate control circuits, and have a modulator bigger than most entire ham rigs. Still, only the five major functions are present.

Since the five major functions are present in all transmitters, from the smallest to the largest, and since no transmitter includes circuits *not* represented by the five major functions, it follows that if we clearly understand the workings of these five functions we must have a clear knowledge of transmitter theory.

We have already examined power supplies in general, and there's not much to add that's unique to the power supplies used in transmitters.

In general, transmitters require larger power supplies than do receivers, because a receiver needs to produce only a few watts of audio power as output (less than that if headphones are used) while transmitters must produce up to several hundred watts of rf output power. Since no transmitter is 100% efficient, production of a 700W output signal (about the best that can be achieved within the ham power limit) requires around 2 kW of dc from the power supplies. The exact ratio between rf output power and dc input power depends largely upon the type of modulation used and is not really important at this stage anyway. The key thing is that transmitter power supplies are just like those used for receivers except that they are heftier and pack much more wallop.

In fact, one of the leading ham radio experimenters of all time (Ross Hull, the discover of VHF tropospheric propagation) met an untimely death from a transmitter power supply which he was using to power a homebrew TV receiver (in 1934). Something shorted out and the full power supply output got into his headphones.

The moral is that safety must be the first watchword in all dealings with transmitter power supplies. Interlock switches to prevent access to the power supply circuits while power is on are a good idea.

Because of the high power levels involved, many transmitter power supplies make use of choke-input filtering, together with full-wave rectification. Mercury-vapor

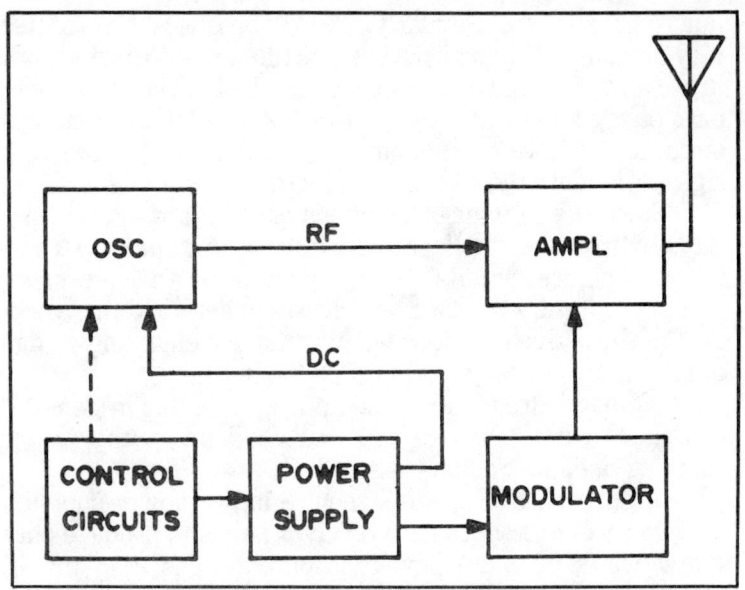

Fig. 8-1. Every radio transmitter contains these five basic functions, as explained in text. Even a CW transmitter has a modulator.

rectifiers are often used in preference to high-vacuum tubes, but high-voltage silicon power diodes are now replacing them. All these steps are taken to reduce power losses in the power supply itself. Yet, even with the most efficient practical power supply circuits, getting 2 kW of dc for the rest of the transmitter often requires the expenditure of 3 to 4 kW of ac from the power line.

The control circuits of a transmitter usually tie in rather closely with the power supply, since they include the off-on switches as well as other operating controls. These circuits differ greatly from transmitter to transmitter, and are often custom designed for each station installation.

One feature frequently found in a transmitter's control circuits is the provision of time-delay relays which delay application of high voltage to the transmitter until after bias voltages are applied, and hold back the grid bias in a tube transmitter until the filaments have warmed up. This greatly reduces stress on the tubes of the transmitter.

On-off switching for the transmitter is usually accomplished by controlling plate voltage to the amplifier. This permits instant changeover from receiving to transmitting, since filament and bias voltages are left on at all times while the station is in use. The same control usually handles antenna switching from receiver to transmitter as well, to permit single-switch operation of the station.

Since the control circuits do vary so greatly from station to station, and are normally just simple power switches in essence (even though they may be hooked together in complicated ways), the license exams do not cover control circuitry—and we'll drop it too.

The modulator is the essential part of the transmitter so far as communication is concerned, but it's a separate subject and will be our target in the next chapter. Its purpose is to put any information to be transmitted by the station onto the *carrier wave* generated by the transmitter, but this can be done in a multitude of ways. Since we'll be looking at this area in detail later, we'll skip it for now.

Which leaves us with only the oscillator and the amplifier to view in the remainder of this chapter.

The purpose of the oscillator is to establish a single specific frequency upon which the transmitter will operate. It accomplishes this purpose by generating an rf signal at this specific frequency, or some other frequency related to it. Output of the oscillator is at very low power level, because any attempt to get high power from an oscillator results in less stable frequency control, and the whole purpose of the oscillator is to provide stable control of the transmitter's frequency. As Robert Heinlein declared in one of his better science-fiction novels, "there ain't no free lunch." We want stability rather than power from the oscillator, and so we must accept low power output as the price of the meal.

However, the transmitter itself must provide power, and so the amplifer comes into the picture. The amplifier takes the minute output of the oscillator, and brings it up to the power level we desire. Rules say that we must never use more power than that "necessary to maintain communications"—but the judgment of what constitutes "necessary" power is left to the operator, and somehow it always seems to work out that "necessary" means "as much power as you can afford." This leads to the rather ridiculous situation of using a 1 kW station to talk across town, during times when 1W transmitters are capable of covering the globe, but that's one of our freedoms as hams.

Since the purpose of the amplifier is merely to boost the power of the oscillator's output, the amplifier must not introduce new frequencies of its own into the signal. This requirement leads to certain complications in the amplifier portion of a radio transmitter which are not found in amplifiers for more general uses.

OSCILLATOR OPERATION

We've said several times here that the oscillator establishes a single specific operating frequency for the transmitter by generating an rf signal. Some time back, though, we discovered that all electrical energy with which we deal in ham radio comes either from the chemical energy in a battery or the physical motion involved in an alternator or generator, and an oscillator has neither of these. How, then, can it "generate" a signal?

Feedback

What really happens is that the oscillator *converts* dc energy from its power supply into rf energy at some specific frequency. Far from actually generating anything, the oscillator dissipates a large part of its input power as heat—but the rest is converted from dc into rf.

To see how this happens, we'll have to back up a bit and think about amplifiers. You'll recall from the previous chapter that the purpose of any amplifier is to boost the power of its input signal to a higher level.

In a normal amplifier, we feed the input signal in from somewhere else, and do as we will with the higher-powered output signal.

But what would happen if we took a little of that output signal and fed it back in as an input, as shown in Fig. 8-2?

Many things might happen. For instance, if the phase relation between input and output is exactly reversed, so that the output signal reaches its positive peak at exactly the same time that the input signal hits negative peak, and if the feedback signal is a small enough fraction of the total output, the amplifier's gain will be reduced, its frequency coverage increased and distortion reduced. This is called *degeneration* or *negative feedback*, and is widely used in hi-fi circuits to minimize distortion of all types. The cathode follower and grounded-grid amplifier circuits are both examples of negative feedback in action.

If the feedback fraction is larger, the apparent behavior of the amplifier will be greatly altered. The differences between cathode followers, grounded-grid circuits, and conventional amplifiers are due entirely to the presence of feedback in large quantities.

On the other hand, if the phase relation between input and output is *not* reversed, but is instead kept in phase so that both input and output signals reach positive peaks together, the picture changes.

With a small feedback fraction, the amplifier's gain is increased, frequency coverage becomes narrower, and distortion rises. This comes about because any small change in input signal will cause a corresponding larger change in output signal, and a part of this change in output signal comes back as additional input to reinforce the original change and make it appear larger. Any distortion introduced by the amplifier is also returned to the input, where it is reamplified—and redistorted.

This type of operation is known as *regeneration* or *positive feedback*. It's the basis for the regenerative receiver (which produces rather outstanding performance from a single tube or transistor) and also for a device called a *Q-multiplier*.

The effects of feedback, both positive and negative, are wrapped up in a simple formula. The effective gain of any amplifier which has feedback connected around it is equal to the *open-loop* gain of that same amplifier (the gain without feedback) divided by what's left when you subtract the product of feedback fraction and open-loop gain from 1. If the feedback is negative, the sign of the feedback fraction is also negative, and the subtraction becomes addition instead. If the feedback is positive, the sign of the feedback fraction is also positive, and the subtraction remains subtraction.

That is, if we had an amplifier with open-loop gain of 100 and connected negative feedback around it with a feedback fraction of 0.5%, the product of feedback fraction and gain would be -0.005 times 100, or -0.5 (remember that negative feedback takes negative sign). The quantity to divide by would then be $1 - (-0.5)$, or $1 + 0.5$, which comes out to 1.5. Effective gain of the amplifier would be 100/1.5 or 66.67X.

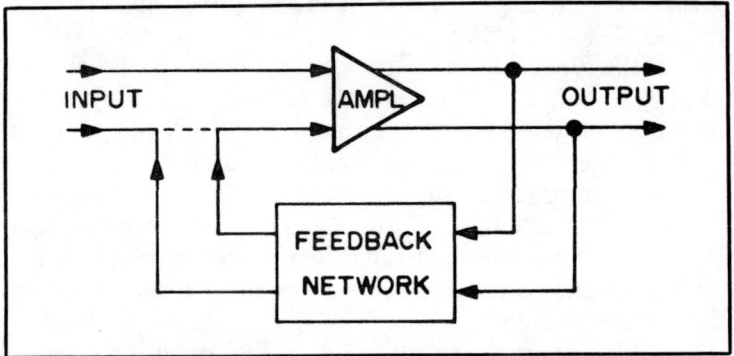

Fig. 8-2. When a part of the output of an amplifier is fed back to the input, the effective characteristics of the amplifier are greatly changed. Normally, only a small fraction of the total output is fed back; the function of the feedback network is to cut total output down to the desired "feedback fraction" for feeding back, and to determine the phase of the feedback signal with respect to that of the input signal.

With the same amplifier, and the same feedback fraction, but using positive feedback, the sign of the fraction would change and the quantity to use for division would become $1 - 0.5$, or 0.5. Effective gain is then 100/0.5, or 200. The gain has doubled—but so has the distortion.

With positive feedback, as the feedback fraction is increased for any specific amplifier, the frequency range becomes even narrower and the gain even higher. Finally a point is reached when the product of feedback fraction and open-loop gain equals 1. Here the bandwidth of the amplifier becomes virtually zero and the gain is effectively infinite.

When this point is reached, the division factor in our formula becomes $1-1$, or zero, and division by zero is "illegal." However, we can see what is happening by looking at the gain when the product is just a tiny bit smaller than 1. For instance, with open-loop gain of 100 and feedback fraction of 0.999%, the product becomes 0.999, and effective gain rises to 100/0.001 or 100,000. With a feedback fraction of 0.999%, the product is 0.9999 and effective gain is one million. With a feedback fraction of 0.999999%, the product is 0.999999 and effective gain is 100 million. You can see that as the feedback fraction increases to drive the product closer and closer to 1, the gain keeps rising at an astronomical rate.

At some point during this rapid rise of gain, no input signal need be supplied to the circuit. Gain is great enough that the random noise caused by motion of the individual electrons within the wires and components of the circuit itself is enough to cause maximum output, and as soon as any output appears that provides all the input necessary to keep things going indefinitely. This self-sustaining circuit is what we call an oscillator, because it oscillates from one state to another continually.

The precise point at which oscillation occurs can be predicted by part of the feedback formula. Whenever the product of gain and feedback fraction become equal to 1, oscillation is sure to result. The oscillation prevents the product from ever becoming greater than 1, incidentally.

This implies that any amplifier can be turned into an oscillator by simply providing enough positive feedback around the amplifier, and the implication is absolutely correct. If the amplifier has no tuned circuits, the oscillation frequency will be determined by the time it takes the amplifier's coupling capacitors to charge and discharge. We call this kind of oscillator a *multivibrator* and it finds wide use in TV, computer, and radar circuitry. We won't go into it here because it's not required for the General class ham ticket.

If the amplifier has neither tuned circuits nor coupling capacitors, then the oscillator has a frequency of zero. This might not seem like much of a circuit to have around, but it's known to the computer industry as a *flip-flop* and provides one of the most popular memory circuits around for information processing. Again, since it's not required for the General class exam, we won't go into this more deeply.

Frequency

The kinds of oscillators we're interested in right now all do have tuned circuits, and are based upon rf amplifiers. The tuned circuits are adjusted to provide maximum gain at some specific frequency, and it's the gain at this frequency (center frequency) which fits into the feedback formula. Thus, when oscillation occurs, the oscillation is at the frequency to which the resonant circuits of the amplifier are tuned.

Because of this, we separate the frequency-determining circuits of an oscillator from the rest of it, and call them the *resonator*. Requirements for a single-frequency oscillator are, then, threefold. We must have an amplifier, a resonator, and a feedback network.

The feedback network provides the essential feedback, the resonator provides frequency control, and the amplifier keeps things going. These three components of every single-frequency oscillator are shown in Fig. 8-3.

The purpose of an oscillator in a radio transmitter is to control that transmitter's frequency, and this means that the frequency of the oscillator must be as stable as possible. This requirement for high frequency stability influences all three parts of the oscillator circuit.

Feedback control is essential, because with too much feedback the gain can be 1 at frequencies which are not exactly at the frequency of maximum gain. This permits the oscillator frequency to wander about, which defeats our purpose. For this reason, no more feedback should ever be used than is required to keep the oscillator going.

The resonator, since it determines the frequency in the first place, must be able to stay tuned to one frequency for an extended time period. This is not so simple as it might sound. For instance, the wire of which a coil is wound expands as it heats, and the passage of current through that wire results in heat being applied to the wire. As the wire expands, the coil dimensions change. Even though the change is microscopic, it will still change the coil's inductance and, consequently change the frequency to which the

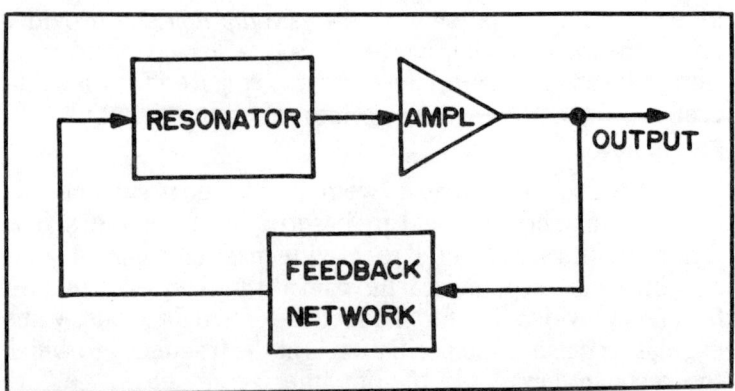

Fig. 8-3. Addition of a resonator to the feedback arrangement of Fig. 8-2, with proper phase of feedback and a feedback fraction large enough to eliminate any need for external input, produces a tuned oscillator. All rf oscillators in common use share these three components, and the differences between one type of oscillator and another all lie in the manner in which one or more of the three basic functions is achieved.

circuit is tuned. In practice, this effect may cause a frequency shift of several kilohertz—enough to take a signal right out of a ham band if you're operating near the edge.

When the utmost in frequency stability is required from an oscillator, then, the resonator is hardly ever a simple tuned circuit. Instead, quartz crystals or *crystal resonators* are used. These are thin plates of quartz, ground to precise sizes and thicknesses, which act exactly like tuned circuits to an oscillator circuit but which are far less effected by the circuit's operation than are ordinary coils and capacitors.

Not all frequency-determining crystals are quartz, but the vast majority are. Quartz is one of the most sturdy of many substances which have a property called *piezoelectricity*; this word means simply that these substances produce electricity when squeezed, pressed, bent, or otherwise mechanically deformed, and conversely can be deformed by application of electrical energy. The crystal in a crystal microphone is another piezoelectric substance.

The crystal resonator gets its frequency stability from the fact that it uses *physical* resonance rather than *electrical* resonance, which eliminates one whole level of physical-to-electrical translations from the process.

When an alternating voltage is applied across the proper faces of a crystal resonator, the crystal will vibrate in step with the voltage. If the frequency of the voltage is such that the vibration is *in tune* with the natural vibration frequency of the crystal, the resulting vibrations will be much larger than if it is not.

In other words, the exchange of energy between the electrical and the mechanical states is much more efficient if the electrical frequency matches the mechanical vibration frequency; this means that very little energy is lost in the transition under such conditions.

If we were to suddenly remove the voltage, the crystal would continue to vibrate for at least a little while, like a Chinese gong which has been struck one time. Each vibration would produce electrical energy which could go back into the external circuit. This is exactly like the energy-swapping role of the LC tuned circuit, and is what makes the crystal resonator act (to the oscillator) exactly like a tuned circuit.

Even though the frequency stability of a crystal resonator is normally much greater than that of an LC circuit, some care is still necessary. Many crystals are capable of acting as resonators at several frequencies, and you have to be certain that you're using the frequency you intended to. The crystal itself can be heated by circuit action, and this will change its dimensions and therefore its frequency. And finally, the dimensions of the crystal itself are always subject to manufacturing and processing tolerances, so that the frequency stamped on the case is not ever exact, but itself has a tolerance. This tolerance must be taken into account whenever you're operating near the edge of a band, or any other time when exact frequency control is important.

Crystal frequency tolerances are usually specified in percent, and a typical tolerance is 0.01%. This means that, in the specified oscillator circuit, the crystal's frequency will be within 0.01% of the marked frequency. The higher the frequency of the crystal, the greater the absolute possible error in hertz. For instance, a 0.01% crystal for 14 MHz might have a true frequency anywhere between 13.9986 MHz (14 − (0.0001 x 14)) and 14.0014 MHz. A crystal to the same tolerance for 7 MHz could have true frequency from 6.993 to 7.0007 MHz. In the first case, absolute possible error is 1.4 kHz; in the second, 700 Hz, just half as much.

This tolerance must be kept in mind when ordering band-edge crystals. The lowest frequency to order, with 0.01% tolerance, for operation in the 7 MHz band, would be 7.00071 MHz. With maximum low-side error, this would still come out as 7.00071−0.0070071, or 7.0000099 MHz, less than 10 hertz inside the band limit.

The way to calculate the tolerance is to add the tolerance to 100%, then multiply the band-edge frequency by the result if you're ordering for the low end of the band, or divide the band-edge frequency by the result if you're going for the upper end (either way, your answer will be farther inside the band than the exact limit). Then round off the answer in the direction which takes you still more inside the band.

While we've discussed tolerance mainly in connection with crystal resonators, an LC resonator has tolerances too, and the same kinds of considerations apply. Any time operation is planned near a band edge, it's necessary to be certain that you know exactly where that band edge is, and keep your signal on the legal side of it.

The major advantage of the crystal resonator is its stability, but at the same time this leads to a drawback. The frequency of a crystal resonator is difficult to change. Because of this, variable frequency oscillators (vfo circuits) using LC resonators are highly popular.

Both crystal and variable frequency oscillators operate in essentially the same way; the major technical difference is the type of resonator used, and the major operational difference is that the vfo is less stable but often more convenient to use than is the crystal.

Basic Oscillator Types

Most of the common oscillator circuits come in both crystal and vfo forms, but sometimes different names are applied to the two versions of the same circuit. Since all oscillators used in radio transmitters must include the three basic components shown in Fig. 8-3, the major differences between different oscillator circuits are the ways in which these components are interconnected. In most cases, but not all, the variations occur in the connection of the feedback network.

The feedback network must couple output back to the input in the proper phase, and with the proper feedback fraction, but so long as it accomplishes these two tasks everything else about it is free to vary without limit. Thus it can be placed at almost any point in an amplifier circuit where it can mesh output with input. Some oscillators (such as the Armstrong and tune-plate-tuned-grid circuits) usually have it connected to the plate, while others (such as the Hartley and Colpitts arrangements) usually have it in the cathode circuit.

Many FCC exam questions ask you to *identify* oscillator circuits such as the Hartley, Colpitts and so forth. In the

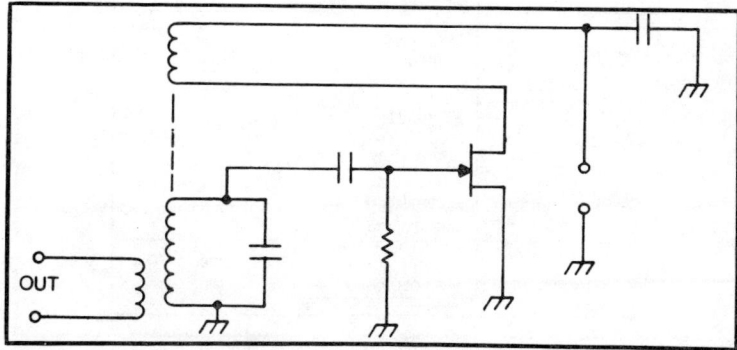

Fig. 8-4. Armstrong oscillator; feedback is via a "tickler" coil.

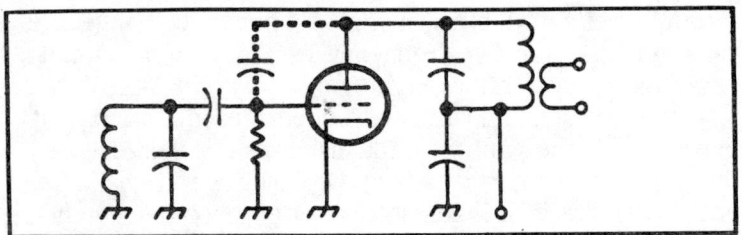

Fig. 8-5. Tuned-plate tuned-grid oscillator feedback path runs through the grid-plate capacitance of tube.

discussions to follow, we will tell you for each case what are the identifying features—all of which are part of the respective feedback network. These features are also summarized in Fig. 8-6.

Figures 8-4 through 8-9 show several oscillator circuits, including those in most common use today. Figure 8-4 is the oldest of all oscillator circuits: the *Armstrong* oscillator. It works only with an LC resonator, which is in the gate circuit, and the feedback is applied by means of a coil in the drain circuit which is coupled to the gate coil. This drain coil

OSCILLATOR TYPE	IDENTIFYING FEATURE	TYPICAL CIRCUIT (BIAS & DC SUPPLIES NOT SHOWN)
ARMSTRONG	*TICKLER COIL* FEEDBACK	
TGTP	TUNED LC CIRCUITS IN BOTH PLATE AND GRID, AND NO NEUTRALIZATION	
MILLER	CRYSTAL IN INPUT CIRCUIT, LC RESONONT CIRCUIT IN OUTPUT	
HARTLEY	*TAPPED COIL* FEEDBACK NETWORK	
COLPITTS (OR HI-C COLPITTS)	TAPPED CAPACITOR FEEDBACK VOLTAGE DIVIDER CONNECTED EMITTER; PARALLEL LC TANK CIRCUIT	
CLAPP	SAME AS COLPITTS, EXCEPT LC TANK IS SERIES CONNECTED	
CRYSTAL COLPITTS	SAME AS HI-C COLPITTS, EXCEPT LC TANK IS REPLACED WITH A PIEZOELECTRIC CRYSTAL	

Fig. 8-6. Basic oscillator circuits.

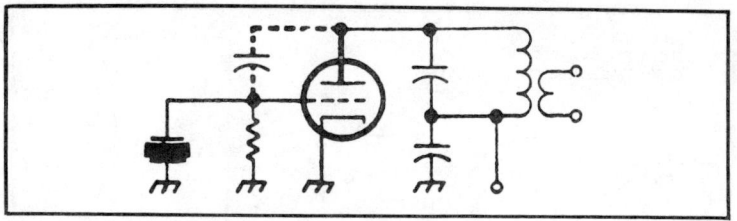

Fig. 8-7. Miller crystal oscillator is simply a crystal version of TPTG circuit (Fig. 8-5), with a crystal replacing the grid tank circuit.

is called the "tickler" and its number of turns is adjusted to vary the feedback fraction. This is the identifying feature of the Armstrong oscillator.

The Armstrong oscillator is virtually obsolete today, although it's still around in some regenerative detector circuits and sometimes finds use in receiver local oscillators. We include it for its historical interest, because it spotlights so clearly the feedback function of all oscillators and because it shows up on FCC exams.

Figure 8-5 is the *tuned-plate-tuned-grid* (TPTG) oscillator. This circuit is identical to a triode amplifier circuit which has not been neutralized, and oscillates when the plate and grid circuits are tuned to *slightly different* frequencies. The feedback path in this one is through the tube, by means of plate-to-grid capacitance.

The *Miller crystal* oscillator (Fig. 8-7), simply substitutes a crystal resonator for the TPTG's tuned-grid circuit. The resulting oscillator is widely used in VHF transmitters.

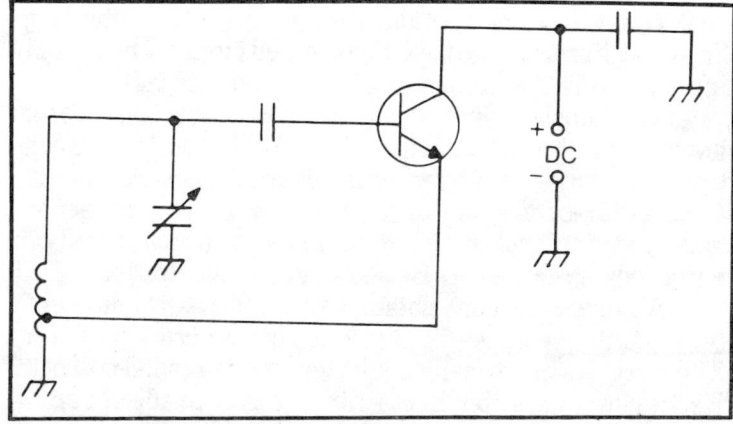

Fig. 8-8. The Hartley oscillator is always identified by a tap on the coil.

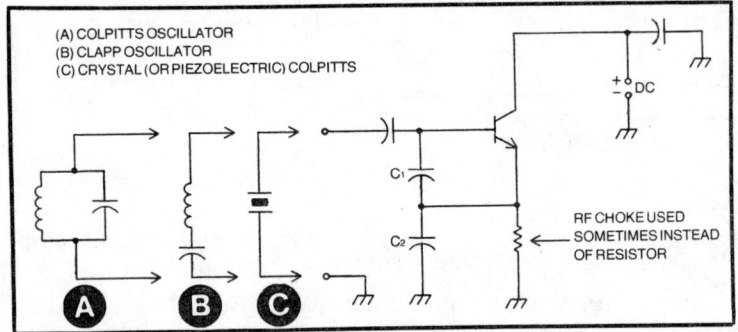

Fig. 8-9. Capacitance feedback circuit goes under various names, depending on the tuned-circuit arrangement.

As in the TPTG, feedback is through the tube or transistor and plate, collector, or drain circuit tuning is critical. For best results, the plate, collector, or drain circuit must be tuned to a frequency slightly different from the crystal. The identifying feature is the crystal at the amplifier input and an LC resonant circuit at the output.

The *Hartley* oscillator (Fig. 8-8) can readily be recognized by its tapped coil and the fact that the emitter returns to the coil tap rather than to ground. Feedback is via the emitter circuit, and is adjusted by varying the tap position (normally by adding or removing turns at the base end of the coil, which effectively moves the tap down or up the total coil). This circuit is almost universally used for receiver local oscillators, and is also found in transmitters both as a vfo and in a crystal version. In the crystal version, the crystal does not replace the LC tuned circuit. Instead, it replaces the coupling capacitor from the base to the tuned circuit. The identifying feature is the tapped coil in the tuned circuit.

The most widely used oscillator circuit at present, however, is the one shown in Fig. 8-9. This one, identifiable by the *two series-connected capacitors* from base to ground and the emitter connection to the junction of these capacitors, goes under at least three different names, which identify the variations shown as A, B, and C in the illustration.

All three versions obtain their feedback from the emitter circuit, by voltage-divider action in the series capacitors. The circuit is electrically equivalent to the tapped coil of the Hartley oscillator, but in practice is easier to adjust because either or both of the capacitors may be an adjustable trimmer, permitting convenient adjustment of feedback fraction.

The differences between the three versions all involve the resonator portion of the circuit. When a high-capacitance, low-inductance LC resonator is used and connected as shown at A, the circuit is called a *high-C Colpitts* oscillator. This version is popular as a vfo. It can be designed to permit extremely precise tuning and high stability.

When a low-capacitance, high-inductance LC resonator is connected as shown at B, the result is the *Clapp* oscillator. At one time this circuit was almost exclusively used for vfo's, but in recent years its sensitivity to small changes of capacitance has caused it to decline somewhat in popularity. It's still around in goodly numbers, though.

When a crystal resonator is used as at C, the circuit is called both a grid-plate or emitter base oscillator and a *crystal Colpitts* circuit. Quite possibly more crystal oscillator circuits in use today are of the Colpitts variety than of any other single type. Since no tuning coil is necessary, the circuit covers a wide frequency range; all that's necessary is to plug in a crystal for the frequency desired. Addition of a tuning coil in the plate circuit makes it easy to pick off any desired multiple of the crystal frequency—a popular feature with VHF operators who often use this circuit with an 8 MHz crystal to get 24 or 25 MHz output direct from the oscillator by taking the third harmonic of the crystal frequency. Stability is excellent, and power output is adequate.

AMPLIFIER OPERATION

We've already examined the subject of amplifiers in general several times so far in this study course, but the amplifier of a transmitter is a bit different from the common run of amplifier circuit. Our question, then, is "how?"

In a transmitter, the amplifier portion serves several purposes. Most obvious is its action in stepping up the relatively feeble output of the oscillator to the power level desired for feeding to the antenna. Not so obvious is its action of *isolating* the oscillator from external influences.

The oscillator, you see, is a rather sensitive circuit. Almost anything—a change in operating voltage, variations in the applied load, or mechanical vibration—can cause its frequency to change. This is something which we do not wish to have happen, and so we connect an amplifier between the oscillator and the antenna even when the oscillator is capable of delivering enough power by itself, in order to provide a constant load on the oscillator and let the variations of operating conditions all be applied to the amplifier.

While the amplifier is performing both these functions, it must of course not introduce any unwanted output frequencies of its own, nor must it influence the oscillator's frequency, itself.

Amplifier Functions

The result of these requirements is that the portion of a radio transmitter which we are here calling "the amplifier" normally is not just a single amplifier stage, but instead is a whole string of amplifiers connected end to end. Some are designed to provide isolation, and some for power handling.

The last stage in the amplifier (the one which feeds the antenna) is called the *final* for reasons which should be apparent. Between the oscillator and the final, we may enconter *buffer* stages which are intended primarily to provide oscillator isolation, *driver* stages which are intended to boost power level up to that required by the final as its input, or both.

The buffer stages accomplish their function of providing isolation in several ways. To begin with, every amplifier stage, whether intended as a buffer or not, provides at least some isolation between the stage which precedes it and the stage which follows. Some buffers, then, are indistinguishable from any other rf power amplifier designs.

Occasionally a circuit designer will set up a buffer stage to operate in class A rather than in class C (the normal operating condition for rf power amplifiers). This is done because a properly operating class A amplifier imposes no load on the stage which precedes it, yet is capable of providing sufficient power output to drive a class C stage behind it. If a class A buffer is used, it normally is driven by the oscillator itself; with no (or little) loading upon the amplifier, frequency stability is increased.

Another trick sometimes used in buffer design is to employ a cathode or emitter follower rather than a normal grounded-cathode or -emitter circuit. The cathode or emitter follower, with its 100% negative feedback, is noted for its isolation-providing capability. While it cannot produce any voltage gain, it can and does provide power amplification.

One of the most popular techniques used in buffering, though, is to operate the oscillator at some submultiple of the desired output frequency, and then use a frequency-multiplying stage or stages as the buffer.

A frequency multiplier looks just like an ordinary amplifier, but is operated with additional bias (deeper into class C), and its input and output circuits are tuned to different frequencies. The class C operation provides current pulses in the output circuit, and if the output tank is tuned to a frequency twice that of the input, these pulses will occur every other cycle of output frequency. That's often enough to keep things going; multiplication of up to five times in one stage is possible.

With the input and output circuits of the multiplier stage operating at different frequencies, isolation between them is naturally better than if they were on the same frequency.

Frequency multipliers are also used as drivers, but their efficiency is much less than that of "straight-through" or "straight" amplifiers (those in which input and output are on the same frequency). A multiplier which doubles its input frequency provides about half the output that the same circuit would give in straight-through operation; a tripler gives about one third, a quadrupler about one fourth, and so fourth.

One great advantage of multiplier stages is that they cannot oscillate because the output signal is different in frequency from the input signal, and so feedback cannot be sustained. A straight-through amplifier, on the other hand, is virtually the same circuit as the TPTG oscillator, and if triodes are used it must be neutralized to prevent self-oscillation. Even with multigrid tubes, which make it possible to operate without neutralization, it's still a good idea to neutralize all straight amplifiers in a transmitter, to keep out of trouble.

Neutralization

Neutralization is the technique of canceling out all positive feedback from an rf amplifier stage, in order to make it impossible for that stage to oscillate. While the amplifier may have been designed to avoid positive feedback, when the thing is actually built it's almost impossible to get rid of all possible feedback sources. Stray capacitance, power wiring, magnetic coupling between coils, and similar factors bring in feedback whether we want it or not. Careful parts layout can minimize the problem, but cannot eliminate it.

Since we cannot eliminate all the positive feedback, we neutralize it instead. We do this by adding negative feedback to the circuit. The negative feedback cancels out the positive feedback, and the net result is (ideally) no feedback at all. In practice, we usually adjust everything to slightly *overneutralize* the stage, in order to have a safety factor against component aging and the like, and also because our means of detecting exact neutralization are not completely accurate.

Some of the ills which can be prevented or eliminated by proper neutralization of the amplifier include self-oscillation, erratic shifts of stage gain with small changes in operating frequency, nonlinear distortion in modulated amplifiers, and splattering in SSB linear circuits.

Several procedures may be followed to properly neutralize an rf amplifier. All involve balancing out the unwanted positive feedback by artificially supplied (and adjustable) negative feedback, and detecting the balance point by means of a sensitive indicator. The negative feedback is usually taken from the plate circuit to the grid circuit.

Figure 8-10 shows a typical neutralization circuit, based on recommendations in the ARRL handbook. Capacitor C1 is the normal bypass capacitor for the grid tank, and C2 is the neutralizing capacitor.

Even though we often consider a bypass capacitor to be a "dead short" for rf energy, actually it must always have at least a little impedance at any frequency lower than infinity. C1, therefore, is not ever a true short for rf, and any signals which are applied across C1 must also appear at the grid end of the circuit. C2 is of much smaller capacitance than C1, which means that it has much greater impedance at any frequency, and so the two capacitors together form a voltage divider between the tube's plate and ground. The small part of the plate signal which appears across C2 is thus fed back into the grid circuit, and the capacitance ratio between C1 and C2 determines the feedback fraction. The feedback is made negative, as required, by the fact that plate and grid voltages in a grounded-cathode amplifier are exactly out of phase with each other.

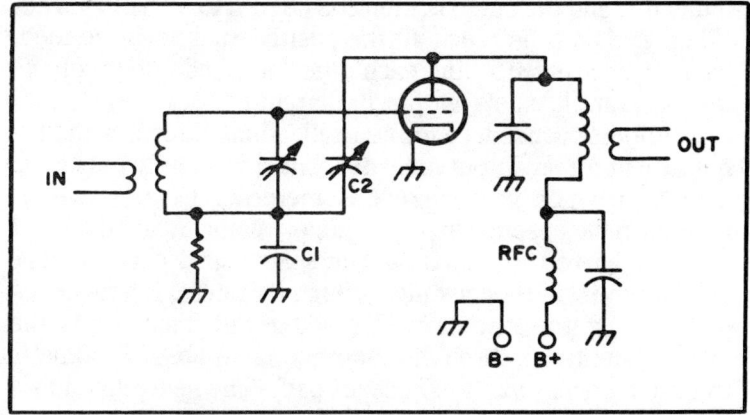

Fig. 8-10. Neutralization circuit for a conventional single-ended rf amplifier makes use of voltage-divider action of capacitors C2 and C1, in series, to provide negative feedback from plate to grid circuit. C1 is a normal grid bypass capacitor and C2 is called the neutralizing capacitor. Normally C1's value is fixed and C2 is variable. By adjustment of C2, negative feedback through the neutralization circuit is made to balance out any positive feedback from the rest of circuit which might cause the amplifier to oscillate.

Adjustment is accomplished by varying the value of either C1 or C2, leaving the other's value fixed, until no feedback exists. In most transmitters, C2 is adjusted and C1 is fixed in value. Special *neutralizing capacitors* with very small capacitance and high-voltage insulation are available for this purpose. Alternatively, C2 can be left fixed, and C1's value adjusted. In this case, high-capacitance trimmers or variable capacitors are required at C1, but the requirement for high-voltage insulation is minimized. Whichever of the two is adjusted, the procedures are the same.

One of the most common procedures for neutralization begins with the removal of all supply voltages except filament power from the stage being neutralized. Then a sensitive rf indicator, such as a vacuum-tube voltmeter with rf probe, is coupled to the output side of the stage, and the stage is driven at full rated input (drive) power. Since no power is being supplied to the stage, any rf which appears at the output indicator must be the result of a feedback path.

While driving the stage and observing the ouput indicator, the neutralization adjustment is then varied slightly. If indicated output increases, the adjustment is moved in the other direction. With careful adjustment, the fed-through power can be reduced to a level too low to be detected by the indicator, and the stage is then considered to be neutralized. When this point is reached, the positive and negative feedback paths are cancelling each other out, and effectively no feedback at all exists within the circuit.

Another procedure for neutralization eliminates the requirement for an output indicator, but applies only to class C stages in which grid current is metered. Drive power is applied in the absence of power-supply voltage, and the input tuning adjusted for maximum indicated grid current. The output tuning is then swung through its range. If a feedback path exists, the grid-current reading will fluctuate as the output tank tunes through resonance, as it absorbs some of the drive energy via the feedback path. The neutralization is then adjusted until this "flicker" of the meter is eliminated.

Either procedure is accurate enough for all practical purposes. The one using an output indicator applies to all kinds of rf amplifiers, while that using the grid-current meter applies only to amplifiers which draw grid current and which have provisions for metering it.

Not all rf amplifiers are single-ended, a circuit called push-pull is popular for rf amplifier use, and its neutralization is handled in a slightly different manner.

With a push-pull amplifier, the neutralizing capacitors are connected from the plate of one tube to the grid of the other as shown in Fig. 8-11. Adjustment is similar to that for single-ended stages, with the added complication that the two separate adjustments interact with each other and both must be checked every time.

POWER MEASUREMENT

It's fairly simple to explain *why* the dc power input to an amateur radio transmitter must be measured. Section 96.67 of the FCC Rules and Regulations declares that "except for power restrictions as set forth in 97.61, each amateur transmitter may be operated with a power input not exceeding 1 kilowatt to the plate circuit..." The restrictions referred to limit power input in the 160-meter (1.8 MHz) band on a state-by-state basis to prevent interference with the LORAN navigation system, and in the 420-450 MHz band on a regional basis to prevent interference with space telemetry and military experimental radar installation.

Additionally, transmitters operated in the Novice bands are limited to a maximum power input of 250 W. In general,

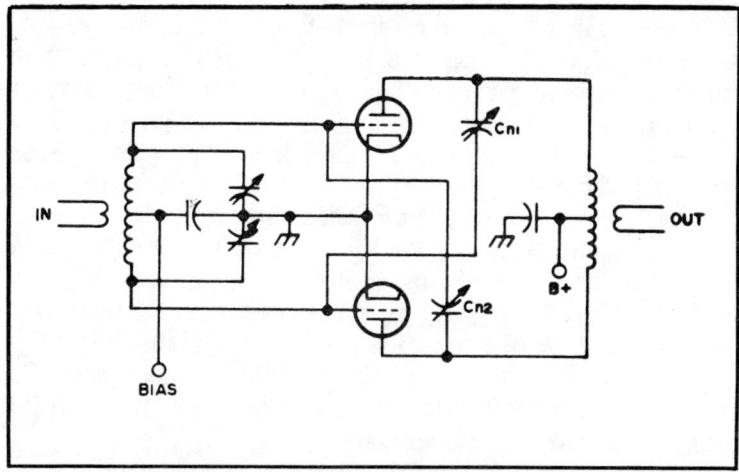

Fig. 8-11. Neutralization techniques for push-pull amplifiers differ somewhat from that shown in Fig. 8-10. Here two neutralizing capacitors are used; they connect from the plate of one tube to the grid of the other and vice versa. Since two tubes always operate out of phase with each other, feedback still works out to be negative. The two neutralizing capacitors interact with each other and adjustment of both must be checked whenever either is adjusted.

though, the power limit for the most popular ham bands is 1 kW. That's what we mean when we say "the legal limit" or "a full gallon."

And even when the limit is lower, such as the 100W permitted for many states in daytime on the 160 meter band, or the 25 to 200W permitted on 160 at night, we still must measure the power to be certain that we never exceed the legal limit for the specific frequency and the specific time at which we are operating. Loran is being phased out, so the FCC may phase out these restrictions in the future.

While we're on the subject of legal limits, it's well to point out that Section 324 of the Communications Act of 1934 includes another legal limit which supersedes that imposed by FCC rules. It's rather specific, too: "In all circumstances, except in case of radio communications or signals relating to vessels in distress, all radio stations, including those owned and operated by the United States, shall use the *minimum amount of power necessary* to carry out the communications desired." (Emphasis supplied.)

That boils down to this: We can never use more power than FCC rules permit, but we cannot use even that much unless it's necessry to carry out the communications.

Unfortunately for the interests of interference reduction, the judgment of how much is "necessary" appears to be entirely subjective, and we have never heard of any operator getting into trouble for using "more power than necessary"—although several have had their licenses suspended and a number of operating awards have been revoked for using more power than FCC rules permit.

The same section of the FCC rules which sets the 1 kW limit goes on to specify just a little bit about how the measurement is to be made. "An amateur transmitter operating with a power input exceeding 900W to the plate circuit," it says, "shall provide means for accurately measuring the plate power input to the vacuum tube(s) or transistor(s) supplying power to the antenna."

Below the 900W level, apparently, an "educated guess" is adequate measurement. In point of fact, the dc plate power input is determined by multiplying plate or collector supply voltage times plate or collector current. Some method of measuring current is necessary in order to properly tune the transmitter, and this is usually a milliammeter. At low to moderate power levels, many operators simply multiply the measured plate or collector current by the calculated plate or collector supply voltage, which is a little better than an "educated guess" but not good enough for high-power use. For 900W or above, it's necessary to have both a voltmeter and a milliammeter in the circuit in order to comply with the rule which requires the means for accurate measurement. It's also a good idea when operating within 10% of the power limits on any of the bands which have lower power limits.

Now that we know why it's necessary to measure power input, let's see how it is done. We've already indicated the actual technique; measure or estimate the plate voltage, and multiply by the measured plate current. But just how should we measure current?

For ordinary AM phone operation, it might not be too confusing. The plate-current meter of a properly operating phone transmitter remains steady, so you have only one reading to concern yourself with.

But, as we shall see in the next chapter, the actual peak power input to the transmitter is at least half again greater than the indicated dc input, and may be twice as great, depending upon your definition for "peak power input."

Should that concern us in our power measurement? This question has not, so far as we know, been officially answered, but in practice the answer is no. The power input referred to by the FCC is the indicated power input, and the additional power involved in AM phone operation is ac power from the modulator.

How about a controlled-carrier AM rig, in which the indicated plate current gyrates with modulation? Or, worse yet, a single sideband transmitter, which has no input power in the absence of speech, yet fluctuates all over the current dial when modulation is applied?

The answer for both these cases is that the measured power, as indicated on the panel meters, must never exceed the legal limit. The FCC recognizes that instantaneous power peaks may easily go above the limit, but by specifying a quarter-second time constant for the meters they assure that this apparent loophole is not really very big, and puts SSB and controlled-carrier operations into the same class as AM (where some of the power, which does not show up on the meters, is "limit-free").

The FM operator doesn't get off so well. His power must be measured on a carrier-level basis, just like the AM operator—but he gets no free-power bonus when he adds modulation.

CW fares poorest of all, because the policy is that power input to a CW transmitter must be measured with the key held firmly down, and this is a condition which is prohibited for any other purpose on the popular CW bands. When CW is being used, the plate current meter fluctuates just as does that of an SSB rig, but the CW operator does not have the freedom to load up to higher power levels and use the "highest flicker" method of reading his power input. He must make his measurement with the key down, and as a result actual measured power during operation seldom exceeds 75% of the "measured" value for official purposes.

For SSB and controlled-carrier AM, plate voltage and current measurements are performed during normal operation, with modulation applied, and the "highest flicker" of each meter is noted. These "highest flicker" readings are used to determine legal power input to the stage, and as in the other case the power input of the transmitter is the total of all stages feeding the antenna. The meter needle never keeps up with the voice peaks, but does record average rather than peak power.

The ratio of peak envelope power to average power in an SSB signal depends primarily on the characteristics of the operator's own voice, and may range from about 1.2 to 1 up to more than 2 to 1. If your voice is such that the ratio is 2 to 1, then you will be able to produce a legal 2000 watts of peak input power while the average upon which the regulations are based remains within the limits. If your voice produces a 1.2-to-1 ratio, you can get only 1200 watts peak input while remaining within the legal kilowatt limit.

This is a notable bonus for SSB operation when compared to either CW or AM operation. An AM transmitter is limited to 1000 watts input in the absence of modulation. Addition of 100% modulation brings the peak input up to 1500 watts (1000W carrier and 500W in the sidebands, contributed by the modulator) but this is still less than the possible 2000 watts with SSB. Especially when you consider that only 250 of the AM rig's 1500 watts are useful and the rest are merely tagging along for the ride.

Despite the possible 4-to-1 advantage in peak power enjoyed by SSB in comparison with CW, the dits and dahs retain the advantage of maximum transmission range. This comes about because CW may be received with only a 50 Hz bandwidth in the receiver, while SSB requires a minimum of 2.7 kHz, some 540 times as great. The 4-to-1 power advantage is canceled out exactly when CW is received with a 675 Hz bandwidth. Cutting that bandwidth in half gives CW a 2-to-1 advantage in effective received signal strength, and each additional halving of bandwidth doubles CW's power advantage. For voice operation, though, no other technique can approach SSB's effectiveness. The nearest competitor is FM.

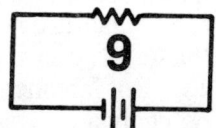

9
Modulation and Modulators

While some of us might find it amusing for a little while to turn on a radio transmitter without transmitting any information, that's not really the purpose of ham radio. In fact, it's so far from the purpose that the broadcast of a raw carrier is prohibited by FCC rules (except briefly for test purposes) at all frequencies below the UHF region.

The real purpose of almost every radio transmitter we're likely to come across or operate is to transmit *information*. So far in this study course directed toward the General class ham ticket, we haven't bothered to look at how this purpose might be accomplished. Now we'll remedy that, and spend this chapter examining rather closely the why, what, and how of modulation.

WHAT IS MODULATION?

If we were to look *modulation* up in a dictionary, we would find that it means (in general) "the act of modulating, or the state of being modulated." To modulate, continuing our search, is to "vary the tone, inflection, or pitch" or "to regulate or adjust, temper, soften." The word comes from the Latin *modulus* or *module*, which in turn comes from *modus* which meant "to measure." None of which, unfortunately, is of much direct help in determining just what modulation amounts to.

233

All these assorted definitions of *modulation* are, however, quite relevant to our purpose here. In communications, modulation is the variation of *any* characteristic of something, in order to convey information or *intelligence*. The white paper upon which these words are printed could be said to be "modulated" by the ink which forms the letters, for the purpose of conveying this information to you.

Modulation Techniques

The characteristic which is modulated may be almost anything capable of being varied, regulated, tempered, or measured. In radio, only three characteristics of a transmitter's signal are normally modulated, and normally only one of these three is modulated in any specific transmitter.

For instance, the *amplitude* or *strength* of the signal may be changed. A simple example occurs in a CW transmitter, where the signal is either full on (with the key down) or turned off (key up). This *is* a form of modulation, since it is a variation of the signal strength caused for the purpose of conveying information. Most folk (including the FCC), however, distinguish between AM and CW despite the fact that CW operation is a simplified form of amplitude modulation.

Amplitude is not the only characteristic available to be varied. We can, if we choose, keep the amplitude constant and vary the *frequency* of our signal. This is just as effective in changing its strength—in many cases, more effective. This technique is known as *frequency modulation* or FM.

The third characteristic which is frequently used for modulation purposes is the *phase* of the signal; the technique is called *phase modulation* or PM, but in practice is virtually indistinguishable from the FM technique since the frequency cannot remain constant as phase changes—which means all FM has some PM and vice versa.

Since all FM does have some PM, and vice versa, these two types of modulation are usually lumped together in engineering textbooks under the heading "angle modulation" (the "angle" referred to here is the "phase angle"). The FCC, however, has very little to say about PM in the amateur license exams; all the questions deal with either AM or FM.

For that reason, in this installment we'll deal with AM and FM. It'll be up to you to remember that in most cases, what we say about FM is also true of PM. In those rare occasions when this is not the case, we'll let you know.

So far, we have defined modulation as the variation of any characteristic of something in order to convey information, and have identified three characteristics of a radio transmitter's signal which may be varied in order to modulate the signal. We then reduced the resulting three types of modulation (AM, FM, and PM) to only two for the purposes of this discussion (AM and FM).

Carrier

That part of a radio transmitter's signal which is sent out in the *absence* of any modulation is called the *carrier wave*. This name, like the British term "wireless," comes to us from the days of early radio when everyone thought that the purpose of the radio signal was to "carry" the information from place to place, just as the wires do in a landline situation. While we now know better (and have since 1927), the old name has persisted.

The carrier wave is the signal generated by the transmitter's oscillator and built up to strength by the amplifier (refer to the previous chapter for discussion of these parts of a transmitter). The modulator then controls either the amplitude or the frequency of this carrier wave, to produce the final radiated signal.

In general, this control process which does the actual modulation of the output signal adds something extra to the signal.

In the case of AM, the "something extra" is usually in the form of added power imparted to the signal. The exact amount of power added by a typical voice signal is almost impossible to estimate; engineers use a steady sine-wave signal instead of voice for measuring the effect, and the FCC expects to see the engineer's answer on the exam. With *100%* (we'll get into percentage of modulation later) *sine-wave modulation of a carrier, one half again is the amount of power added*. That is, a 500W AM transmitter would be operating with 750W input during this type of modulation. The added power, being ac, does not show up on your input-power measurements, nor need it be included in your records.

With FM, the "something extra" takes the form of a frequency change in the signal sometimes called "vanishing carrier." At certain specific levels of modulation with FM, the carrier wave literally disappears and all the power shows up at other frequencies in the immediate neighborhood. This vanishing point for the carrier is sometimes used to measure modulation levels for FM transmitters.

Sidebands

With any form of modulation of a carrier wave, sidebands are created. The unmodulated carrier is, just as closely as we can make it so, a single spot-frequency signal. That is, if we are operating at a frequency of 3.735 kHz, *all* of our output energy is at 3.735 MHz; none at all is at 3.735000001 MHz, or at 3.734999999 MHz, even though these frequencies are only 1/1000 of 1 Hz away from our chosen frequency. Of course, no known measuring technique can prove that this is the case, but it's what we are trying to achieve.

When we modulate with a single-frequency tone, we introduce additional frequencies into the output signal which are known as side frequencies. If we apply AM, using a 1 kHz tone, one side frequency will be 1000 Hz higher than the carrier, and the other will be 1000 Hz lower. The first of these is known as the upper side frequency and the other is the *lower side frequency*. They are also called sum and difference frequencies, because the upper side frequency is the *sum* of the frequencies of the carrier wave and the modulating signal, while the lower side frequency is the *difference* between the frequencies of the carrier wave and the modulating tone.

A man named John Carson developed the theory of side frequencies and *sidebands* (with voice rather than single tones, a whole band of frequencies is involved in the modulating process, and the side frequencies smear out to become sidebands) in 1927, and obtained a patent in that year for a system of radiotelephone transmission which made use of only one of the two mirror-imaged sidebands. Even though his technique worked, and was used for many years by the Bell System for transatlantic conversations, most people felt that it was a mathematical fiction and refused to believe that the side frequencies really existed. Development of accurate measuring apparatus in connection with radar's development in the years 1940-45 made it possible to actually see the sidebands in a scope display of a radio signal and proved their existence. Since 1948, single sideband has been an important technique of ham radio voice communications. While SSB is a special form of AM, it differs rather drastically from ordinary AM and so we won't talk about it any more in this chapter. Here, we'll look only at ordinary AM and at FM.

The sidebands do a bit more than merely making SSB possible. They cause the signal to occupy more spectrum when it's modulated than it does when it's not. That, however, is getting ahead of our subject.

MODULATION AND BANDWIDTH

We observed, a few paragraphs back, that the *unmodulated* output of a radio transmitter is (to the limit of our ability to achieve the goal) a single, spot-frequency signal. Strange as this may sound, such a signal would occupy no space at all in the radio spectrum, because it would be present only at one single frequency!

The space in the spectrum occupied by an actual signal, on the other hand, is known as the *bandwidth* of that signal, and is measured in hertz or kHz just as is the signal's frequency. The bandwidth is determined by subtracting the lowest frequency in the signal from the highest frequency present; the resulting difference is the bandwidth of that signal.

Our example spot-frequency signal at 3.735 MHz would have both its highest and lowest frequencies equal, and their difference would be zero. That's why we say that such an ideal signal would take up no space at all.

However, one of the lesser-publicized researchers into communications discovered that such a signal, while it might occupy no space at all, would be rather useless because it could not convey any information either! The man who made this discovery was also involved in circuit design, and his name is more familiar to most hams in connection with an oscillator circut he derived, but Hartley's law relating information rate and bandwidth is probably more important because it sets an absolute limit on the amount of information which may be sent in any specific portion of the rf spectrum.

The law itself is simple: the *bandwidth* required to transmit information is *directly proportional* to the *information transmission rate*. To see why this is so, it's easiest to look at a simple case involving only one item of information such as presence or absence of a carrier wave. This is known in *information theory* as one *bit* of information.

If we need not know whether the carrier is on or not more often than once every second, and look at it no more frequently than this, then it would be sufficient to turn it on for a full second and then leave it off for the next second. This would be an information rate of one bit per second, or one *baud*. The baud is a unit of information transmission speed, named for Georges Baudot, the French telegrapher who invented the 5-unit teleprinter code used in radioteletypes.

Going further, we could represent this one-baud information rate by a sine wave with a frequency of 0.5 Hz because one bit of information would correspond to each half-cycle of the sine wave.

This sine wave could not, however, tell us anything about events happening more often than once per second. If we wanted to know conditions 20 times a second (an information rate of 20 baud), we would have to look at things 20 times more frequently, and the lowest-frequency sine wave which could carry this information rate would be one of 10 Hz.

To carry information at a rate of 200 baud, we would have to increase the frequency of the sine wave to 100 Hz, and to get up to 2000 baud, we would require a frequency of 1000 Hz. That is, the frequency is always half the information rate, according to Hartley's Law.

This is, however, the theoretical limit, and has never yet been achieved in a practical system. Even more frequent sampling than Hartley's Law would indicate is required; to carry an information rate of 2000 baud in practice requires a sampling rate of 5000 Hz.

Now let's go back to sidebands and see what happens when we attempt to transmit our information, using the corresponding sine wave (and the frequencies determined by Hartley's Law). Even with our theoretically perfect zero-bandwidth carrier wave, as soon as we put even a one-baud information rate onto it we create side frequencies 0.5 Hz either side of the original, so it now has a bandwidth of 1 Hz—just the same as the baud rate of the information. If we attempt to transmit information at the rate of 2000 baud, we will have a signal with a bandwidth of 2000 Hz, 1000 either side of the carrier.

What this means in practice is simply that even were we able to achieve a "perfect" zero-bandwidth carrier, just as soon as we add modulation to it to convey information it will require sidebands. These sidebands must be large enough (wide enough) to carry information at the rate we are pumping it in.

And if you think this is just an exercise in doubletalk, then fire up a supersharp receiver, crank its selectivity down to a needle-sharp 10 Hz or so. (It can be done, with a good crystal filter together with a Q-multiplier and a sharp audio filter—ask any serious VHF operator.) Then try to copy some of the 100 wpm CW that shatters the airwaves down around the low end of 40 meters. You'll find that the dits and dahs all run together into a solid tone. We normally say that the filter is "ringing" in such a situation, but what's really happening is that the sharp filtering is shaving off the sidebands which carry the 100 wpm information, leaving you only the carrier.

When we deal with phone signals and ordinary voice modulation we don't speak of baud rate or bits; these terms are more often encountered in commercial telegraphy and the computer industry (where they are standard) than in radio to begin with. The principle of Hartley's Law still applies to voice communications, however. Through experiment (mostly by the telephone people over the years) it has been determined that the normal frequency range used for voice communication is from 300 to 3000 Hz, which means a bandwidth of 2700 Hz.

This means that for normal communication, each of our sidebands extends from 300 to 3000 Hz away from our carrier frequency, giving us a total bandwidth for normal AM of 6 kHz (from -3000 to $+3000$ Hz).

We don't get this automatically, though. The voice contains frequency components ranging up to 15 kHz, and if we go for "hi-fi" or "broadcast quality" communications, we will take up 30 kHz of the spectrum instead of 6. That's five times as much space as we need, and means that we stand to create five times as much interference to other stations. For this reason, it's considered good operating practice to restrict the frequency range of a phone transmitter to 3 kHz at the upper end, thus limiting its bandwidth requirement to 6 kHz.

While courtesy and sharing are important, there are more selfish reasons also for limiting the frequency range of a phone signal. The phone company studies have shown that voice components outside the 3 kHz "communications" region are not worth the effort required to transmit them. By restricting the bandwidth, then, you can make all the sideband power into *effective* power and not waste any of it. This means a stronger signal at no extra cost, or, to put it another way, less cost per unit of signal strength! If courtesy doesn't convince you, maybe economy will.

If we're using FM rather than AM, then the relation between modulating signal and bandwidth is a bit different. We'll go into that in more detail when we look at how AM and FM carry voice. For right now, the key difference is that with AM, it's the frequency range or bandwidth of the audio which determines the bandwidth of the transmitted signal, while with FM, it's the amplitude or intensity of the audio which sets bandwidth of the transmitted signal.

Even with AM, though, the amplitude of the signal has some effect upon bandwidth. Specifically, if the audio signal is too strong for the carrier, it will tend to cut off all carrier power on negative-going peaks. This causes a clipping of the audio, which creates an extremely wide band signal. The "splatter" which results can interfere with all the signals in a single ham band, and is prohibited by FCC regulations. So long as signal amplitude is kept below this limit, however, only the frequency of the audio can affect bandwidth of an AM signal.

The important points to remember out of what we've covered so far, then, are these: Any modulation of a signal must increase its bandwidth, with the bandwidth increase being determined by the rate or speed at which information is conveyed. Slow-speed CW has the narrowest bandwidth, and TV video (which must carry the equivalent of millions of bits of information per second) the widest. For any one type of modulation, the bandwidth is primarily dependent upon a single characteristic of the modulating signal. And finally, signals with greater bandwidth than that required to carry the desired information are to be avoided for many reasons.

Now that we've learned a little more about just what modulation amounts to and how it's closely interwoven with signal bandwidth, let's see just how voice signals are carried by AM and FM.

AM AND FM MODULATION

As we have already seen, the principal types of modulation used for ham phone transmission are AM and FM. The oversimplified view of how is far too condensed to be precise, but it is useful in getting a starting point: AM varies the strength of the signal (leaving the frequency constant), and FM varies the frequency (leaving the strength constant). This is often shown by *envelope waveforms* which are supposed to look something like those in Fig. 9-1, and *spectrum distribution* charts such as in Fig. 9-2.

As you can see, these show no significant frequency distribution for the AM signal, and no significant change in amplitude for the FM signal. And since you can actually see such displays on the face of a scope properly connected to the output of an FM or an AM transmitter, it's sometimes difficult to see just why these ideas should be too simple.

The problem is that any scope hookup, and for that matter any receiver, cannot examine a single spot-frequency

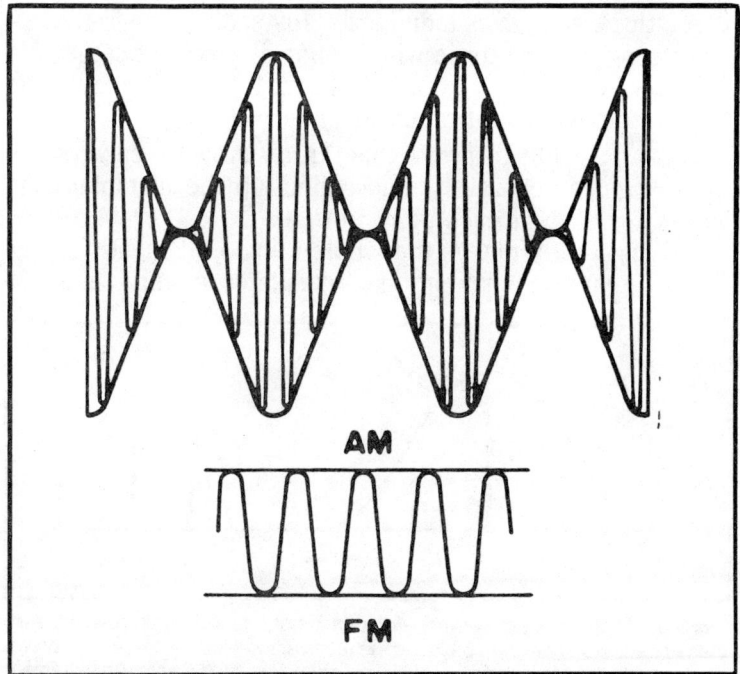

Fig. 9-1. Envelope waveforms of AM (left) and FM signals show that amplitude varies in AM, but remains steady with FM.

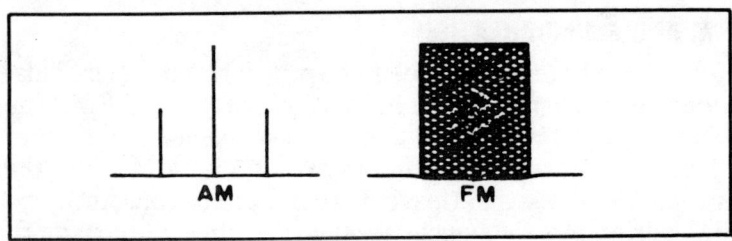

Fig. 9-2. Spectrum distribution charts for AM and FM show that AM occupies fixed positions in the rf spectrum, while an FM signal appears to vary in frequency over a band of closely related frequencies. With most receivers, the AM spectrum distribution would appear to be a single point rather than the three points shown.

signal. Instead, it must act on a whole band of signals which make it through the *selectivity curve* of the device, and the result is influenced by all the signals in that band.

Thus, when you have a 1000 kHz carrier wave, together with two side-frequency signals at 1000.5 and 999.5 kHz (each with the proper phase relationship to the carrier), neither the scope nor the receiver will actually show you these three signals as individuals. Instead, the overall *average* strength of the three will be shown—and that comes out as the typical AM envelope waveform. Figure 9-3 shows this by comparing two different spectal charts.

It's possible to prove this out by drawing each of the three signals separately, and adding up the instantaneous values for each at each point in time.

What really makes the whole thing a bit mystifying at first is that the conventional modulator circuits used to

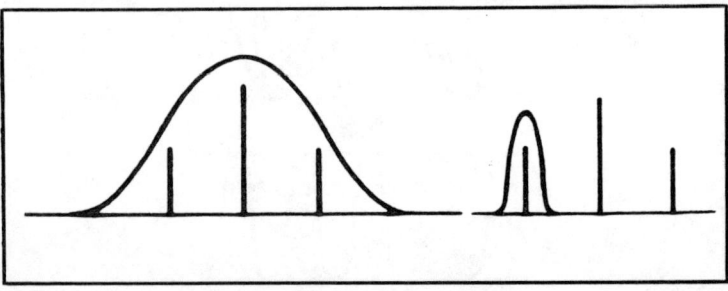

Fig. 9-3. AM appears to be a single-frequency signal to almost all receivers, and many spectrum analyzers, because the receiver's selectivity curve is so broad as to include the carrier and all sidebands as shown at left above. An extremely selective receiver (right), or analyzer, is capable of selecting any one component of the complete signal and isolating it.

achieve AM really look as if all they do is control the amplitude of the signal. Take, for instance, the high-level plate-modulation circuit shown in Fig. 9-4. This circuit is used in AM broadcast transmitters, and the few amateur AM transmitters still in use. It is shown here to illustrate the process of modulation.

Nothing would be easier than to look at this circuit as a means of controlling power output of the rf amplifier by adjusting its plate voltage with the audio signal to be trans-

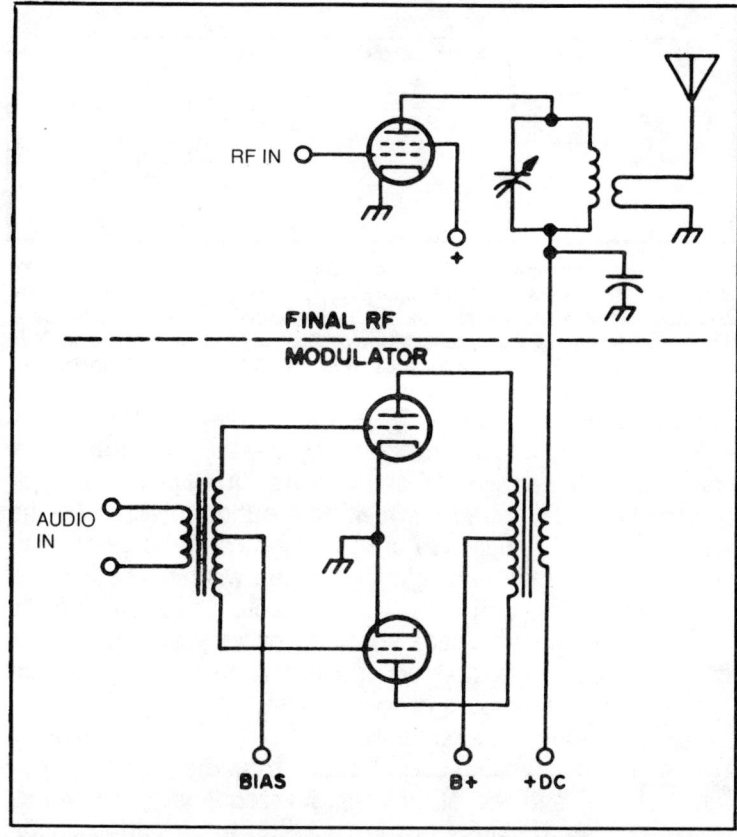

Fig. 9-4. This high-level plate-modulation circuit, which is typical of the most popular type of AM modulator, was wrongly explained for many years as a type of "valving" action in which the high-power audio signal from the modulator's output transformer or modulation transformer alternately add power to, or canceled power out of, the steady dc supply level, and thus caused the rf amplifier stage to produce more or less output power accordingly. The explanation is highly plausible, but fails to account for the presence of sidebands which are present in every AM signal.

243

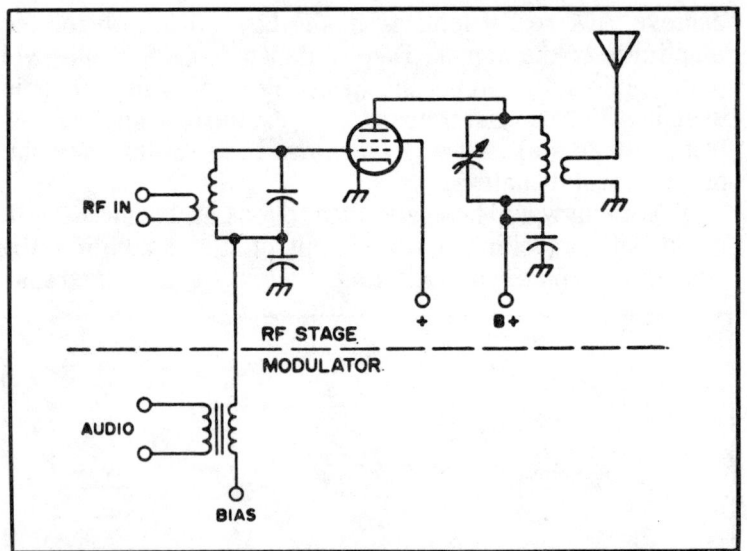

Fig. 9-5. Grid-modulator circuit for AM could be viewed as valving action by way of grid bias, but it's much easier to take the correct viewpoint and see that the rf and audio input signals mix together in the amplifier stage to produce a modulated signal as the output. Mixer circuits used in receivers and SSB exciters, which we'll examine later, are very similar to this grid modulator circuit.

mitted. In fact, a couple of generations or so of hams learned solidly that this was just how it worked—the rf amplifier was said to act "like a resistor" and vary its output power according to the plate voltage, which was either lowered (if the audio was going negative) or raised (when audio went positive). This is simple and clear cut—and wrong. Those hams who learned it so solidly had a really rough time adapting to SSB when it came in, because SSB forces you to know how it really does work instead of a might-be explanation that isn't correct regardless of its glibness.

What really happens is this: The rf amplifier receives two separate signals, one at radio frequency (the normal input signal) and the other at audio frequency (through its power leads). Because of the variation in operating conditions, and more to the point, because the rf amplifier distorts its signals (and therefore is not a linear device), these two separate signals interact with each other and the result is not just one, or even two, but four signals at the output. These four output signals consist of the two input signals and two new signals. One of the new signals has a frequency equal to the sum of the frequencies of the two input signals, and the other has a frequency equal to the difference.

When the two input signals are far apart in the rf spectrum, as they are in this case, we call this process *amplitude modulation*. If the input signals are closer together, we call it *mixing*, and we'll meet it again several times as we progress through this course. No matter what we call it, it's the same process every time—any time that two signals meet in a nonlinear device (it doesn't even have to be an amplifier), we're going to get four signals out.

You may have noticed that these "new" signals which appear in the output of our modulated amplifier are identical with the side-frequency signals we met earlier.

The appearance of a change in signal strength is, as we mentioned, because at any one instant the net energy in a circuit depends on *all* the signals present, and the phase relation between carrier and sidebands is such that the sidebands alternately add to and subtract from the carrier-wave signal's power to give the appearance (and practical effect) of varying signal amplitude.

Once the basic idea of what really happens during modulation takes hold, it's much easier to see how such circuits as the grid-modulated amplifier (Fig. 9-5) can work. These are a bit difficult to comprehend on a straight amplitude-control basis, but when you look at modulation as being essentially a mixing process, things come through more clearly. The major difference between the different amplitude modulators then boils down to where the audio is fed in.

The mixing-action principle also explains why the bandwidth of an AM signal depends upon the bandwidth of the audio signal applied to the modulator. The higher the maximum frequency in the audio signal, the higher will be the sum or upper sideband limit, and the lower will be the lower sideband limit. Since signal bandwidth is the difference between upper and lower sideband limits, this means that high-frequency audio means wider bandwidth.

With signal sideband signals, incidentally, both the carrier and one sideband are normally suppressed, and the bandwidth is thus cut to be roughly equal to that of the audio signal fed in (or half that of the corresponding AM or DSB signal).

The basic principle of FM is, unfortunately, not so easy to see accurately. The simplified principle (which is no more accurate than the amplitude-control idea of how AM works) is typified in Fig. 9-6, where a capacitor microphone is used to vary the frequency of an oscillator, directly controlling the frequency of the transmitted signal. This was the old standard explanation, but as we said it's not accurate. The fact is that FM signals have sidebands just like AM signals do, and what's more the only difference between (ham style) FM and AM in the sidebands is that FM's sidebands are phased differently with respect to the carrier. (Wideband broadcast FM has many more pairs of sidebands than does AM; that's where much of its improved audio quality comes from.)

The big difference in sideband phase is shown in Fig. 9-7. You can see that where both the AM sidebands are going up, one of the FM sidebands goes down. Thus the two FM sidebands appear to cancel each other so far as amplitude is concerned, giving rise to the steady envelope we saw in Fig. 9-1. However, this imbalance between the sidebands makes that effect on apparent carrier frequency much greater, so that the frequency of the composite signal appears to swing with FM, where it remains constant with AM. In fact, as the

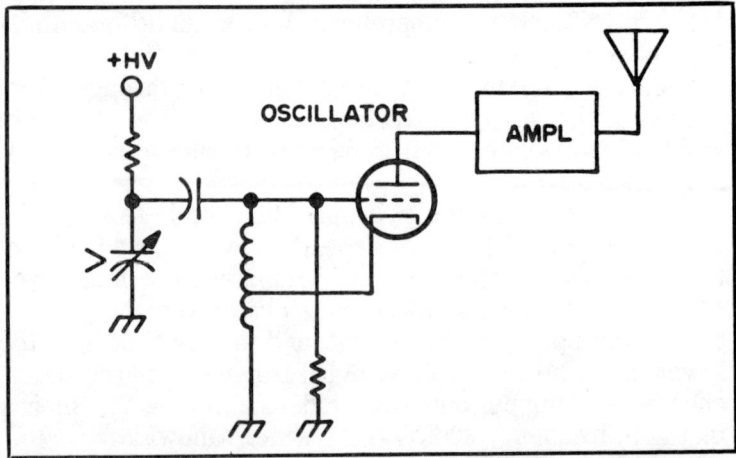

Fig. 9-6. Oversimplified view of how a frequency modulator works is provided here, where a capacitor microphone (in which capacitance is changed by sound waves striking it) serves as a tuning capacitor for the oscillator in the transmitter. Speech causes the capacitance to vary, which changes the transmitter's frequency. In practice, a voltage-variable capacitor is used and the voltage applied to it is varied by the audio signal.

modulation level is increased, you can reach a point with FM where the effect of the two sidebands is to completely cancel the carrier wave. This is called the point of *vanishing carrier* and can be used to measure modulation level.

The fact that sideband phase makes the difference between AM and FM is the basis of the Armstrong system for frequency modulation, which made commercial FM practical. This system used conventional AM techniques, but took the carrier out of the signal and shifted its phase by 90 degrees, then put it back. The result was FM.

A not-so-widely known fact, suggested by the sideband relationships shown in Fig. 9-7 and proved by experiments a number of years ago, is that you can combine AM and FM of the same signal to produce an AM output signal with only one sideband.

This is not SSB as we know it, because the carrier is not suppressed, but by proper adjustment of the levels of both types of modulation you can make the upside-down FM sideband cancel the corresponding AM sideband, while the carrier and the other sideband reinforce each other. The technique has been used on SWBC stations, but it's illegal for hams—because FCC regulations prohibit simultaneous FM and AM of the same signal.

Practical ham FM modulator circuits usually involve phase modulation of some type which is made to look like FM by adjustment of the audio frequency response (that's the only effective difference between FM and PM). The phase modulation is achieved by electronically detuning a tank circuit, to cause the tank's phase shift to vary in step with the audio signal.

MODULATION MEASUREMENT

To carry information of any sort, a radio signal must be modulated. Modulation consists of varying the characteristics of the signal in order to transmit the information. Naturally, when we vary any characteristic we can either vary it only slightly, or we can carry the variation to extremes—and so the degree of modulation applied to the signal can vary from almost none up to some limiting point.

The measurement of modulation is the determination of the degree of modulation applied to a signal, and it's necessary to measure modulation of any modulated signal for a number of reasons.

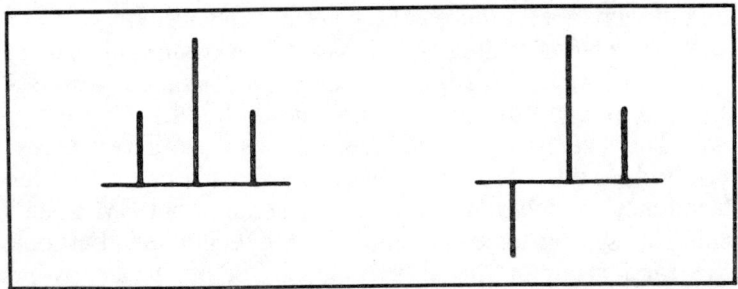

Fig. 9-7. Essential difference between AM sidebands (left) and FM sidebands (right) is in their phase with respect to the carrier signal. AM sidebands are completely symmetrical about the carrier; FM sidebands are offset in phase and so pull the frequency to one side or the other instead of affecting the envelope amplitude. This is shown in these plots by the reversed direction of the lower FM sideband.

Most of these reasons boil down to the basic factor of complying with rules and regulations. For instance, we've already mentioned that it's illegal to apply both AM and FM to the same signal at the same time in the ham bands. To be sure that we are complying with this rule, we have to be able to measure both the AM and the FM of the signal, and be certain that whenever one is present, the other is not.

We've also seen that the bandwidth of the modulated signal is determined by its modulation. FCC rules limit the bandwidth of a ham signal, and so we must perform modulation measurements to be certain that we remain within the bandwidth limit.

When too much AM is applied to a signal, we saw that the result is splatter, which interferes with other communications over a wide band of the spectrum. Because of this, the level of AM permitted is limited. Again, we must make measurements to assure ourselves that we are complying with the rules.

Before we can examine the how of modulation measurement, we must meet some of the words and phrases we'll be using. The quantity most often measured is known as *percentage of modulation*, and is derived from the instantaneous rf voltages present in the modulated signal as shown in Fig. 9-8 when we're talking about AM. An unmodulated carrier has a modulation percentage of zero, while a fully modulated signal is 100% modulated. While it's possible to exceed 100% modulation, doing so is forbidden by FCC rules

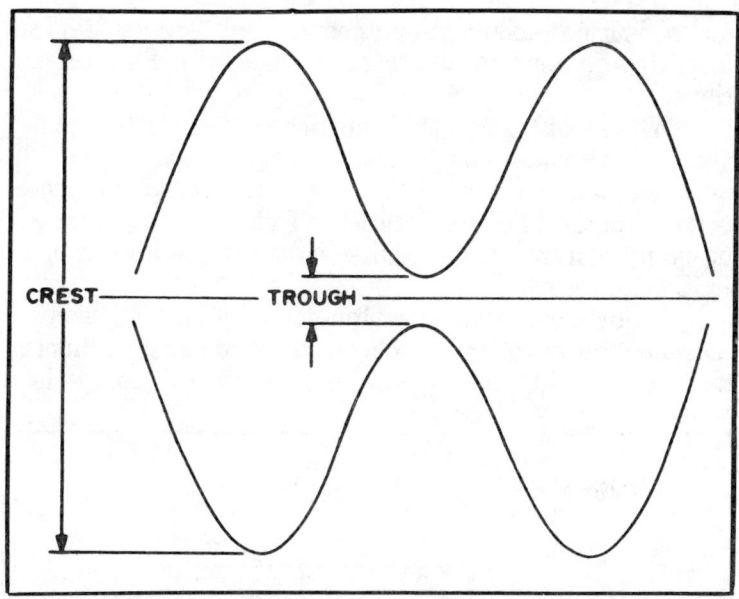

Fig. 9-8. For an AM signal, modulation percentage is a function of crests and troughs on a sum and difference basis. The resulting fraction is multiplied by 100 to change it into a percentage. Both crest and trough quantities are usually measured as voltage. In this example, the crest is 86 units and the trough is 12 units, so modulation percentage is (86−12)/(86+12) or 74/98, which comes out to 0.755, or 75.5%. Modulation percentage as defined here is meaningful only for AM; since FM has no amplitude variation, it would always come out to 0%.

because of splatter, and such a condition is called *overmodulation*. Most good AM signals average about 80% modulation; some safety factor is necessary to keep from going over 100% with audio peaks.

Since the amplitude of an FM signal does not vary when it is modulated, modulation percentage as defined in Fig. 9-8 is meaningless if applied to FM; all signals would be 0% modulated. The corresponding term for FM is *modulation index*, which is the ratio of signal frequency deviation to modulating signal frequency. For broadcast FM, the modulation index is usually 1.0 or less.

Now we're ready to look at the techniques of modulation measurement. We'll take up AM measurements first, since most ham work on the HF bands makes use of AM, and then turn our attention to FM.

For AM, the most useful measurement is that of modulation percentage. The oscilloscope is the best instrument for measuring modulation percentage, but simpler devices may also be used to assure compliance with FCC regulations.

With most types of AM, the plate current of the final rf amplifier remains constant with proper modulation, but overmodulation (more than 100%) will cause the plate-current meter's needle to flicker. This is the simplest go/no-go measurement of modulation, and will suffice to meet FCC requirements.

A high-resistance ac voltmeter may also be used to measure the ac voltage produced by the modulator; this can be calibrated in percentage of modulation (by comparison

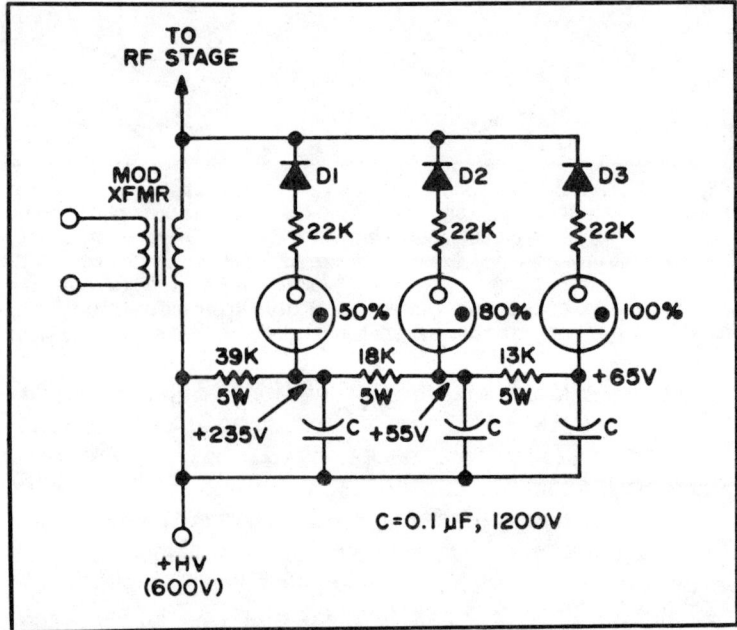

Fig. 9-9. Simple devices can be built to indicate when defined modulation percentages are achieved. This arrangement uses neon bulbs and switching diodes to fire the bulbs when the negative-pack modulation reaches 50%, 80%, and 100% respectively. Bottom ends of the neon bulbs return to voltages which are uniformly 65V more negative than the voltage marking the corresponding modulation percentage. These voltages, and the resistance values in the divider which establishes them, depend upon the voltage supplied to the transmitter (here assumed to be 600V). The 65V offset is the firing voltage of the neon bulb. When the upper end goes 65V more negative than the lower end, the corresponding bulb can fire. In use, the object is to keep the 50% bulb on all the time, the 80% bulb on as much of the time as possible, but never permit the 100% bulb to flash.

with scope measurements, one time) and is more accurate than the plate-current indication. The calibration will be correct only so long as the dc voltage to the modulated stage remains unchanged.

Neon tubes may be used as visual indicators of modulation percentage, by connecting them through diodes so that they compare the instantaneous supply voltage of the modulated stage to the unmodulated voltage and thus flash whenever a specified percentage of modulation is exceeded. Figure 9-9 shows such a circuit, set up with three bulbs which glow 50%, 80%, and 100% modulation. Similar circuits may be set up with "magic-eye" tubes as indicators.

To measure modulation percentage with an oscillscope, the modulated rf signal is applied directly to the vertical plates and the modulating audio signal is applied to the horizontal plates. In the absence of modulation, the display is a single vertical line in the center of the screen, because no horizontal deflection voltage is available. This vertical line is a picture of the carrier, and its height is proportional to the carrier's intensity.

When modulation is applied, the modulating signal provides horizontal deflection to the scope and at the same time causes the carrier's intensity to vary. At the positive peak of the modulating signal, carrier intensity will be maximum, and the display will be at one limit of its horizontal deflection. At the negative peak of the modulating signal, carrier intensity will be minimum, and the display will be at its other horizontal limit.

With 100% modulation, in which the maximum carrier intensity is twice that of the unmodulated carrier and the minimum carrier intensity is zero, the resulting display is a triangle. With less than 100% modulation, it is a trapezoidal shape, and for this reason the scope measurement method is often called a *trapezoid pattern* modulation measurement. With more than 100% modulation, the triangle develops a horizontal line at its tip, representing the excess modulating signal (the carrier cannot reach less than zero intensity). Figure 9-10 shows these patterns, and Fig. 9-11 shows the typical test hookup for the average ham transmitter.

The value of the scope method over all simpler techniques is that the two vertical edges of the display are direct representations of the quantities which define modulation percentage, and so you can calculate modulation percentage accurately from the scope display by the formula shown in Fig. 9-10. In addition, the slanting edges of the display are drawn by the two signals involved, and should be straight at all times. Any kinks indicate distortion being produced by the modulator, which could result in interference to other signals and even at best would mean a sloppy output signal.

Scope measurements can be made on any AM transmitter, since they simply compare the signal doing the modulating to the signal being modulated. The simpler techniques

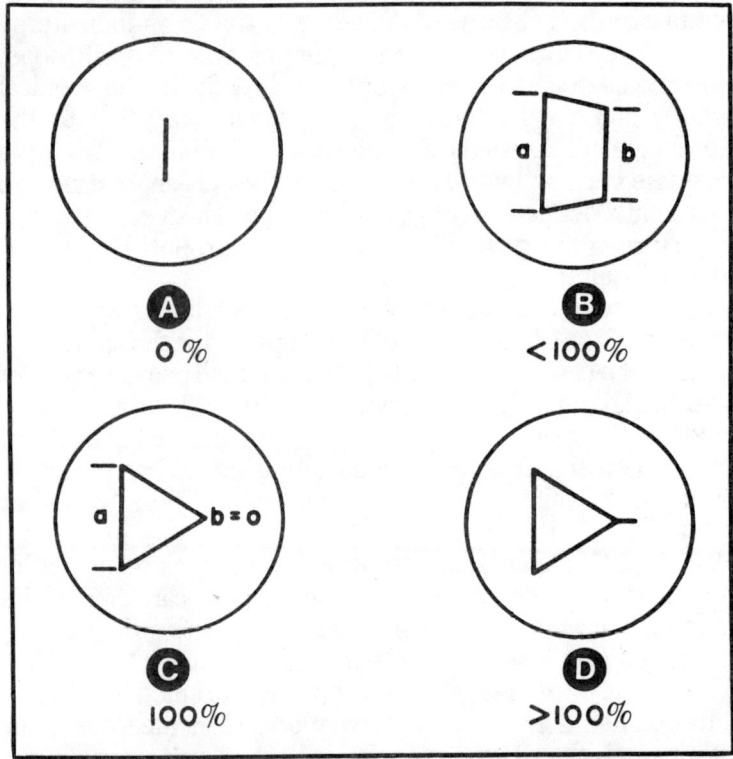

Fig. 9-10. Trapezoid patterns for 0%, less than 100%, 100%, and greater than 100% (overmodulated) modulation percentages are shown here in ideal form. Modulation percentage can be measured directly from the scope screen by taking the fraction $(a-b)/(a+b)$ and multiplying by 100 (patterns B and C). When $b=0$ as in C, this becomes a/a or 1 time 100, for 100% modulation. In addition to indicating modulation percentage, this measurement tells whether the modulator is distorting the audio signal. Clean signal is indicated by straight sides. Any kinks or curvature of the slanting parts of the display mean trouble in the transmitter.

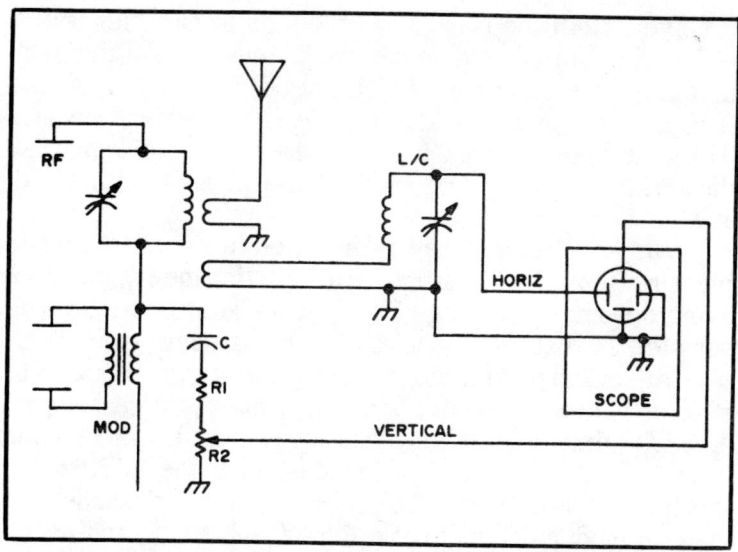

Fig. 9-11. This is how to hook up a scope to get the trapezoid pattern of Fig. 9-10. Rf is sampled from the final tank circuit by means of a link coil and tuned circuit L/C, and fed to the vertical plates of the scope through short, direct connections. Audio is sampled from the secondary of the modulation transformer and applied to the horizontal plates. Capacitor C is a safety blocking capacitor to prevent supply voltage from getting into the scope; resistors R1 and R2 cut down the audio signal level as required, and may not be necessary for low-power rigs. Size the display is adjusted by the tuning of L/C and the setting of R2.

apply primarily to high-level plate-modulated transmitters (the most common type), and must be used with caution if any other modulating technique (screen, grid, or cathode modulation, for instance) is employed.

When it's FM instead of AM you're measuring, the situation is somewhat different. Most of the common modulation measurement techniques apply only to AM, and cannot be used for FM.

The accurate technique for measuring FM involves a device known as a spectrum analyzer, which is not normally found in a ham shack. The normal ham technique is to locate the point at which the first pair of extra sidebands is produced. This requires a 3 kHz oscillator and a sharp-tuning receiver capable of detecting one signal which is only 3kHz from another, stronger signal.

The transmitter is modulated with the output of the 3 kHz oscillator, and the output signal is examined with the receiver. Both the carrier and the side frequencies 3 kHz either side of it should be detected. Audio gain of the modulator is now slowly increased until a second pair of side frequencies, this time 6 kHz away from the carrier, appears. The gain should be backed off until the second sidebands just disappear, and this point marked as a modulation index of 1.0.

An ac voltmeter may be installed in the modulator to measure signal level for the 1.0 modulation index, and when a microphone is substituted for the oscillator the gain can be adjusted so that this modulation index is never exceeded.

At certain points, the carrier of an FM signal appears to vanish. This occurs as the modulation index is increased past 1.0. The first point of vanishing carrier is at a modulation index of 2.4; the next one is at a modulation index of 5.52, and the third at 8.65. Similarly, the first pair of sidebands vanishes at modulation indexes of 3.83, 7.02, 10.17, and 13.32. Modulation indexes greater than about 8.0 are too wide for even VHF FM bands, but the points of vanishing carrier and vanishing sidebands can be used to set certain indexes with certainty.

The technique is similar to detecting the second sideband pair; the difference is that the receiver is left tuned to the carrier, and the audio gain increased until the carrier vanishes. This represents a modulation index of 2.4. The receiver is then tuned to the first pair of sidebands, and gain increased until they vanish. This is a modulation index of 3.83. Back to the carrier, and increase modulation until it vanishes again at 5.52, and so forth. If two separate audio gain controls are provided on the modulator, one can be used to set modulation level and it can be calibrated by this technique; the other can then adjust for differences between microphones.

VHF FM enthusiasts sometimes use *frequency deviation meters* to adjust or measure the modulation level of their transmitters. These meters are similar to FM receivers, and provide a direct indication of the bandwidth of the modulated signal.

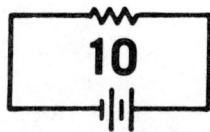

The Listening Post: Radio Receivers

One of the oldest sayings in radio (attributed by some to Signor G. Marconi in the years preceding 1900) is, "You can't work 'em if you can't hear 'em." The direct implication of this maxim is that a good receiver is essential to any ham station.

Not only is a good receiver essential, but the FCC requires that any aspirant for the General class license have some knowledge of what goes on inside that receiver. While only three questions on the official study list of questions to help in preparation for the license exam involve receivers, those three between them manage to require a fairly complete knowledge of receiver theory.

In this chapter we're going to tackle those receiver questions, and while we're at it we'll take up SSB (single sideband). This isn't as farfetched as it might sound, because the generation of an SSB signal is mighty like the reception of just about any kind of signal.

RECEIVER TYPES

The earliest kind of radio receiver on record was a device which today would startle almost anyone involved with electronics. It was "a circle which could be rotated within itself . . . made of copper wire 1 mm thick, and had a diameter of only 7.5 cm. One end of the wire carried a polished brass sphere a few millimeters in diameter; the other end was pointed and could be brought up, by means of a fine screw insulated from the wire, to within an exceedingly short distance from the brass sphere."

This receiver was used in 1888 by 31 year-old Heinrich Rudolph Hertz (the man for whom the unit of frequency is named) to demonstrate the existence of "distinct rays of electric force" carried without wires, through space. The quotations are from his own description of the experiments, published in December 1888. The copper-wire loop was the complete receiver; the accompanying transmitter was a sparkgap arrangement, and the maximum distance between transmitter and receiver was limited to about 5 ft (when everything was working perfectly, amazing DX of 6 ft could be achieved).

After Hertz' experiments demonstrated the existence of radio waves, Marconi developed a communication system around them. The Marconi system used a "coherer" as the main part of the receiver; this was much more sensitive than Hertz' loop-and-sparkgap, but still nothing compared to today's apparatus.

By the time amateur radio began to become popular, the coherer was on its way out, having been replaced by the much-more-sensitive galena crystal. This was a first cousin to today's semiconductor crystal diodes, and operated in essentially the same fashion (but for more than 20 years no one suspected as much). Finally, during World War I, the vacuum tube came into use for radio reception, and a young Signal Corps major named E. H. Armstrong invented, in order, the regenerative receiver, the superregenerative receiver, and finally the superheterodyne circuit. The superhet is still the standard receiver circuit, although transistors and integrated-circuit chips are replacing the vacuum tubes Armstrong knew.

Despite the almost universal use of the superhet for serious communications today, the diode, regenerative, and superregen circuits are still very much with us. The popular "Twoer" and "Sixer" transceivers from Heath used superregen receivers, and most superhets include at their core at least one diode detector which is a direct descendant of the ancient "crystal set." Regenerative circuits are popular for beginners' construction projects and can deliver amazing performance with a minimum of components.

All these receiver types so far are for AM reception. For CW, the regenerative and superhet circuits work nicely. The diode circuit and the superregen do not produce easily audible output from CW signals; the crystal set worked on the CW of its day only because the output of a sparkgap transmitter was not a clean CW signal, but instead was a raucous buzz (or smooth whine if the gap was driven by an alternator or used a rotary gap arrangement, as some of the large commercial installations did).

Almost all FM receivers are of the superhet variety, although a superregen will operate nicely on FM and even a diode circuit can perform creditably.

The modern standard, as we said, is the superhet. These come in several types, and we'll find out more about them in the next section.

THE SUPERHET

Before we can get very far in finding out what goes into a superhet circuit, we must look at receiver circuits in general. This, in turn, is going to take us back in places to the subject matter of our previous chapter, modulation.

For now, incidentally, we'll only be examining receivers for AM signals. Once we have them down cold we can proceed to the features required for other types of modulation.

The simplest practical receiver for today's signals is the diode detector, shown stripped to its basics in Fig. 10-1. It will work just this way, too, if you care to try it. Incoming rf picked up by the antenna must pass through the diode and the

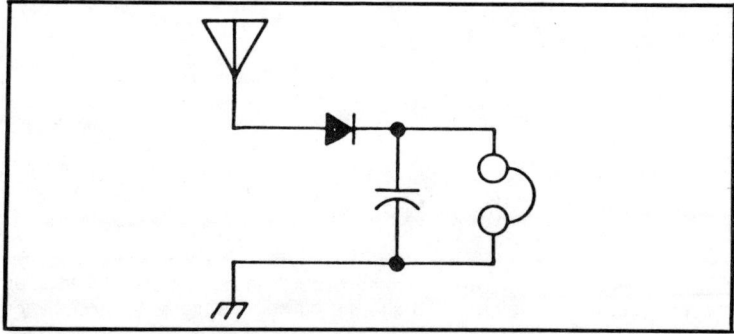

Fig. 10-1. Simplest radio receiver consists only of antenna, ground, diode, capacitor, and earphone connected as shown.

capacitor (the inductance of the headphones acts as a choke) to reach ground. The diode, however, permits current flow in only one direction and blocks any reverse current flow. This results in production of a charge across the capacitor which follows the *peak* values of the radio-signal envelope waveform. The headphones are operated by this voltage. Figure 10-2 shows the waveforms and brings out how the rectifying action of the diode causes the modulation envelope to be reproduced across the capacitor.

Such a receiver as this will have little or no selectivity. That is, it will receive *all* radio signals which reach it. Most will not produce enough power across the capacitor to operate the phones, so if you try it you probably will get only the

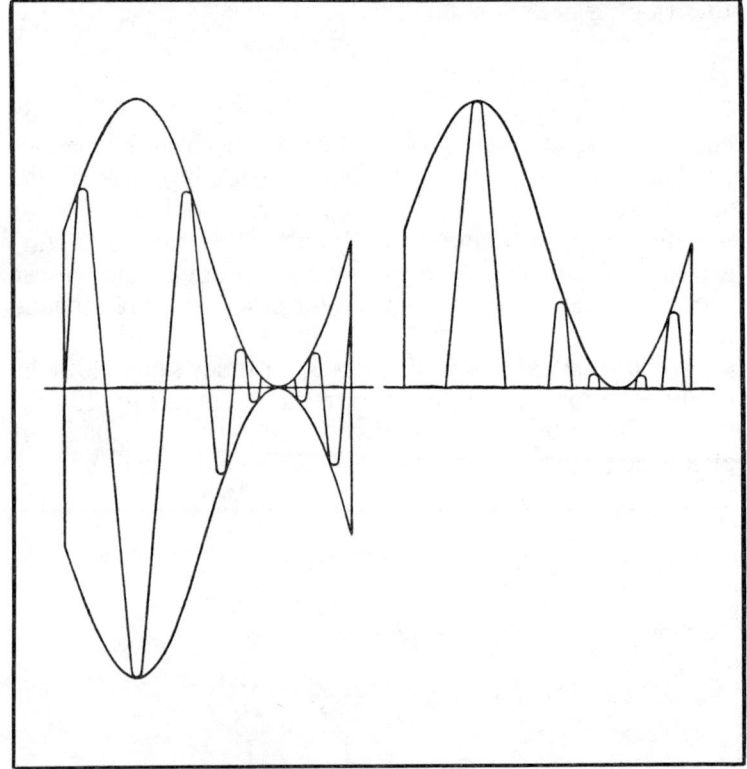

Fig. 10-2. Receiver of Fig. 10-1 operates by rectifying an rf signal picked up by the antenna (left waveform) and shaving off half the signal to produce the pulsating-dc output shown at right. This signal charges the capacitor to the peak voltage reached by each half-cycle of rf, and earphones respond to the audio envelope thus reproduced.

strongest broadcast station in your area (and if you live more than 15 to 20 miles from the transmitter, you may not get even that). Some degree of tuning is provided by the antenna, but that's the only control over what you get.

Obviously, this simple circuit suffers from three major problems. It is not very sensitive (cannot pick up weak signals), has poor selectivity (cannot separate a single signal from all the rest), and has low output (earphones only, and no control of volume).

Selectivity was improved in crystal-set days by adding tuning circuits between antenna and detector. As many as three or four separate tuned circuits were found in the fancier sets, and they helped to shave reception down to the desired signal. However, they could not produce enough selectivity to separate two signals only a few kHz apart—and they did nothing for sensitivity or output.

It took the vacuum tube to do much for receiver performance in the areas of sensitivity and output volume. The amplification made possible by the tube could be used at either radio frequency, to provide added sensitivity by producing a stronger signal to the detector than the antenna alone could produce, or audio frequency following the detector, to provide louder output. Normally it was used both places—and still is.

Early vacuum tubes were expensive, and experimenters were hard-pressed to obtain more than one at a time. Ingenious circuits were devised to permit this one tube to amplify the signal at rf, then detect it, and amplify the resulting audio some more, all in the same stage.

Then Major Armstrong came up with the regenerative detector, which made use of positive feedback to produce both exceptionally high gain and very sharp selectivity within a one-tube stage. For years, the regenerative circuit was the standard for both amateurs who could afford the tube, and commercial installations.

By this time researchers had realized that the particular set of problems represented by a radio receiver had no simple direct solution. Radio communications require high frequencies, but amplifiers give greatest gain at low frequencies. Similarly, selectivity requires sharp tuned circuits, but the higher the frequency the less sharp any tuned circuit becomes. These problems, as much as anything else, limited early radio to frequencies below the upper end of today's broadcast band. Available receivers simply would not operate properly at higher frequencies.

The regenerative receiver amounted to an oscillator with not quite enough feedback provided to sustain oscillation. This gave extreme gain and excellent selectivity. By this time, CW transmitters were replacing the noisy sparkgaps, and the regenerative receiver also turned out to be excellent for receiving CW signals (which the diode detector simply could not detect). All that was necessary was to push the receiver just over the line into oscillation; the oscillating receiver was then tuned to a frequency just a few hundred cycles off from that of the transmitter. The receiver's own rf mixed with the incoming rf from the transmitter to give a *beat note* or *heterodyne* signal equal to the difference between receiver and transmitter frequencies, and this pure tone was easy copy. This technique was known as *heterodyne reception*.

While serving overseas during World War I, Major Armstrong had what was probably his most important of many new ideas. He saw how to eliminate all the problems of reception by applying the heterodyne principle in a new way. Rather than offsetting local and received frequencies by an amount in the audio range as was done for heterodyne reception, why not make the difference come out in the lower end of the rf range, at a point where good gain and the desired selectivity could easily be achieved. The resulting *intermediate frequency* signal could then be redetected into audio, after being amplified and shaved down as desired.

That was the *superheterodyne* idea, and it's still in use in essentially the same form its inventor first conceived. A superhet today will have, as a minimum, a local oscillator to provide the offsetting frequency, a mixer stage to combine the local and signal frequencies into the intermediate frequency or i-f, and i-f amplifier, a second detector, and audio stages. Most will also have one or more stages of rf amplification preceding the mixer (sometimes called the first detector), and if intended for communications use, will provide heterodyne action for the second detector.

Figure 10-3 shows the stages of a single-conversion superhet communications receiver. The purpose served by each stage is described in the following paragraphs.

The rf amplifier's primary purpose in superhets operating below the VHF range is to isolate the local oscillator from the antenna and thus prevent interference by the receiver to other receivers. At VHF, the rf amplifier serves primarily to boost the signal strength and thus help overcome noise problems, but at HF and lower frequencies ample gain is available in later stages.

The local oscillator provides the *offset frequency* signal and thus controls the receiver's tuning. The receiver will pick up only signals which are separated from the local-oscillator frequency by the amount of the i-f frequency, and so changing the frequency of the local oscillator changes the frequency to which the receiver is tuned. The local oscillator must be stable; that is, its frequency must remain constant once set. Otherwise the receiver will *drift* off frequency and will require frequent retuning.

The mixer or *first detector* operates in the same manner as the amplitude modulator we examined in the last chapter, combining two input signals at different frequencies to produce four output signals which are at the two original frequencies, their sum, and their difference. Most often only the difference frequency is of interest. The mixer actually gets many more signals in, and produces many more out, because all the signals which make it through the rf amplifier stage are acted upon by the mixer. The relative frequency difference between different signals is much greater at the output than at the input, however, because the mixer essentially subtracts a constant (the local oscillator frequency) from each of its input signals.

The i-f strip is the heart of the superhet receiver. This block consists of one or more stages of rf amplification, operating at a fixed and relatively low frequency. For many purposes, the standard intermediate frequency is 455 kHz. At this low frequency, very sharp selectivity may be attained, yet good gain is still possible. The i-f amplifiers are set to one frequency and left there (a process called *alignment*), and thus the receiver's gain and selectivity are the same at any frequency within its tuning range. Because of its selectivity, the i-f strip accepts only one of the many signals appearing at the mixer output; all the rest are eliminated, but the one signal accepted goes on through and is amplified to respectable strength before reaching the second detector.

The second detector is essentially the same as the simple diode detector of Fig. 10-1; since both sensitivity and selectivity have been provided by the preceding stages, the only function of this stage is to convert the i-f signal into audio.

For reception of CW and SSB signals, a *beat frequency oscillator* (bfo) may be provided. This oscillator produces a signal offset from the i-f by a difference within the audio range; normally, it includes a *pitch control* which adjusts its output frequency to fall either side of the i-f. The bfo output is fed to the second detector along with the i-f when heterodyne reception is desired. In more recent designs, the bfo is sometimes included as part of a separate *product detector* circuit which is switched in to replace the second detector for heterodyne reception. We'll look at this more closely in the next section.

Once the incoming signal is converted to audio by the second detector or its equivalent, the audio is then amplified to drive a loudspeaker (if desired) by the audio section.

Everything has a price, and the performance of the superhet is no exception. Since the basic principle is that of converting an incoming signal to an intermediate frequency by mixing action, which produces both a sum and a difference signal, it's possible to receive two different signals simultaneously with a superhet. Such action is called *image* reception.

An example may make it clearer. Broadcast-band receivers usually have the local-oscillator frequency higher

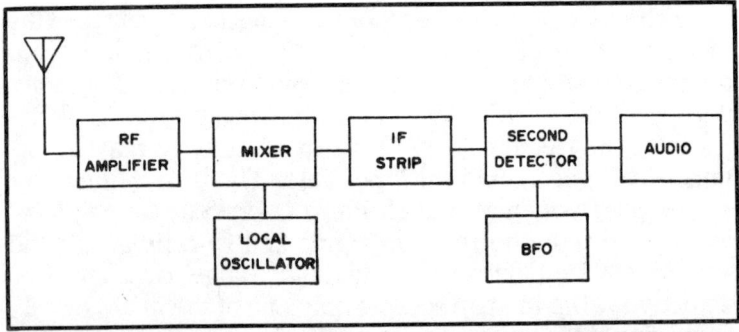

Fig. 10-3. Basic stages of a conventional superheterodyne receiver are shown here and described in the text. Rf amplifier and bfo are not normally included in non-communications superhets for entertainment use, but are considered essential in quality communications receivers.

than that of the signal to be received (partly to minimize the image problem), and normally use a 455 kHz i-f. Thus, when one is set to receive a station at 610 kHz (the low end of the band), the local oscillator will be set to 1065 kHz, 455 kHz higher, in order to produce a difference frequency of 455 kHz.

Unfortunately, with the oscillator at 1065 kHz, any incoming signal at 1520 kHz which happens to reach the mixer stage will also produce a 455 kHz output (1520—1065), which the i-f stage will not be able to distinguish from the desired signal. Thus, the 1520 kHz signal can and will interfere with the desired one at 610 kHz.

The characteristic of image reception is that the undesired signal is always separated from the desired one by *twice* the *i-f* and is on the *other side of the local-oscillator frequency*.

Another aspect of the image problem, which shows up on many inexpensive ham receivers, is the possibility of tuning the same signal in at two places on the dial. To reverse our previous example, you could tune in the 1520 kHz station either at its true frequency of 1520 kHz, or at the image point 910 kHz lower, 610.

As the signal frequency gets higher, with a constant i-f, the separation between true and image points becomes proportionately smaller. In the 40 meter band, for instance, 8 MHz commercial signals show up on top of 7 MHz ham signals, and the ham signals themselves can also be tuned in around 6 MHz if you like. By the time you get to the 10 meter region, ham signals at 29.5 MHz can be causing interference with other ham signals around 28.59 MHz.

Since the separation between true and image points is always twice the i-f, one solution to the image problem is simply to use a higher i-f. If the i-f is raised to 1600 kHz, for instance, the true and image points will be separated by 3.2 MHz, and then a single rf stage ahead of the mixer normally will reduce the image response to the point that it is not objectionable.

Unfortunately, the higher the i-f, the poorer the sensitivity and selectivity of the receiver.

One of the more ingenious solutions to this problem was the invention of the *double superhet* or *double conversion* system. This is essentially two superhets end to end, with the mixer of the second taking the place of the second detector of the first as shown in Fig. 10-4.

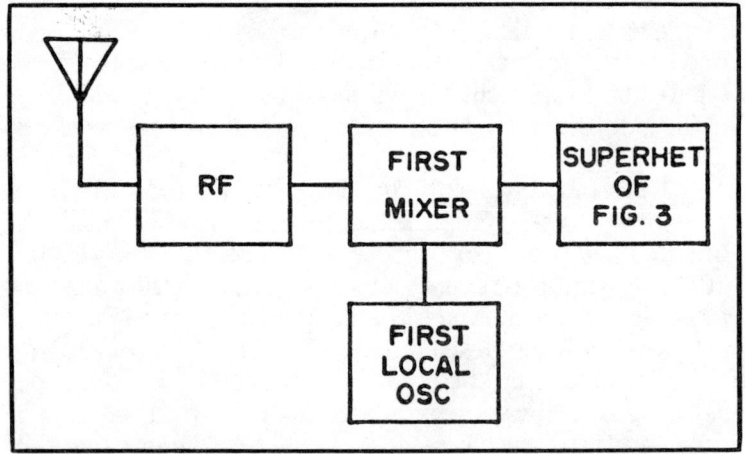

Fig. 10-4. Double-conversion superhet duplicates the "front-end" circuitry of a single-conversion superhet, then uses the entire single-conversion superhet at the output of the first-mixer. Resulting composite circuit has two local oscillators, two mixers, and two i-f strips (rf stage of single-conversion receiver becomes first i-f of a double-conversion circuit). Advantage is improved image rejection because the first i-f can be a relatively high frequency, while maintaining selectivity and gain with a low-frequency second i-f.

The first i-f is picked for best image rejection, while the second is chosen for gain and selectivity. One popular receiver using the double-conversion principle used 1825 kHz as the first i-f, and converted this down to 85 kHz for the second to achieve high selectivity.

When double conversion is used, either the first or the second conversion oscillators may be tuned. Use of a VHF converter ahead of a tunable communications receiver is an example of double conversion with the second oscillator being tuned; the converter produces an i-f between 14 and 18 MHz, for instance, and the receiver then tunes over this band. Many high-performance receivers now make use of double conversion with tunable second oscillators in order to achieve high stability in frequency, and to get the same tuning rate on each band covered (since the tuning is always the same).

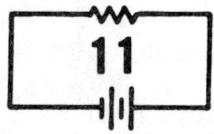

Single Sideband

In a preceding chapter we made the acquaintance of sidebands and found that sidebands are necessary in any kind of modulated signal—in fact, that sidebands carry the information which composes the modulation of that signal.

We also saw that any modulated signal has two sidebands, one on either side of the carrier, which are mirror images of each other.

It follows from this that only one of the two sidebands is really necessary, and that the carrier isn't necessary at all. That is, the same amount of information could be transmitted if we simply didn't bother to transmit either the carrier or one sideband, just so long as long as we did transmit one of the two sidebands accurately.

Since with amplitude modulation the limit of 100 percent modulation is reached (with sine-wave signals) when the modulating signal has half as much power as the carrier wave, and since this modulating signal's energy is split between the two sidebands, that means that 5/6 of the power transmitted from an AM station is useless; all of the information can be recovered properly from one sideband, which represents only 250W of the power transmitted from a 1 kW transmitter (250W out of a total of 1500W: 1000 + 250 + 250).

That's what SSB is all about—the technique of taking advantage of these facts, to achieve that 4-to-1 power advantage by concentrating all the transmitted power in the only meaningful part of the signal, and simultaneously to conserve space in the rf spectrum by holding the bandwidth of the signal down to half that of a conventional AM transmission.

Unfortunately, doing so is not so simple as we may have just implied. Getting rid of the carrier and one sideband requires special circuitry in the transmitter, and, in addition, SSB signals cannot be received as simply as can those from ordinary AM stations. The missing carrier must be put back into the signal at the receiver in order to recover the audio, and that's a bit of a job.

SSB has been in use since 1927, but until 1948 was applied only to commercial telephone communications. The first ham station to use SSB was apparently W6YT, the Stanford University club station. But within a matter of weeks after W6YT appeared with SSB, a number of other stations followed suit. Today, SSB is the standard means of voice communication on HF ham bands.

A number of factors combined to produce the 21-year delay between first use of SSB and hams' acceptance of the technique. Most of them boiled down to the fact that at first, it was simply not simple enough to be practical for the vast majority of hams. Costly installations and special equipment were required. The advances in radio technique during World War II changed this situation and made it practical for hams to take up SSB operation.

Still, SSB is nowhere near as simple as ordinary AM, either to generate for transmission, or to receive. In addition to all the functions of a normal AM transmitter, the SSB transmitter must include facilities for removing the carrier and unwanted sideband from the signal. These normally operate properly only at low signal levels, and once the signal is converted into SSB it cannot be amplified by normal rf power amplifiers (which would introduce distortion). This introduces the requirement for linear amplifiers, which are more difficult and critical to adjust than are ordinary class C stages.

SSB RECEPTION

Once the signal is generated and transmitted, it must be received. Frequency stability is necessary in an SSB receiver, because a drift in receiver tuning of as little as 20 Hz is noticeable, and a 100 Hz drift renders the signal unreadable in most cases. The receiver must, in addition to having high stability, include provisions for putting the carrier back into the signal for detection, and this means that heterodyne reception is necessary.

Let's look at the requirements for both generation and reception of SSB signals, now that we have a general idea of how they differ from conventional AM techniques. We'll take the transmitter first.

SSB TRANSMITTER CIRCUITS

Figure 11-1 shows a block diagram of a typical SSB transmitter, contrasted with that of a typical AM transmitter. The difference in complexity is obvious.

In the SSB transmitter, the audio is applied to a sideband generator stage where it is converted to an rf signal which reproduces the audio at some radio frequency. This is done by a variation of the normal AM process, but the carrier and unwanted sideband are removed before the signal gets out of the sideband generator. Output of the sideband generator is an SSB signal, but is not necessarily at the desired signal frequency because it turns out to be much simpler to always generate the SSB signal at the same frequency, and then translate it by a mixing operation to the desired frequency for transmission.

Sideband generator output, therefore, is applied to a mixer stage, and an output-frequency tuning signal is applied to the mixer's other input. This output tuning signal may be obtained either from a fixed-frequency oscillator such as a

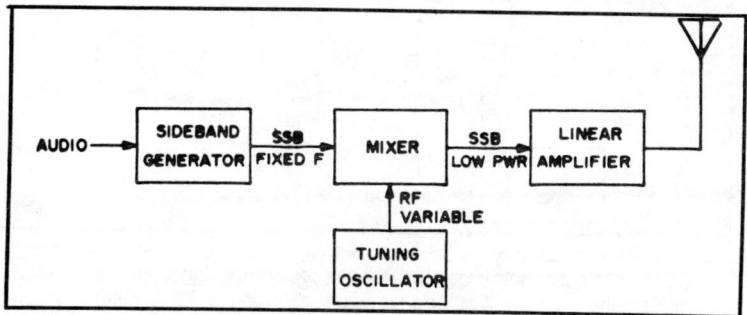

Fig. 11-1. Typical SSB transmitter consists of a sideband generator, provisions for changing the output frequency, and linear amplifiers.

crystal circuit, or from a stable vfo. The mixer's output includes an SSB signal at the desired output frequency, which is selected and amplified to the output power level by means of linear amplifiers.

The sideband generator stage is the heart of the SSB transmitter, and warrants additional examination. At least three different techniques for sideband generation are known, but only two of these are used in practice. They are known as the *filter method* and the *phasing method*.

Both the filter and phasing methods make use of *balanced modulator* circuits, which are special types of mixers which eliminate one of the two input signals from their outputs. These balanced mixers are usually set up to eliminate the carrier signal, because the audio signal (at a far different frequency) can easily be rejected by tuned circuits. Output of an SSB balanced modulator, then, usually consists only of the two mirror-imaged sidebands which result from the mixing process.

Figure 11-2 shows a block diagram of a filter-method sideband generator. Incoming audio and the rf carrier signal are applied to a balanced modulator where the original signals disappear. Output of the balanced modulator is a double-sideband (DSB) signal with both carrier and audio suppressed. This DSB signal is then passed through an extremely sharp filter, which shaves off one of the two sidebands, and passes the other relatively unaffected. Output of the filter, therefore, is the desired single sideband signal.

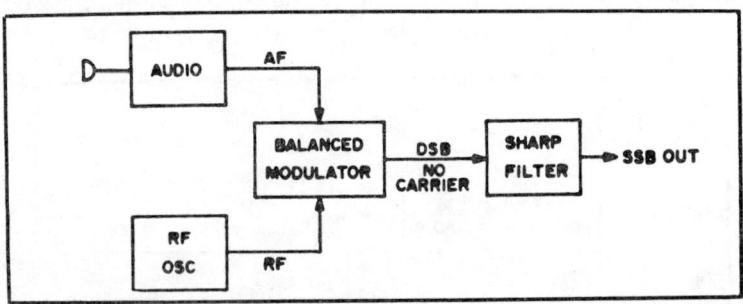

Fig. 11-2. Filter-method sideband generator converts audio and fixed-frequency rf into a double-sideband suppressed carrier signal using a balanced modulator, then filters away one sideband to produce SSB. Filter may use crystals or a mechanical technique, and may operate at frequencies from 17 kHz up to 9 MHz.

which shaves off one of the two sidebands, and passes the other relatively unaffected. Output of the filter, therefore, is the desired single sideband signal.

The phasing method (Fig. 11-3) is more complex, but eliminates the need for the sharp (and costly) filter. Incoming signals, both audio and rf, are applied to phase-shifting networks. The phase shift networks produce pairs of outputs, which are identical except that they differ from each other in phase by 90 degrees.

One audio-rf pair is applied to once balanced modulator, and the other pair goes to a second balanced modulator. Outputs of these two balanced modulators are double sideband signals.

The balanced-modulator outputs differ from each other in phase, however. Because of the phase shift introduced into the signals before mixing, one sideband (either the upper or the lower one) will have the same phase in both modulator outputs, while the other sideband will have a 180-degree phase difference between one modulator output and the other.

The two outputs are combined in a summing network, where the sideband with no phase difference survives while

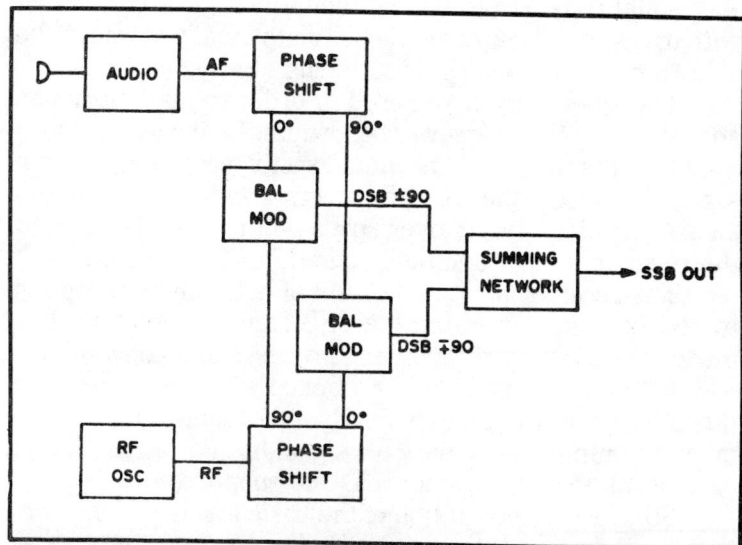

Fig. 11-3. Phasing method of SSB generation makes use of balanced modulators operating with different combinations of input signals, forcing one sideband to cancel out in the summing network. See text.

the one with 180-degree phase difference cancels itself out. The output of the summing network is, therefore, the desired single sideband signal.

Both the filter method and the phasing method have advantages. The filter is simpler in concept, but requires components which are usually more costly. On the other hand, phasing is less expensive but requires more critical adjustments. Either technique is capable of producing outstanding results—or garbage, depending upon the skill and care of the operator.

Once the sideband generator produces an SSB signal, regardless of the method used to obtain it, the mixer (Fig. 11-1) translates it to the desired output frequency and its power level is determined by linear amplifiers.

SSB RECEIVER CIRCUITS

Now that we understand how SSB is generated and transmitted, let's look at how it is received:

Almost any receiver capable of copying CW signals *can* be used for SSB reception, but for generally satisfactory results an SSB receiver should have high frequency stability, a slow tuning rate, and sharp selectivity. We've already seen why stability is required. The tuning rate goes right along with it; when a 20 Hz error causes distortion, you want to be able to make fine tuning adjustments slowly.

The selectivity is required in order to take maximum advantage of SSB's narrower bandwidth. Best use is made of a signal if the receiver bandwidth exactly matches that of the signal, because if the receiver covers a wider band than the signal occupies, the leftover space will contribute noise to the receiver's input and thus degrade total performance.

The receiver must be capable of heterodyne reception to recover the audio from the SSB signal. Detection, like modulation, is actually a mixing process. In a conventional AM detector, the sidebands are mixed with the carrier, and the difference frequency (which is the audio involved) is taken as output. Since the SSB signal has no carrier to mix against, an external carrier must be supplied.

SSB forced recognition of the basic identity of modulation, detection, and mixing—and led to improved techniques for detection when the principles long established for mixer operation were applied. The resulting circuits are now

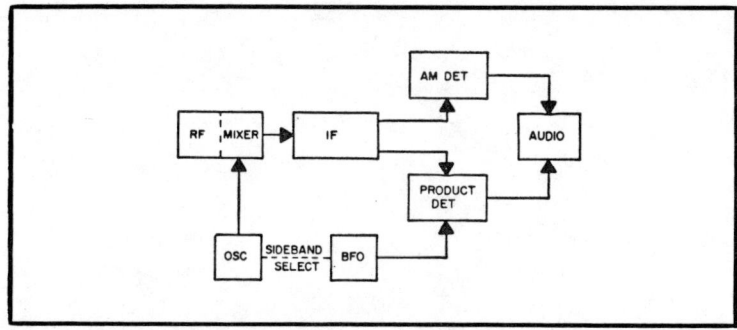

Fig. 11-4. Selectable sideband receiver shows great similarity to an ordinary receiver (Fig. 10-3), but with added extras to facilitate SSB reception. Actually, any receiver capable of receiving CW can tune in SSB, but extra features shown here are needed for extended use.

known as *product detectors* and provide outstanding performance for both CW and SSB signals.

A good agc system is convenient to have in an SSB receiver, because the signal strength varies widely and rapidly. This creates some problems, because conventional AM receivers drive their agc (sometimes called avc) systems with a signal derived from the carrier strength (the same signal drives the S-meter), and SSB has no carrier. The problems have been solved in many different ways, and again CW operators as well as SSB users have benefited.

The block diagram of a typical single-conversion selectable sideband receiver intended for AM/SSB/CW operation is shown in Fig. 11-4. The circuit may be compared with Fig. 10-3, which shows the conventional version of a similar receiver. You can see that the sideband receiver contains all the functions required for ordinary reception, and merely adds a few new ones to handle SSB.

The purposes served by these added functions have been indicated in the preceding paragraphs. Naturally, many more functions *could* be added. For instance, most selectable sideband receivers involve double-conversion techniques to achieve sideband selection. For this, add the features of Fig. 10-4.

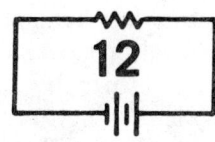

Antennas: The Common Denominator

One item which every radio station, be it for two-way communications or merely a listening post, has in common is the *antenna*. By definition, the antenna is the part of an installation which couples the transmitter or receiver to that little-understood medium in which radio signals travel, so without one neither transmission nor reception could be possible.

Because of its central importance, the antenna is involved in a number of questions in the FCC examination for the General class ticket.

Since these questions cover not only antenna operation and characteristics, but those of antenna feedlines as well, we're going to have to look at everything between the transmitter output connector and the input connector of the distant receiver to get a full grasp of them.

ANTENNA OPERATION

In the beginning of this study course we made the acquaintance of the electric and magnetic fields which are inseparably associated with the flow of electrical current or the motion of a magnetized object. At that time, we observed that while the electric and magnetic fields were alternately swapping their energy content to provide motion of the current, some of the energy was lost to the mysterious surrounding medium, and this loss was known as *radiation*.

An antenna is simply a device intended to make this "loss" of electromagnetic energy easy; its whole purpose is to couple energy from a normal wire conductor into space, or from space back into a normal conductor.

Since any flow of current within a conductor involves at least some radiation, almost any conducting material must act as an antenna of sorts, coupling energy into space and vice versa. That's why portable radios operate so well with self-contained antennas. They have high sensitivity, and the tiny internal antennas extract enough energy from the powerful broadcast station signals to do a good enough job.

Virtually all ham antennas, though, are conducting surfaces which are large in comparison with the wavelength involved. When the conductor is long in comparison to the wavelength, the field strength at any one point on it will not be balanced out by similar strengths at many other nearby points. Instead, the strength of both the electric and the magnetic field around any one point on the conductor will vary, and the conductor will vary, and the strength from one point to another will vary depending upon the distance between the points.

To see how this works, let's look first at a half-wave dipole antenna which is fed at its center. This antenna (Fig. 12-1) is one of the most commonly used varieties, and is also one of the easiest to visualize in operation.

First, let's assume that we have neither voltage nor current associated with the antenna. With no voltage, there's no electric field, and without current, no magnetic field. The antenna is just so much wire, hanging in the breeze.

Now let's apply a voltage suddenly at the feedpoint, say 100V between the wires. That would be +50V to one wire and −50V to the other, for instance.

These voltages cannot appear instantly at the ends. Instead, they must travel along the wire (at approximately the speed of light) to get there.

At each instant during this rather brief trip, there's a point on one wire at which voltage is just changing from 0 to +50V, and a corresponding point on the other at which it's just going from 0 to −50V. As the *wavefront* moves toward the ends of the antenna, these two points are getting farther apart from each other.

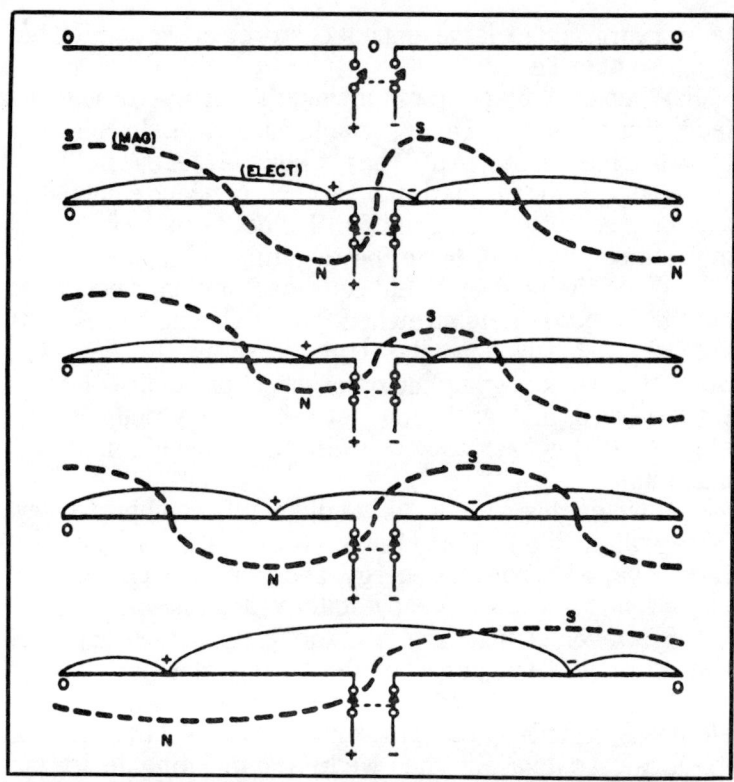

Fig. 12-1. Half-wave dipole provides an example of how a resonant antenna operates. At the top, the antenna is disconnected from any energy source and so is inactive. When energy is supplied, fields begin to travel toward the tips of the antenna. As they travel, radio waves are emitted into space. Each cycle of rf repeats this action to provide continuous radiation.

Since these two points are at different voltages, an electric field will exist between them. And since they're moving away from each other, the field is expanding rapidly. The energy which enables this field to expand is extracted from that which we fed into the wires to cause the initial change.

But whenever an electric field moves, a magnetic field moves right along with it. That voltage change involved a current flow, and the current flow is greatest back at the feedpoint because that's where all the energy must pass.

Eventually (rather rapidly, in fact) the voltage change reaches the ends of the antennas. Now it has no place to go. The field, however, is expanded to a large region, and contains a surprisingly large amount of stored energy which has to go someplace.

When the voltage stops moving, the current through the feedpoint stops also. The associated magnetic field collapses (because motion stopped). Thus, when the electric field is at its strongest, the magnetic field is at zero, which establishes a 90-degree-out-of-phase relationship between the fields and imparts outward motion to the resulting radio wave.

When the magnetic field collapses, nothing remains to hold the electric field stretched, and it also collapses. The voltages rush back down the elements toward the feedpoint. Since they're now going the other direction, their polarities are reversed, and when they get back to the feedpoint the electric field is back to zero while the magnetic field is at maximum.

If we do this with dc, as we did in this example, everything will stabilize after a few such cycles. The fields will carry away any transient energy, and when the full length of the antenna reaches a steady state, radiation will cease.

However, if instead of dc at the feedpoint, we apply ac, and adjust the frequency of this ac so that each time a reflected wavefront comes back from the tips of the antenna elements, it finds a new "push" just ready for it to send it on its outward path again, then we have a radiating device par excellence. It will accept just enough energy from the feedpoint to balance out that which it "loses" to space, and thus maintain what is known as a *standing wave* of rf energy on the antenna. During each cycle of applied rf, this standing wave will launch an infinite number of traveling radio waves into space.

In practice, we cannot adjust the frequency to suit our antenna, because ham frequencies are assigned within relatively narrow bands by the FCC. What we do instead is to adjust our antenna length to fit the frequency at which we desire to operate.

FREQUENCY AND WAVELENGTH

Frequency and wavelength are closely related, since frequency is the number of complete cycles which occur in a specified period of time, and wavelength is the distance between corresponding points of two successive cycles. The speed at which radiation travels is the relating element;

wavelength is equal to speed of travel divided by frequency. If wavelength is in meters and frequency in megahertz, the formula comes out to be $\lambda = 300/f$.

The half-wave antenna we've been using as an example, then, should have a length approximately half of one wavelength. To operate at 7 MHz, its length should be about 22½ meters.

This relationship is only approximate, though, because every antenna has some capacitance to ground, which makes it appear to be a little longer (to the fields) than it really is. The difference usually amounts to about 5%; the formula usually used to determine length of a half-wave antenna takes this into account, and also comes out in feet rather than meters. It's length = $468/f$, where f is frequency in MHz.

Figure 12-2 shows what happens when an antenna is either too short or too long for the operating frequency, as well as the desired result when the length and frequency match.

However, one-half wavelength is not the only practical length for antennas. Almost any multiple of quarter-wavelengths (with the 5% correction factor applied to the one at the far end, only) can be used. Such antennas are

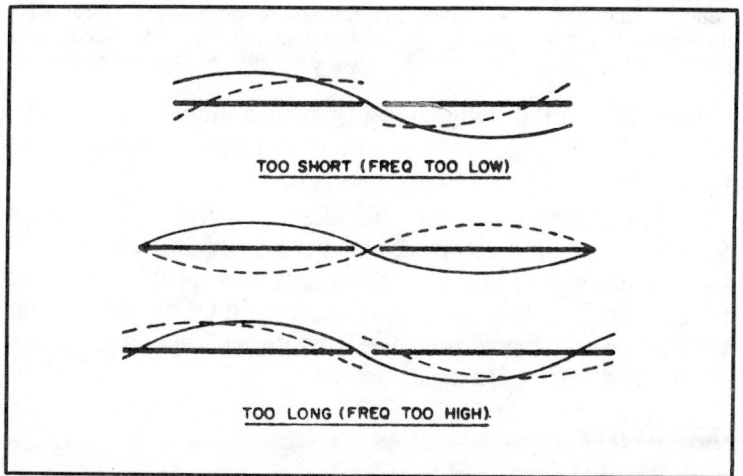

Fig. 12-2. When an antenna is too short for the operating frequency, the reflected electric field is out of phase with the driving energy (top) and part of the driving energy is cancelled. If the antenna is too long for the frequency (bottom), the same thing happens in reversed phase. Only if antenna length and operating frequency match (center) does all energy radiate, and the antenna's impedance has no reactive component.

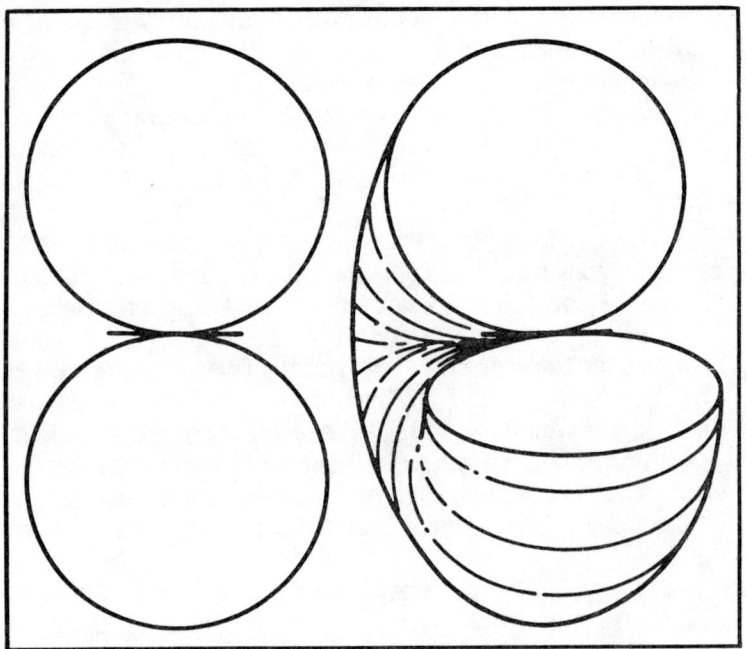

Fig. 12-3. Radiation pattern of a half-wave dipole is shown here both in cross-section (left) as generally drawn in pattern diagrams, and three dimensionally (right). Resemblance to a donut strung on the wire is easy to see. This pattern occurs only if the antenna is in "free space" with nothing affecting it, a condition which never actually occurs. Actual antennas are influenced by the presence of ground, which reflects energy and modifies the radiation patterns.

sometimes called *long-wire* antennas, although some hams use the term long wire for a random-length antenna which operates in spite of rather than because of its length.

Any antenna has two characteristics by which its performance can be measured, and a third which is essential to making use of it. The *rating* characteristics are its directional pattern or *directivity*, and its *efficiency*, and the third characteristic is its *impedance*. We'll examine impedance later. Right now let's take up directivity.

DIRECTIVITY

When the antenna launches its radio waves into space, each tiny part of the conductor launches its own collection of rays. That is, each infinitesimal portion of the antenna acts like a tiny isolated antenna, and each radiates its energy equally in all directions like an expanding sphere.

In some directions, however, the traveling waves from various parts of the overall antenna structure have phase relationships which cause them to cancel each other, while in other directions these phase relationships cause the individual rays to reinforce each other and build up added strength.

For the half-wave dipole, the resulting pattern of *far-field signal intensity* is something like a donut strung on the antenna wire (Fig. 12-3), with the strongest part of the pattern at right angles to the wire itself, and no field strength at all off either end.

Long wires modify this pattern to produce a cone of field strength (Fig. 11-4), aimed in the same general direction as the wire is pointing, but covering a region at an angle to the wire itself.

Combinations of long wires can be put together as in the rhombic antenna (Fig. 12-5), to give almost no response at all except in the two directions in which all the combinations add together. Such antennas are favorites of commercial installations, as well as of those hams who can afford enough space to string them. Most of us, though, must settle for much less.

In fact, the differences between the various types of antennas are largely differences in directional patterns, or in different types of construction in order to achieve equivalent directional patterns.

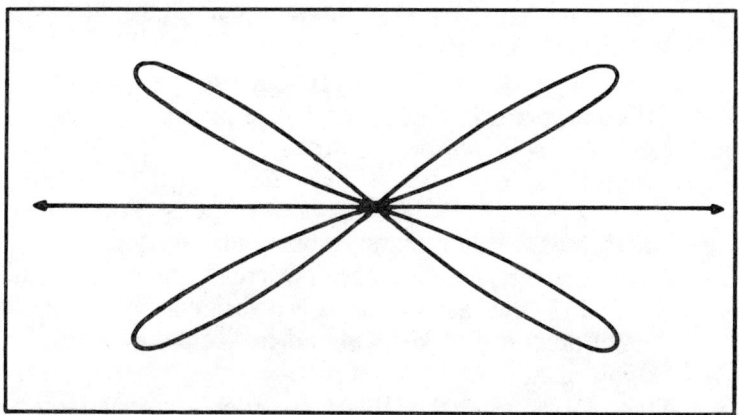

Fig. 12-4. Long-wire antenna's radiation pattern stretches the figure-8 pattern of a dipole into a pair of lobes. The longer the wire, the closer the lobes approach the wire direction. In three dimensions, the pattern would resemble two cones, point-to-point, strung on wire.

ANTENNA EFFICIENCY

The efficiency of an antenna is difficult to measure. In general, resonant antennas (those which are multiples of quarter wavelengths long) have the highest absolute efficiency, and directive antennas (those with sharp directional patterns which concentrate the radiated energy within relatively small volumes of space) have higher apparent efficiency than do nondirectional antennas. One rule of thumb left over from the earliest days still is surprisingly valid—the more wire, the better.

In addition to the characteristics of directivity, efficiency, and impedance, which we've mentioned, every antenna has one other characteristic called *polarization*.

POLARIZATION

While it's a bit hard to visualize clearly, every radio wave is composed of two related fields in motion—the *electric* field and the *magnetic* field. Each of these fields has its own plane, and the two planes cross each other at right angles to establish the line along which the wave travels. If it helps to think of the two walls of a room meeting at the corner to establish the position of the corner, do so.

The polarization of this radio wave is simply the direction in which the electric field's plane extends. If the electric-field plane is vertical, the wave is said to be vertically polarized, and if the electric-field plane is horizontal, the wave is horizontally polarized.

Since it's impossible to separate out a single radio wave in practice, a vertical or horizontal polarization of a signal must refer to the polarization of the majority of the individual waves which make up that signal. As a result all actual signals contain portions of both polarizations. A vertical signal will have some horizontal components, and vice versa.

A signal which contains both vertical and horizontal components is known as an *elliptically* polarized signal, and if it has equal amounts of vertical and horizontal, we call it *circular* polarization.

What determines the polarization of a signal is the antenna that transmits it—but polarization of any individual wave will change every time that wave is reflected from anything on its path from transmitter to receiver.

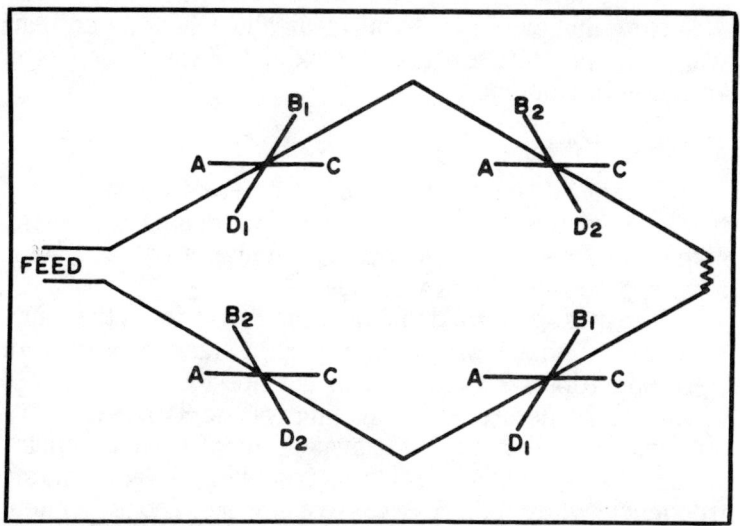

Fig. 12-5. Rhombic antenna is derived from the long-wire (Fig. 11-4). Each leg of the rhombic has a long-wire pattern symbolized here by lines rather than lobes. Radiation arriving from the left, however, goes to the terminating resistor and does not affect the feedline; thus lobes A, B2, and D1 have no effect on antenna operation. That arriving from the right can reach the feedline. Of this, only that in lobe C of each leg adds up. B1 lobes tend to cancel each other out, as do D2 lobes. Result is a highly directional single-lobe pattern, but the antenna is too large for most ham installations.

The plane in which the transmitting antenna polarizes the signal is the same as that containing the wire of the antenna, for the most common antenna types. Thus, a horizontally stretched half-wave dipole will transmit a horizontally polarized signal, while a vertical antenna will transmit a signal having vertical polarization. If the signal is transmited at the same time from two antennas, one vertical and one horizontal, the result is an elliptically polarized signal.

In the normal HF region (3 to 30 MHz), the polarization of the signal doesn't mean too much because most communications in these bands are by means of the ionized layers of the atmosphere, and the polarization is rotated when the signal reflects from these layers. At VHF, however, polarization is important. VHF work in this country now seems to be about equally divided between vertical and horizontal polarization. Elliptical polarization is seldom used except in the UHF regions and above, where the helical antennas which produce it most simply are small enough to be convenient.

Now that we have some ideas about how an antenna works in general, let's take a look at the various types of antennas in common use.

ANTENNA TYPES

Antennas are classified in many different ways. Some of them are named for the people who first popularized them, some are named for their major characteristics, and some have apparently arbitrary names.

The most general division of antennas breaks them into two groups: those which depend upon standing waves for their operation (as described in the previous section) and those which operate with traveling waves. Traveling-wave antennas are sometimes known as *terminated* antennas. Hams make very little use of this type, because at moderate frequencies they require excessive acreage, and at frequencies where their size is practical, they are outperformed by other types. The rhombic is virtually the only terminated antenna used by hams, and we won't go into any of the others in this group.

The antennas using standing waves can be subdivided again into those which are *resonant* and those which are not. A resonant antenna is one which is cut to a length which is a multiple of ¼ wavelength, or made to appear to be such a length by electrical tricks such as loading coils, capacitance, or tuned traps. Most ham antennas are resonant.

A nonresonant antenna is simply a random length of wire. Normally such an antenna does not operate well unless it's tuned by an antenna tuner, which turns it into a resonant antenna so far as the electrical properties are concerned.

Among resonant standing-wave antennas, though, there are still an amazing number of types to be examined. The most common is the half-wave dipole which we used to examine antenna operation. In its purest form, this consists of a length of wire one-half wavelength from end to end (less the correction factor we mentioned before, for end capacitance), with an insulator in the middle. This divides the antenna, both physically and electrically, into two equal quarter wavelengths of wire set end to end. These are fed with a *balanced signal* (one which is balanced with respect to ground, so that when the signal on one wire is at its positive peak that on the other is at negative peak).

The dipole continues to be a good performer even when its two wires are not stretched in a straight line. Such variants have many names.

For instance, the popular *inverted vee* antenna is merely a half-wave dipole suspended from a single pole at the center, with the ends allowed to drop down to much shorter supports. Its behavior is much the same as that of the dipole.

A half-wave dipole performs well not only at its fundamental frequency but at the third harmonic of this frequency, where each wire is ¾ wave long instead of ¼ wave. That is, a dipole cut for 7 MHz also does well at 21 MHz.

However, at even harmonics the dipole does not do so well. The feedpoint impedance becomes high rather than low, (voltage fed rather than current fed). And most ham bands are in an even-harmonic relation to each other.

Operators who want to use all bands thus find themselves in need of several antennas if they intend to use the dipole. One solution to the problem which finds wide use is to place all the antennas together (Fig. 12-6). This is sometimes called a *parallel dipole* antenna. Only the pair of conductors which are resonant at the particular frequency in use are effective, and they operate as a half-wave dipole. The rest of the conductors merely go along for the ride.

Another solution to the problem is the *trap* antenna. This is essentially a half-wave dipole of the lowest frequency to be used, with tuned parallel-resonant traps inserted in the wires. At the lowest frequency, the trap circuits show up as inductance which electrically lengthens the wire, so that the physical length of the antenna is less than its electrical length. At the next higher band, the outermost traps are resonant and effectively disconnect the ends of the wire, leaving a half-wave dipole of the higher frequency. In a

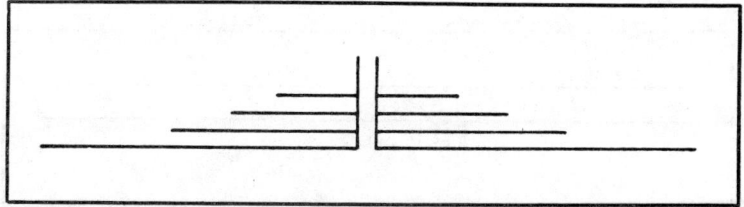

Fig. 12-6. Parallel-dipole antenna consists of several half-wave dipoles all strung together and connected at the feed-point. Only that which is resonant in any band is effective; the rest disconnect themselves.

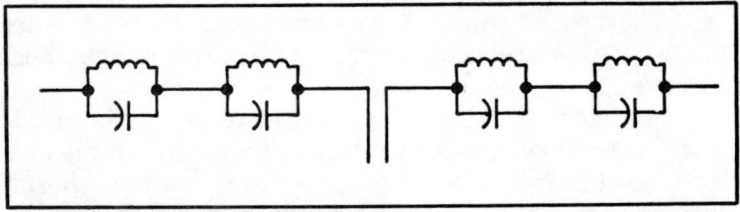

Fig. 12-7. Trap dipole makes use of parallel-resonant trap circuits in the wire to alter the length of the antenna depending upon frequency. See text for explanation of the antenna's action.

5-band trap antenna (Fig. 12-7), this dipole also is loaded by the inner traps.

At the next higher band, the next set of traps is resonant and so cuts off all the ends (including the outer traps). The fourth band of the 5-band trap is 21 MHz, which takes advantage of the third-harmonic operation and uses the 7 MHz dipole. Finally, at the highest band, the innermost traps isolate the outer portions of the antenna to again produce a half-wave dipole.

The trap antenna is a bit more difficult to adjust properly than is the parallel dipole, because the wire length, trap inductance, and trap capacitance all interact with each other at every operating frequency. However, these multiple interactions permit fine adjustments which permit improved performance if you have enough patience.

Still another solution to the all-band antenna situation is the *Windom* antenna. The Windom is a pseudo-balanced antenna similar in appearance to a half-wave dipole, but with the two conductors being of unequal length (Fig. 12-8). This raises the feedpoint impedance to approximately 300 ohms, permitting use of common TV twinlead. Interaction between the shorter wire and the longer wire permits resonance on several ham bands simultaneously.

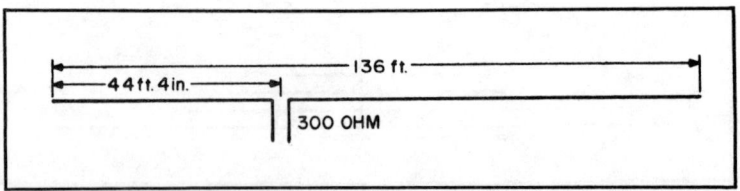

Fig. 12-8. Windom antenna provides a match for a 300-ohm feedline and has been popular since mid-1930s. Dimensions are critical in order to approximate a 300-ohm impedance at the feedpoint on all hf ham bands.

The "L" antenna is similar to the Windom, but the two legs of the antenna extend at right angle to each other rather than in straight line. This tends to broaden the directional pattern in comparison to the two-lobed effect common to almost all dipoles.

All of the antennas we've examined so far have been of the *balanced* variety, which are normally stretched in the horizontal plane and so produce horizontally polarized signals (they don't have to be; at higher frequencies vertical dipoles are often used).

However, radio waves will reflect from any conducting surface, and the reflection produces the same effect as a mirror does for light rays. That is, if a half-wave dipole were to be cut in two at its midpoint and a huge reflector properly placed (Fig. 12-9), the effect would be the same as if the other half of the dipole (the *image* antenna) were actually still present.

This is the basis for the quarter-wave grounded antennas often known as *Marconi* antennas. They are, in effect, half of a half-wave dipole, using the ground as the reflector to provide the image of the rest of the antenna.

Such an antenna must be vertical in order to use the ground as the reflector, and so this type of antenna is often called simply the vertical antenna.

Traps can be used in a vertical, for the same effects obtained in a trap dipole. Similarly, several verticals can be erected in parallel with each other and sharing a common

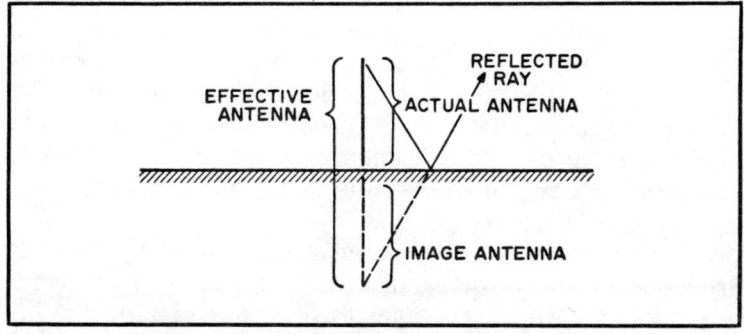

Fig. 12-9. Image principle is the key to the operation of a quarter-wave vertical antenna. Reflected ray shown actually originates at the tip of the actual antenna, but apparently originates at the tip of the image antenna beneath the ground surface. Effectively, then, the antenna is a half-wave dipole standing on end, while the actual structure is only quarter-wave high.

feedpoint to produce an all-band vertical operating much like the parallel dipole.

The major differences between the vertical and the dipole stem from the different orientation. Where the directional pattern of the dipole (on the earth's surface) has two lobes of major response, with nulls off the ends of the wire, the vertical is omnidirectional. Its single null is off the end of the wire still, but now that points straight up. The three-dimensional directional pattern is still the same, but since it intersects the earth's surface differently, the practical effects are greatly changed.

The second major difference comes from the change from balanced (two-wire) to unbalanced (one wire against ground) operation. This changes the feedpoint impedance and the type of feedline required.

ANTENNA ARRAYS

It may seem as if we've omitted several common antenna types from this listing. For instance, we have not yet mentioned the folded dipole or the monopole. However, while these are often spoken of as if they were distinct antenna types, they are in fact only variations of other types which we have already described, which have different impedances. We'll get to them in the next section.

So far we've discussed only the simpler types of antennas. Long ago people found that the directional properties of an antenna could be modified by adding more antennas to it, forming an antenna array. The added antennas can be parallel to the original but separated by some fraction of a wavelength (broadside array), in a straight line with the original (collinear array), or a combination of these (Franklin array). The directivity can be controlled by proper phasing of currents in the individual elements of the array, to give broadside or end-fire patterns, or to vary the direction of maximum response (beam-steered array).

Most of the different antenna types used by hams at frequencies above 14 MHz involve some type of array. Among them are such designs as the *sterba curtain*, the *ZL special*, the *8JK*, the *flattop*, and so forth. A special group of array designs involves *parasitic* elements which are not driven by feedlines, but reradiate energy absorbed from that radiated from the driven element. The most common of these *beam* antennas is the yagi, but again many variations have been designed and each has its own name.

One of the more recent types of antenna arrays is the *log periodic* antenna, which involves exotic mathematics in its design. Such an antenna can cover a bandwidth of 10 to 1 while maintaining a good beam pattern and a low swr, but it's a bit beyond the scope of our discussion here.

When you reach the VHF bands, the number of antenna types takes another sudden jump. There you may encounter such things as the discone, the helical, or the reflecting dish (parabolic antenna).

However, for the General class exam, it's enough to know the simpler types such as the dipole and the vertical, as well as having some idea of the different types in use. We've got enough for that now, so we'll continue and examine impedance and feedlines.

IMPEDANCE

Whenever you discuss antennas, you can't escape *impedance*. It's one of the major characteristics of any antenna, and of any line used to feed the antenna with energy. What we hope to do here is to clarify just what it amounts to, and why it's so important to antenna operation.

Back in one of the early chapters of this study course we made the acquaintance of impedance in its most general form, and discovered that impedance amounts to "any quan-

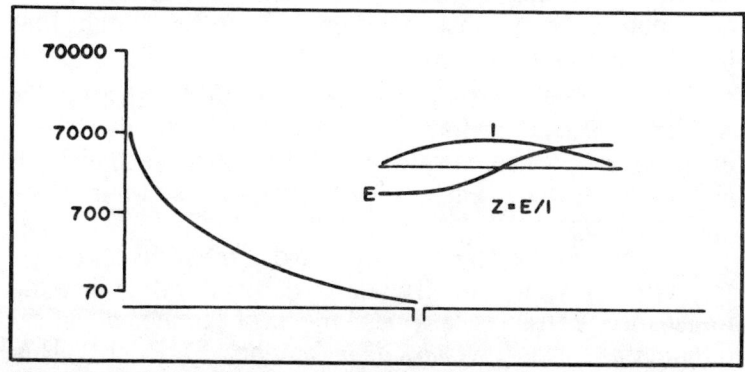

Fig. 12-10. Variation of impedance along a half-wave dipole from center to ends is shown here. Impedance is the ratio of voltage to current, both of which vary along the antenna's length as shown in the inset, and so ranges from very low at center to very high at the ends. Because of radiation, impedance never reaches zero, but has a minimum of about 70 ohms for a half-wave dipole in free space.

tity measured in ohms" while the ohm is simply the ratio of voltage to current at any point in a circuit.

Since both voltage and current are present in an operating antenna at every point along the conductor, it's only natural that every point on an antenna should have an impedance of some type.

But because of the way in which an antenna operates, both the voltage and current are varying, and where voltage is high, current is low and vice versa. Therefore, the impedance must change from point to point along the antenna. Where voltage is high, so is the impedance; where voltage is low and current high, the impedance is low.

Figure 12-10 shows how impedance varies along a half-wave dipole. At the ends, it's very high (the capacitive end effect which makes it necessary to shorten the wire by 5% from an actual half wavelength keeps some current flowing even at the end, and so keeps impedance from becoming infinite). In the center, it is low.

You might expect the impedance to drop to zero at the center of an ideal dipole, but it doesn't because the antenna is radiating power and so there's always a little more power going in than there is reflecting back from the ends. In free space separated from all other conductors, a dipole would have an impedance of about 73 ohms at its center.

Any actual antenna's impedance will be modified by the energy which it receives from nearby reflecting surfaces such as the ground. Figure 12-11 shows how the impedance of a dipole varies because of this ground effect, assuming that the ground is a perfect reflector.

When several antennas are used together in an array (or when parasitic elements are placed near a driven element as in a yagi or other beam antenna), the power in each of them affects the power in each of the others—and so their impedances change.

Because impedance is a measure of the ratio of voltage to current, anything which modifies this ratio will change the impedance of an antenna. One of the most common modifications of this sort is the *folded dipole* antenna (Fig. 12-12). This is an ordinary half-wave dipole, with a second half-wave of wire strung alongside and connected to it at the ends. Now, when the voltage peaks reach the ends of the antenna and reflect back, they have not one but two paths available—one

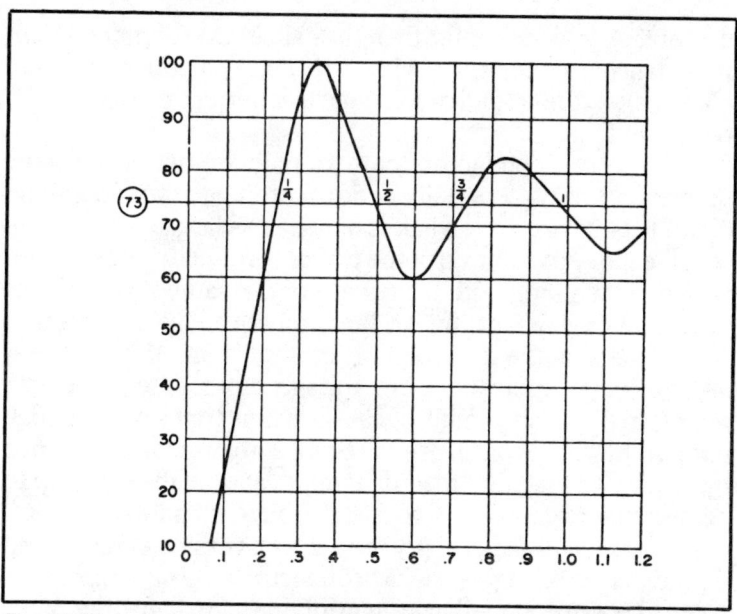

Fig. 12-11. Impedance of a half-wave dipole illustrates the action of reflected energy from ground in changing the antenna's actual impedance. Vertical scale is impedance in ohms; horizontal is height above electrical ground in wavelengths. At multiples of one-quarter wave, conditions are similar to those of free space. At other distances, impedance may be higher or lower than the anticipated value.

back down the original wire, and the other down the second wire alongside it. This splits the current between the two wires, and at the same time doubles the voltage for the same power (to keep power constant). The result is that the impedance is multipled by four, two times being for the doubling of voltage and the other two for the halving of

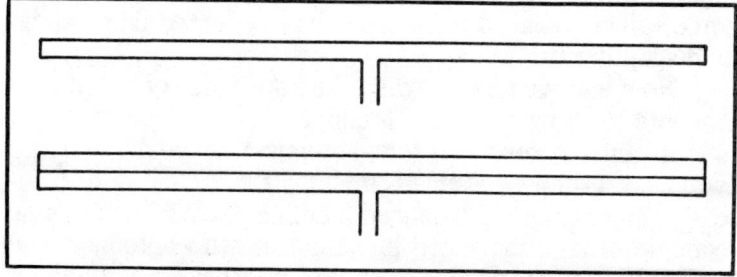

Fig. 12-12. Folded dipole multiplies the antenna's impedance by providing alternate paths for current and so raising the voltage-to-current ratio. Two-wire version (top) multiplies by 4; 3-wire version by 9.

current. A folded dipole, then, has about 300 ohm feedpoint impedance in comparison with the ordinary dipole's 73 ohms. This makes the folded dipole a nearly perfect match for 300 ohm twinlead.

If the two conductors of the dipole are not of the same diameter, the current will divide according to the ratio of the diameters. If the driven half is only half as large a wire as the passive side, two-thirds of the current will flow in the driven portion, assuming both halves are of the same material. To keep power constant, the voltages must change by the same ratio, so that at the feedpoint the current is one-third as great as in a plain dipole, and the voltage three times as large, giving an impedance multiplication of nine times or about 650 ohms. This fact is often used in beam antennas for impedance matching, to raise the feedpoint impedance of the array up to something large enough to match conventional feedlines.

The *monopole* is simply the vertical version of the folded dipole, in which the passive conductor is grouped. It provides the same impedance-multiplying effect, for the same reasons.

What makes impedance so important in an antenna and feedline is the fact that rf energy flows smoothly only when the impedance of the circuit in which it is flowing does not change. In fact, the reason an antenna radiates is directly related to the fact that its impedance changes, as we saw earlier in this chapter. So long as circuit impedance remains constant, the rf energy finds a smooth path to follow and like all kinds of energy, flows along the path of least resistance. When impedance changes, this produces a "hump" in the path, and energy goes in all directions. If the hump is small, most of the energy keeps going in its original direction, but some will be radiated and some will be reflected at any hump or discontinuity.

Now that we've seen what the impedance of an antenna amounts to, how about the feedline?

In general, feedlines (like antennas) can be divided into two major groups: balanced and unbalanced. 300-ohm twinlead is an example of balanced feedline. Coaxial cable is an example of an unbalanced line. Just as with antennas, the "balance" is with respect to ground. In a balanced feedline, any voltage in one wire with respect to ground is "balanced out" by an equal voltage of opposite polarity in the other wire. In an unbalanced line, one conductor is always at ground potential and the other carries the full energy flow.

Most balanced feedlines are composed of two parallel conductors, and this type of line is often called parallel-conductor transmission line. It may, however, have more. In some commercial installations, four wires are used in a balanced configuration to carry higher power. 300-ohm line is a parallel-conductor transmission line.

In a parallel-conductor line, the two conductor are kept a fixed distance apart by means of some type of insulating material. In the 300-ohm line, the insulating material is typically polyethylene. At higher power levels, air is often the insulating medium. In an air-insulated parallel-conductor line, conductor spacing is fixed by means of insulating spreaders at regular intervals.

While both balanced and unbalanced feedlines have a *characteristic* impedance, it's easier to visualize in the case of he parallel conductor lines, and so we'll look at them first.

Whenever such a line is carrying power, a voltage exists between the two conductors at all times, and current is flowing in each. Since every conductor has inductance, each of the line's two conductors will have inductance. Since the two conductors are separated by an insulating material, capacitance will exist between them. Both the inductance and the capacitance are "distributed" along the full length of the line rather than being "lumped" into separate inductors and capacitors. The inductance will restrict current flow, while the capacitance will inhibit voltage change. The net result of the inductance and capacitance is that each tiny incremental portion of the line is equivalent to the circuit shown in Fig. (12-13).

Since both resistance and reactance are present in this circuit, it must have some value of impedance associated with it. As it happens, the impedance turns out to be constant with frequency, and is equal to the square root of the ratio of inductance to capacitance.

The inductance is established by the diameter of the conductors, and the capacitance depends primarily upon the spacing between conductors (although the type of insulating material also has a major effect). The larger the conductor, the lower its inductance and therefore the smaller will be the line's characteristic impedance. Similarly, the closer together the two conductors are, the greater will be the capacitance, and again the value of the impedance will decrease.

The key point here is not so much the formula which determines impedance, because you can look that up in a handbook whenever you might need it (most of us purchase feedline ready-made, anyway). Rather, it's the fact that the characteristic impedance of a parallel-conductor line depends entirely upon physical factors: the diameters of the conductors, the spacing between conductors, and the type of insulating material.

In the balanced line, any radiation which might occur from one conductor is balanced out by equal and opposite radiation from the other, and so long as conductor spacing is very small (as compared to wavelength) the line has no effective radiation.

However, keeping such a line perfectly balanced is a bit tricky. Anything that might affect one conductor must be made to have exactly the same effect on the other. Because of this, a balanced line should be kept clear of all surrounding objects.

The coaxial line bears the same relationship to the balanced line as the vertical antenna does to the dipole. It's effectively just half of the balanced line.

Coax consists of a single conductor running down the middle of a hollow conducting tube. Both conductors share the same axis, and that's where the name "coaxial" comes from. Most commonly used coax consists of a center conductor surrounded by foam insulation, with a flexible metal shield braid providing the outer conductor, and the whole business enclosed in a weatherproof jacket.

In coax, the center conductor is free to radiate in all directions, but the outer shielding conductor confines all this radiation to the inside of the line. Thus, such a line does not depend upon tricky balancing to prevent radiation, and can be run almost anywhere.

Like parallel-conductor line, coax has both inductance and capacitance along its length, and so has a value of characteristic impedance fixed by these L and C values. Again, the impedance of any line is fixed by physical factors: the diameter of conductor (inner conductor), spacing between conductors (ratio of conductor diameters), and insulating material between them. Most common coax has either 52 or 75-ohm impedance.

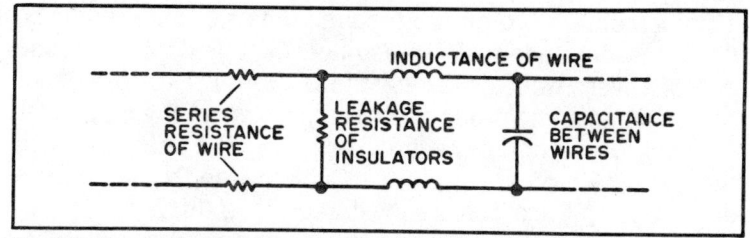

Fig. 12-13. Equivalent circuit of a transmission line includes the series resistance and inductance of each conductor, shunt resistance of insulators, and capacitance between wires. Inductance and capacitance here determine the line's characteristic impedance.

Now that we've met impedance in both the antenna and the feedline, it's time to put the two together. We've already observed that rf energy flows smoothly only when it encounters no "humps" or impedance discontinuities in its path. Therefore, if both the antenna feedpoint and the feedline connected to it have exactly the same impedance, the energy cannot tell the difference between them. There will be no "bump" to cause reflection of energy, and everything will flow smoothly. Such a condition is known as a *match* because the impedances match perfectly, and that's the goal of most antenna and feedline adjustments.

Since antenna impedance depends upon so many variable factors, however, while feedline impedance is fixed by a few physical constants, the matched condition seldom happens by accident. If the impedance of the antenna feedpoint and the feedline differ, you have a mismatch which reflects energy back toward the transmitter. This mismatch creates a standing wave on the feedline in just the same way that the ends of the antenna create a standing wave on the antenna, and the feedline then can radiate energy—which is normally not a desirable situation, because this radiation may interfere with the desired pattern from the antenna.

In addition, the energy reflected down the line may cause the impedance at the transmitter end of the line to be either too high or too low for proper operation. That's why most hams strive to match their antennas and feedlines.

Note that while a standing wave is necessary for operation of the antenna, it's not desirable on the feedline. In fact, the strength of the standing waves on the feedline (voltage standing wave ratio, often abbreviated merely swr) is used as a measure of antenna system performance.

This standing wave on the feedline is not considered to be desirable, because from an engineering standpoint any one part of a system should do only the job it's there to do, and not do the job of some other part. The purpose of the antenna is to radiate, that of the feedline is to carry the energy—all of the energy—from transmitter to antenna, and if the feedline does some of the antenna's job of radiating, this is not good.

Of course, you might connect the transmitter directly to the antenna with no intervening feedline, and that's exactly what connecting a random-length, nonresonant hunk of wire to a transmitter amounts to. If it's necessary to get a signal out at any cost, this is an acceptable solution, but the directive pattern of such an antenna is unknown, and transmitters sometimes turn out to be difficult to operate with six-inch arcs of rf energy leaping from every sharp corner!

It's important to note clearly at this point that the existence of standing waves on the feedline will *not* make it impossible to radiate energy. They may make it impossible to get the energy out of the transmitter, by causing changes in effective feedline impedance, but that can be remedied by adjusting the length of the feedline. The primary purpose of getting a good match and resulting freedom from feedline standing waves is to let the antenna operate as it was designed to, and to prevent undesired radiation.

The most commonly used measure of feedline performance is the swr (or vswr). The term derives from the earliest means of measurement. You'll recall that the standing wave is essentially stationary; that is, on the antenna the voltage is always maximum at the ends and minimum in the middle (for the half-wave centerfed dipole we used as an example). It works the same way on the feedline. The standing wave on the feedline will produce a voltage maximum at some points, and at other points a quarter wavelength away from these maxima, voltage will be minimum. The swr is simply the ratio between the maximum rf voltage and the minimum.

Thus, if there's no standing wave at all, the maximum and minimum will be the same, and the ratio will come out to be 1.0. Therefore an swr of 1 (or 1:1) means that no standing wave exists, and any *smaller* value is simply impossible, by definition.

If the standing wave were using all the energy in the feedline, the ratio would be nearly infinite, as it is on the antenna.

In practice, the swr usually ranges from 1.05 or so up to possibly 10. The higher the swr, the greater the magnitude of the standing wave on the feedline.

We saw earlier that the ratio of voltage to current created an impedance for the antenna, which varies as we move along the conductor. We also saw that the characteristic impedance of a feedline is determined by physical rather than electrical quantities.

The swr on a feedline modifies the *effective* impedance of that line, although its characteristic impedance remains unchanged. An swr of 2 means that the voltage swings through a 2-to-1 range of variation. The minimum will only half the voltage that is present at a maximum.

Since with power remaining constant, voltage and current must have a constant product, the current goes through the same range of variation. This must, then, mean that effective impedance goes through a 4-to-1 variation. When current in the line is maximum and voltage minimum, impedance is minimum. When current is at its highest value and voltage at its lowest, impedance is maximum.

As it happens, this variation in effective impedance swings around the characteristic impedance as a center point. With 75-ohm line and an swr of 2, the effective impedance would range from 37.5 ohms minimum to 150 ohms maximum. (Actually, since it's impedance, there's also a reactive component, but right now we'll ignore it. This is oversimplification, but with a purpose.)

If the swr were to climb to 10, then the impedance swing would be from 7.5 ohms to 750 ohms.

One of the effects of swr on a feedline, then, in addition to making possible unwanted radiation, is to make the effective impedance at the transmitter end of the line become something other than what you thought it was. This is what makes a line with high swr difficult to work with at times. Transmitters are designed to feed specified feedline impedances; if swr puts the actual effective impedance out of that range, things don't work right.

The multiplying effect upon voltage (maximum-voltage levels) is another problem of swr. It's easy for the voltage at a maximum to be greater than the feedline or the transmitter insulation is designed to handle. Even a 10W transmitter can develop quite respectable voltages when the impedance is high enough. Similarly, the current maximums may cause the feedline conductors to heat up enough to melt the insulation.

So now that we know some of the problems, what can we do about it?

The cause of standing waves on feedlines is, as we saw, the presence of impedance "humps" in the path of the rf energy. With no humps, no standing waves are created. The cure for unwanted standing waves, then, is to remove the humps.

The easiest way to do this in theory is to adjust the antenna impedance until it's a perfect match for the feedline. Then there will be no hump, and no problem. The swr will be 1.0.

However, antenna impedance is subject to many variables, and feedlines come in only a few impedance levels. What if they don't match?

The answer in this case is to introduce a *matching network*, which is a fancy name for "anything that will make a smooth match." A tuned circuit makes an excellent matching network; when it's resonant, you can get almost any impedance levels you like by just tapping the coil at the proper number of turns. These are often used, and known as *antenna tuners*, but usually a long run of mismatched feedline separates the antenna and the tuner, with a short run of line from transmitter to tuner, with a short run of line from transmitter to tuner. The swr is low only over the short line.

Many other types of matching networks are possible. Impedance of a parallel-conductor line depends upon its conductor spacing, in part, so a *tapered* section can be built which changes smoothly from one impedance level to another, and used as a transformer.

It's also possible to use a quarter-wave section of feedline and a transformer to step impedance up or down, and to cascade several such sections if necessary to get the proper ratio.

Any antenna handbook includes quite a bit of material on impedance matching networks, because they are the heart of practical antenna construction.

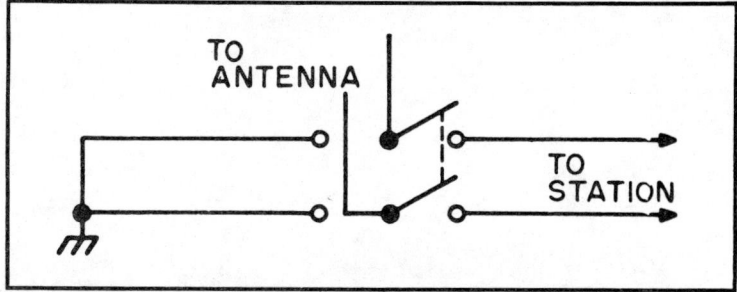

Fig. 12-14. Most certain protection for amateur equipment against the danger of damage by lightning involves a large DPDT switch connected as shown here, to remove the antenna connections from the station's equipment and connect the antenna instead to ground by heavy, short leads. This effectively converts the antenna installation into a lightning rod structure. Even this, though, may not protect against a direct hit.

LIGHTNING PROTECTION

One point mentioned in the FCC exam has not yet been discussed, and it fits into a discussion of matching networks as well as it does anywhere else. That's the subject of protection against lightning.

Obviously, any large metallic structure such as an antenna is a tempting target for lightning, and since the laws of nature have neither conscience nor memory, it's a fairly good bet that unless equipment is protected, it can be damaged.

The most certain protection is a large knife switch, of the sort you see in old monster movies where the mad scientist shoves home the switch to jolt life into his creation (Fig. 12-14). A hefty switch of this nature, with the arms connected to the antenna, one pair of poles to the equipment, and the other pair directly to ground with a short, heavy braided strap, provides almost certain lightning protection. The energy need not jump the gap to get to the equipment, when it has a direct path to ground.

Such a switch upsets the swr on coax lines, though, and isn't the most attractive item in any station. An alternate but less sure means of protection is to disconnect the antenna cable from the equipment and connect it to a direct ground whenever a thunderstorm threatens, and whenever the station is not in use.

Lightning arrestors are sold which provide a spark gap and are said to help prevent lightning damage. While they work as claimed, they are much less sure protection than is physical disconnection of antenna leads.

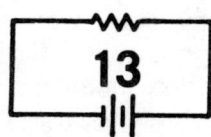

13

Radio Propagation Phenomena

We've come a long way in our study course for the General class amateur radio exam, but there is still one area we haven't examined in any great amount of detail yet—the subject of propagation and the manner in which radio waves of different frequencies and wavelengths act in air and space. The study of radio waves encompasses more than mere propagation, though, and so this chapter will go beyond that subject area as well. We'll look at some specific frequency bands to see how their characteristics differ from one another, we'll examine a few of the FCC rules, and we'll touch on the subject of interference and how to get rid of it. Finally, we'll look at some operating practices to determine how we can improve conditions within the bands themselves.

HAM BAND DIFFERENCES

It takes only a receiver and a little listening to discover that each different ham band has a "flavor" all its own. That's one of the reasons why many hams settle down to operate on only one of the many available bands, and become known as 40-meter or 20 meter operators or what have you.

One of the major factors contributing to this unique identity for each of our bands is that no two ham bands have identical transmission or propagation characteristics. Some are almost useless for short-range operation but perform spectacularly for long-distance contacts. Others are limited to line of sight. Operators who prefer DX (long distance) gravitate to the DX bands, while those who just like to chew the rag tend to stick to short-range bands where they get a chance to become personally acquainted with their fellow rag-chewers, as well as by radio.

WAVE PROPAGATION

Before we look in detail at the characteristics of each of the popular bands, let's see how radio waves in general are propagated. Earlier, we saw how waves are launched into space in all directions from an antenna, and are reflected from any large conducting surface. This is the heart of all radio propagation, and what makes the difference between one band and another is the difference in what will reflect the wave.

When radio waves leave the antenna, they go in all directions. Some travel directly to the receiving antenna without reflecting from anything on the way. Radio engineers call these *direct waves*, while hams usually call them *ground waves*. Some travel along the surface of the ground, if the radio frequency is low enough. Almost all the ham bands are high enough in frequency that this wave is ignored by hams; engineers call this the ground wave (which sometimes leads to confusion when hams and engineers talk with each other). Most signals, however, radiate out into space.

If conditions are right, some of this *space wave* will be reflected by ionized layers in the upper atmosphere, from 10 to 200 miles above the surface of the earth, and will return to earth at far distant points. These signals are known as *sky wave* signals since they appear to come from the sky (because of the reflection), and are the basis of almost all shortwave communication except for the line-of-sight ground wave operations conducted on low-frequency bands during daylight hours and on VHF bands around the clock.

The higher the frequency of the signal, the less effective will be this reflecting action. Balancing this, however, is the fact that other ionized layers tend to absorb lower-frequency signals, so that for any specific conditions in the ionsphere, both lowest and highest useful frequencies exist. The lower end of the range of useful frequencies provides shorter range, because the more effective reflecting action will permit the signal to go nearly straight up and bounce right back down. The upper end provides greater range, because only those signals which hit the ionized layers at a relatively shallow angle (which provides a longer *skip distance*) can be reflected.

Figures 13-1 through 13-3 compare the different kinds of waves used by hams. Figure 13-1 shows the line-of-sight conditions for UHF operation. In this direct-wave type of operation, the key factor is antenna height above ground. The higher the antenna, the greater the range—for exactly the same reason that you can see farther from the roof of a skyscraper than from ground level.

At all except the very highest radio frequencies, a bending of the radio waves around the curvature of the earth occurs as shown in Fig. 13-2. The waves are still direct waves, in that they do not reflect from anything, but they go a short distance past the visible horizon (about the same as if the earth's radius was 4/3 of its actual value).

Atmospheric effects can create *ducting* or a sort of waveguide action between the earth's surface and an air stratum, or between two different air strata, which gives the same sort of action as the refraction shown in Fig. 13-2, but actually depends upon reflection of the waves from the invisible "walls" of the waveguide. This type of action occurs reasonably frequently on VHF bands, where it is known as *tropospheric propagation* and leads to exciting DX.

The sky-wave propagation which accounts for the majority of ham long-distance operation is illustrated in Fig. 13-3. When a signal is propagated by the sky wave, it may leave the transmitting antenna at any angle from one which causes it to just graze the horizon up to straight vertical.

A signal leaving the transmitting antenna vertically will return to earth very close to the originating station, which does not give exceptional range but makes possible good coverage of moderate-range areas.

One which leaves at the grazing angle, however, gets the maximum range. It skims the surface of the earth at the horizon, then travels on until it hits the ionosphere at a distance from the transmitter somewhat greater than twice the distance to the horizon (because the reflecting layer is far above the surface of the earth), then is reflected at a shallow

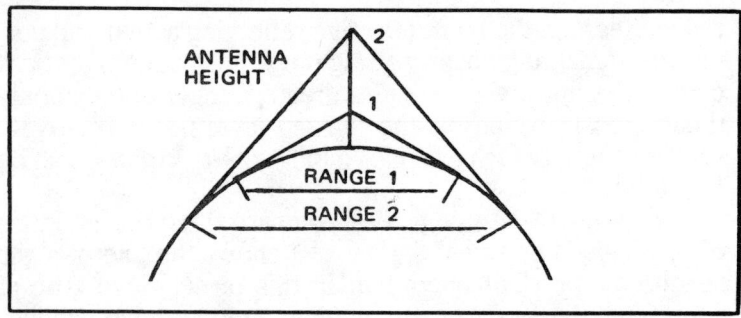

Fig. 13-1. Line-of-sight and horizon range of UHF radio waves are shown here. Curvature of the earth has been exaggerated for emphasis. Key point is that antenna height above ground is virtually the only factor determining the range of line-of-sight signals.

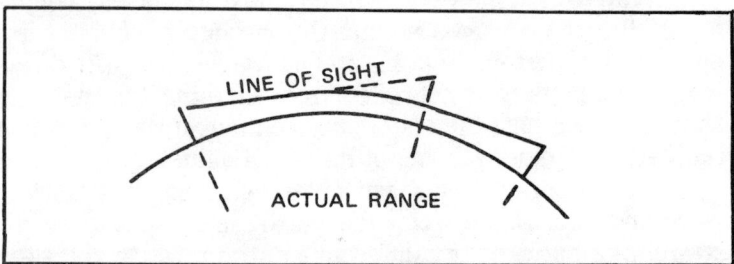

Fig. 13-2. Refraction of radio signals around the earth's curvature as shown here extends line-of-sight distances in many cases. In almost all cases the actual effective range is the same as it would be if the earth's radius was 1/3 larger than it actually is. Weather conditions (tropo bending and waveguide effect) can extend the range amazingly at times. 2-meter transmissions from Hawaii to California have been accomplished by this method.

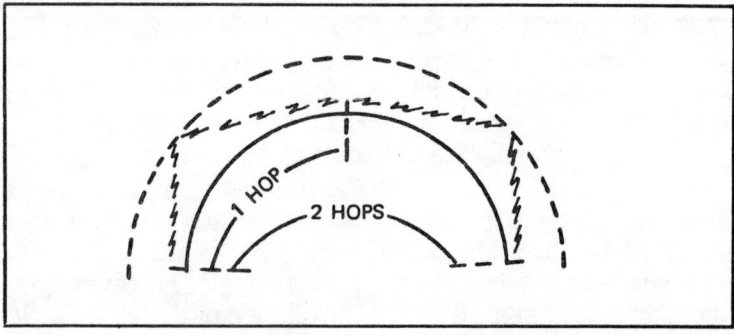

Fig. 13-3. Most ham DX depends upon reflection of radio signals from ionized layers in the upper atmosphere. This ionospheric reflection permits radio transmission from almost any point on earth to any other point, by proper choice of frequency and beam direction. Reflection is most effective in HF bands between 3 and 30 MHz, but sometimes occurs in the 6-meter band as well.

angle, and travels an extended distance before returning to earth. When it returns to the surface, it may be reflected back up for another hop, or may graze the surface and thus begin a second hop without surface reflection. Multiple-hop transmission makes it possible to reach almost any spot on the planet from any other spot by proper choice of frequency to suit the prevailing ionospheric conditions.

ATMOSPHERIC LAYERS

At least four different layers in the ionsphere (Fig. 13-4) have been identified. They have been assigned letter designations. The D layer is the lowest of those which seriously affect ham radio signals. It's about 35 miles above the surface of the earth, and absorbs low-frequency signals somewhat like the way in which fog swallows up light. The D layer is present only during daylight hours, and makes the lower-frequency ham bands almost useless for sky-wave propagation in the daytime.

Some 25 miles above the D layer is the E layer, which is also present only during daylight (but during the winter, it may persist for several hours after sunset). This layer reflects signals which reach it, and causes short skip. *Sporadic E* clouds are patches of extremely dense ionization within the E layer which are capable of reflecting VHF signals, and provide skip signals on 6 meters (and on rare occasions, on 2 meters as well).

Far above the E layer, at an altitude of about 120 miles, is the lower of the two F layers, called F1 to distinguish it from the 200-mile-high F2 layer. Like D and E, F1 is present only during daylight and for a short time after sunset. This layer reflects signals at a higher frequency than does the E layer, and provides the majority of the intercontinental DX worked by hams during daylight.

At night, the F1 layer either vanishes or rises to merge with the F2 layer above it. This uppermost of the reflecting layers provides the longest range for signals, but is less effective at higher frequencies than are the lower E and F1 layers.

Both the frequency limits and the altitudes of these layers are influenced by the intensity of the ionization, which in turn depends upon many things. The major factor affecting the ionosphere is solar activity. During years of high activity

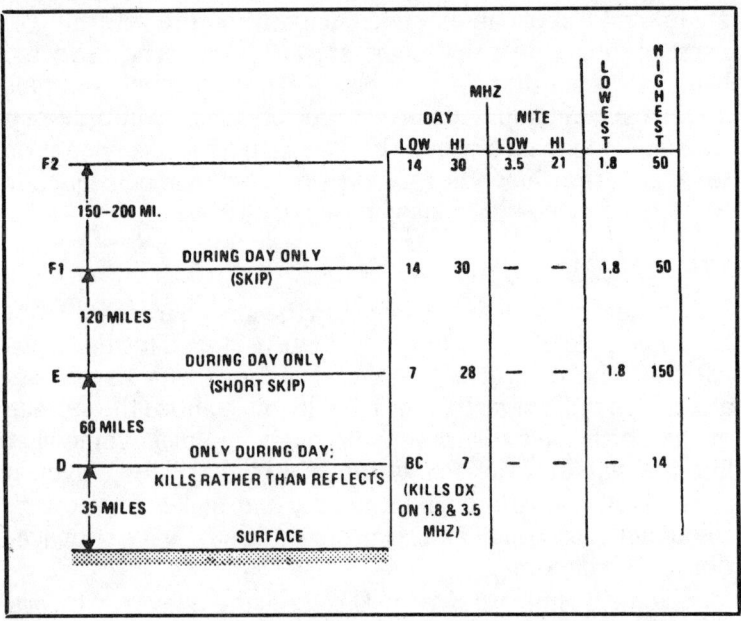

Fig. 13-4. Various layers of the ionosphere and their characteristics are shown here. Both the upper and lower frequency limits for each layer, day and night, are listed. Seasonal variations also occur, and the entire range is shifted by the 11-year sunpot cycle. These variations are shown as lowest and highest columns in the listing.

on the part of the sun (sunspot maxima), ionization is more intense and the maximum usable frequencies are higher, than in years of low solar activity. The seasonal variation between winter and summer performance is due to the change in the angle at which solar radiation reaches the earth, which affects the height of the layers. For serious DXers, intense study of the ionsphere is indicated.

TYPICAL HAM BAND PROPAGATION

Because of all the variable factors affecting the ionosphere and thus changing the way in which signals are reflected, any generalizations about the various ham bands' typical characteristics must be subject to rather large amounts of error. The following summary is intended to portray "typical" conditions which, unfortunately, are hardly ever met in actual practice.

The lowest frequency ham band, 160 meters, has only limited usefulness for any purpose during daylight because of the strong absorption by the D layer. Even at night when the D layer is no longer present, the range of top band operations is limited, but transoceanic DX has been achieved and many operators take a special interest in activity on this band, traditionally the oldest of them all (because original ham activity was restricted to shortwaves, and these wavelengths are the longest of the original ham region).

The next higher ham band, from 3.5 to 4.0 MHz, is known as 80 meters to CW and novice operators and as 75 meters to phone enthusiasts. It is one of the most popular bands for domestic activity. During the day its range is restricted by D-layer absorption, but this restriction usually vanishes near sunset and the band then opens up to permit transcontinental contacts, with little or no dead zone between. It could easily be possible to assemble a roundtable network on 75 or 80 meters with stations from every one of the 48 contiguous states, with every station hearing each of the others, were it not for the congestion due to the band's popularity.

Above 75 meters, the next ham band is 40 meters. This band sometimes suffers from D-layer range restrictions during daylight, but even during the day ranges up to 1000 miles or so are not uncommon. The maximum usable frequency at any time is usually well above 7.3 MHz, so that 40-meter signals may reflect nearly vertically from the E and F layers to provide good coverage of extended areas. At night, the band may open up for intercontinental operation, and nationwide night coverage here is the rule rather than the exception.

The 20-meter band, from 14 to 14.3 MHz, is traditionally the DX band. It is seldom used for domestic purposes, other than coast-to-coast contacts via single-hop F2 skip, but provides worldwide coverage capability day or night. Sometimes the maximum usable frequency (MUF) drops too low at night to permit use of 20 meters, but this occurs only near the periods of minimum solar activity. At most times, this band is always open to somewhere far away.

On 15 meters, DX is also the normal condition, but by this time we are into the upper frequency region (above 21 MHz) where sky-wave operation is not always possible. When the ionosphere will not reflect 15-meter signals, this band is virtually dead. On the other hand, when the band is open it frequently offers a longer operating range (more distant DX) than does 20.

The 10-meter band, from 28 to 29.7 MHz, is like 15 only more so. During sunspot maxima, the band is often open around the clock, and intercontinental contacts are there for the working. For a large part of the time, though, the band is dead and is used only for local mobile operation such as is found in the VHF region.

The 10-meter band is the upper limit of HF ham bands; the boundary between HF and VHF lies at 30 MHz. The lowest VHF ham band is 6 meters (50-54 MHz), just below TV channel 2 in the spectrum. This band is an experimenter's delight, because unusual propagation techniques such as scatter, tropo propagation, etc. become easy to handle at this frequency. Normal sky-wave operation so common to "dc bands" (the VHF operators' derisive label for the HF region) is rare on 6 meters; sporadic-E brings 1500-mile ranges occasionally, and about one year out of eleven at sunspot maximum the F2 reflecting capability may climb up to the lower part of this band, but for the rest of the time most use of this band is for local communication and mobile operators.

The 2-meter band is like 6 only more so, somewhat as 10-meter operation resembles that on 15. Sky-wave propagation is almost unknown at 2 meters and above, but the exotic propagation techniques such as meteor trail reflection, moonbounce, and satellite relay begin to become practical. The dedicated 2-meter operator usually concentrates on this type of operation, while the casual user and the public-service-oriented operator make use of mobile installations and FM repeaters.

The uppermost VHF ham band is 220 MHz or 1¼ meters. Its characteristics are very similar to those of 2 meters, but because of the increased difficulty in making circuits operate easily this band is not so popular with most operators. It may be lost before long because of its relative unpopularity, as was the old 11-meter band (which became the CB)!

FCC RULES

The final authority concerning what a licensed ham may and may not do with his station if Part 97, FCC Rules and Regulations, published by the Government Printing Office and available from the Superintendent of Documents. Every ham or would-be ham should have an up-to-date copy, but the sad fact is that few really do.

The pertinent portions of Part 97 are published as an appendix to this book, but you are advised to check with the FCC for any updated versions issued after the date of this publication.

The rules themselves divide into several categories. Some govern the physical characteristics of an amateur station; this is particularly directed at antenna height, and the idea is to be certain that no ham antenna is high enough to be a menace to airplanes. At this writing, the limit was 170 ft, or 1 ft above ground for every 200 ft range from an airport, whichever is less. Exceptions may be granted upon written application, which means in effect that if it's all right with the FAA, it suits the FCC also. Also in this group are the rules which require every amateur station to have a *fixed* transmitter location, which means a mailing address at which the licensee can always be reached.

Another group of regulations within Part 97 establishes technical standards for amateur stations. These rules establish the authorized frequencies for each class of licenses, the power limits, the type of signals permitted, and the quality of signals required.

A third category of the rules sets up operating requirements and practices required of all licensees. One key point which the FCC has always felt it necessary to explicitly state, and which many hams often violate, is that in all situations not specifically covered by these regulations, each amateur station shall be operated "in accordance with good engineering and good amateur practice."

Among the rules in this category are those requiring a licensed amateur to control the station at all times when it is operating, prescribing the method for identifying the stations in a conversation, permitting certain kinds of one-way transmissions, and detailing procedures for portable and mobile operation. Log-keeping and emergency operation are also covered in this section, as are "permissible communications."

Not all the "permissible communications" rules binding the U.S. ham are listed in Part 97, however. International regulations also apply, and these international regulations restrict the kind of communication permissible between two amateur stations to "messages of a technical nature relating to tests" being carried out, and "remarks of a personal character for which, by reason of their unimportance, recourse to the public" communication services is not justified. They continue to spell out that it is "absolutely forbidden" for amateur stations to be used for transmitting international communications on behalf of "third parties" (meaning someone other than the two hams involved in the contact). The term "third party traffic" has come to signify this "absolutely forbidden" type of message.

What makes third party traffic legal at all is the fact that it is not forbidden in domestic communication, only internationally; and even in international situations, the prohibition "may be modified by special arrangements between the administrations of the countries concerned." The United States has made such special arrangements with a number of countries. The exact list of nations with which third-party traffic is legal varies from time to time, and may be obtained directly from the FCC.

The same international agreement also makes it illegal for hams of one country to contact those in any country which has filed formal objection to such international contacts. Several countries are currently on the forbidden list. Again, contact the FCC for the up-to-date listing.

Back to Part 97, a fourth category of FCC rules specifies "prohibited practices and administrative sanctions." Prohibited practices include broadcasting, accepting any form of compensation for use of a ham station, transmitting music or secret codes, use of "obscene, indecent, or profane" language or expressions on the air (which, like many present laws in this area, may not be enforceable, and which the FCC has indicated it will not enforce in all cases), transmitting false or unidentified signals, willful or malicious interference with any radio communications or signal (whether a legal signal or not), assisting anyone to obtain a license by fraud, and willful damage to any radio apparatus. If you get mad at your equipment, don't smash it.

Administrative sanctions include "quiet hours" or restricted operation, in case a specific ham is causing interference to other services. Several levels of restricted operation are prescribed; the most restrictive permits the ham to operate only between the hours of midnight and 8 a.m.

A fifth category of rules sets up the Radio Amateur Civil Emergency Service and governs its operation, while a sixth category prescribes requirements for ham station operation in this country by aliens who are licensed in their own countries. Only the first four categories are likely to be covered in detail on the General class exam.

Because the exact provisions of the rules are subject to change much more frequently than are the technical matters which we have been discussing in most of this course, we won't attempt to go into more detail concerning the rules. We recommend that you obtain a copy of Part 97 shortly before taking the exam, and studying the four categories of rules we've discussed here. Particular points to note are the frequency limits for the various classes of license and types of emissions, the corresponding power limits, and the section dealing with technical standards.

ELIMINATING TVI

Of all the "alphabet soup" combinations of initials which have hovered around our culture for the past 40 years or so, probably none has caused so much agony to so many hams as TVI.

Those three letters stand for television interference—specifically the type resulting from ham operation. In the early days of TV, TVI was the rule rather than the exception for an amateur station, and many hams simply shut down their operations for considerable periods of time until they learned how to overcome the problem.

They had good reason to do so, too. A woman scorned *may* equal the fury of a televiewer whose picture is suffering interference, but it would be difficult to convince those hardy hams who have faced both furies that the one resulting from TVI wasn't the worst of the pair.

History is full of instances of antennas being cut down, tires being slashed, antenna towers toppled, and hams being threatened with assault (and worse) because of TVI.

One factor which made the problem so difficult in the early days was that almost all early TV receivers used an i-f of 21 MHz, smack in the middle of one ham band, at the third harmonic of another, and the sixth harmonic of still a third band. In addition, the 6-meter ham band and TV channel 2 are adjacent to each other. No matter what band a ham used, he could hardly help getting into one of those old TV receivers. If he didn't make it in through the tuner, he would come in through the i-f section. On the rare occasions that both these failed, the audio sections would pick up his energy and detect it, and there he would be in the sound channel.

For several years, it seemed almost as if TV designers went out of their way to create the probability of ham interference. One ham thought he had escaped from TVI when he abandoned his favorite HF bands and went to 2 meters for all operation. He reasoned, correctly, that at this high frequency he couldn't possibly transmit any harmonics which could get into either the front end or the i-f of a TV receiver, and since he was an apartment dweller surrounded by televiewers this was important.

He learned, rapidly and to his utter dismay, that most of the TV sets in his apartment building were of a specific make and model which had the volume control on the front panel, and the audio section at the rear of the chassis, with a pair of 19 in. wires connecting the volume control to the audio section. As it happens, 19 in. is a quarter-wavelength at 144 MHz; almost every one of his neighbors' TV sets had a very good quarter-wave whip antenna sucking in his 2-meter signal and spewing in into the audio section!

From experiences such as these, the subject of TVI gathered about itself a mystique and books full of exotic cures. One, for instance, which would cure such cases as the 19 in. leads (if you could convince the televiewer to do it), was to wrap the TV set completely in aluminum foil, and bond the foil to a water pipe. This made it a bit difficult to see the picture, but they solved that by putting screen wire over the picture tube face. Naturally enough, this approach did not prove popular with the public. They had spent much money for those sets, and obviously it had to be the hams who were at fault.

Not all the complaints were so serious, though. Occasionally a viewer would call a ham to mention that he was hearing the ham's side of the conversation, and far from complaining, was interested enough to want to hear the rest of the conversation as well. This offered a source of new converts to ham radio—and it was sorely needed in those trying times.

Along the way, of course, those hams who stuck to it managed to learn how to clean up their transmitters and receivers (yes, even receivers created TVI) so that the only troubles left were due to faults in the TV sets themselves (and in the whole nature of radio—a class of problem exists which just happens, and we'll get to it shortly). These lessons were passed on through the ham magazines and by discussions, reaching the designers of commercial ham equipment, and the status of being "TVI-proofed" rapidly became a key sales feature for a commercial rig.

What really made the problem manageable, though, was the TV industry's realization (with a bit of government prodding) that it wasn't really good engineering to select ham bands for the i-f, and the subsequent switch to 41 MHz as a standard i-f for TV receivers. Once the obsolete sets with 21 MHz i-f strips became extinct, the number of TVI complaints dropped noticeably. And the existence of TVI became a sales pitch for pushing those new TV designs as well.

Nevertheless, the problem is still with us (although at reduced magnitude), and probably always will be. So, far that matter, is the inverse problem—interference to ham stations by TV receivers. The 15.734 kHz horizontal sweep frequency of a TV set is rich in harmonics, and on occasion makes the lower frequency ham bands unusable. Here, however, we'll concentrate on TVI rather than ITV (interference by television).

A TVI problem normally will fall into one of three categories. Either the ham station is at fault, the TV receiver is to blame, or it's the result of a law of nature about which no one can do anything.

Faults at the ham station usually boil down to the fact that the transmitter is letting unwanted harmonic energy get out, and some of this energy falls within the TV channel someone wants to watch.

The cure for this class of problem is to prevent the transmitter from letting the unwanted harmonics out. Careful tuning and operation can reduce the amount of harmonic energy generated, but a certain unavoidable percentage of harmonics are inevitable when class C amplifiers are used. A single-band antenna rather than a multiband design can help prevent the radiation of harmonics, but the accepted cure for the problem is installation of low-pass or TVI filter in the feedline between transmitter and antenna. This filter will have little effect upon normal transmitter operation, assuming that the feedline is matched so that the filter sees its

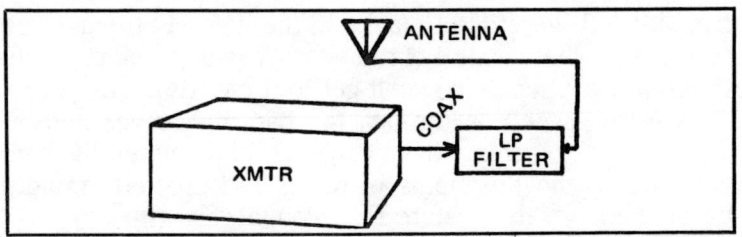

Fig. 13-5. Low-pass filter connected between transmitter and antenna as shown will help control TVI if harmonics are getting out through the feedline. For this to be effective, the transmitter itself must be properly shielded and all power and control leads filtered, so that no harmonic energy can escape except through the feedline. Antenna turner serves just as well as does special TVI filter, and also permits accurate matching to the antenna.

design impedance levels at input and output, but will block the path for the higher frequency interference-creating harmonics. Figure 13-5 shows the hookup.

Faults at the TV receiver may be insufficient selectivity, which permits the ham signal to get into the receiver front end despite the wide difference in frequency and thus over-

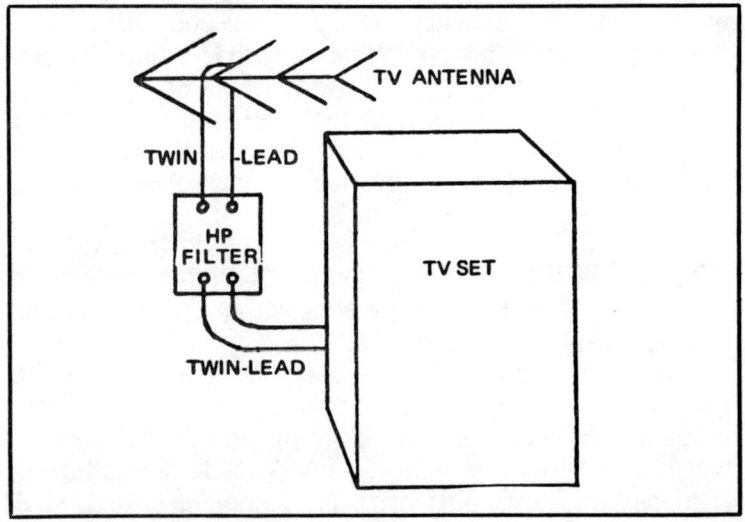

Fig. 13-6. When a ham station is operating properly and TVI is due to the TV set's failure to reject ham signals, a high-pass or TVI filter between the TV antenna and TV receiver as shown here may prove helpful. Most TV manufacturers will provide such filters free of charge upon request by the TV set owner, in compliance with FCC requirements that sets not be capable of radiating energy (the filters serve to prevent radiation of signals from the set, as well as reception of undesired signals by the set).

load the set, or inadequate shielding, which may lead to the audio pickup problem mentioned earlier.

If the trouble is due to the ham signal getting in through the front end, a high-pass filter in the feedline may help. This won't affect the high-frequency TV signal but will cut back on the amount of ham signal which gets through. Figure 13-6 shows how such a filter is installed. In weak-signal areas, it may be possible to readjust the TV antenna to put the ham station's signal in a null without significantly reducing the signal from the TV station, also.

If audio pickup is createing the problem, the surgery indicated by Fig. 13-7 is an almost-guaranteed cure. This may be applied to any kind of audio equipment which is bothered by ham interference, such as BC radios, record players, tape units, and so forth. It simply prevents the ham signal from being detected by the audio stage, without harming the normal function of the audio stage.

The most difficult class of problem to deal with is that which is nobody's fault. The most common such problem is one in which two radio signals, each faultless in itself, mix in some accidental circuit (such as a corroded rain gutter or a rusty metal fence) to create either a sum or a difference product which comes out in a valid TV channel and causes interference. Sometimes one of the original signals is itself a TV signal.

For instance, if a city has stations on channels 4 and 13, it's easy for a ham operating on 144 MHz or a frequency close to 144 to create interference through no fault of his own. This comes about because TV channels 4 and 13 are exactly 144 MHz apart, and the ham's 144 MHz signal can mix with the channel 4 signal to produce interference on 13, and at the same time mix with the channel 13 signal to produce interference on 4. While all this is going on, the two TV signals are mixing and wiping out the ham's receiver on 144 MHz.

When this happens, you can do only one thing: find a new operating frequency which does not cause interference. There are just too many possible sources of mixing action to hope to track them all down, and even if you could, new ones would develop naturally within a few days.

Most cases of TVI, however, do not involve this kind of mixing. The majority result from improperly operated ham transmitters, and misused TV receivers. Usually , tack and

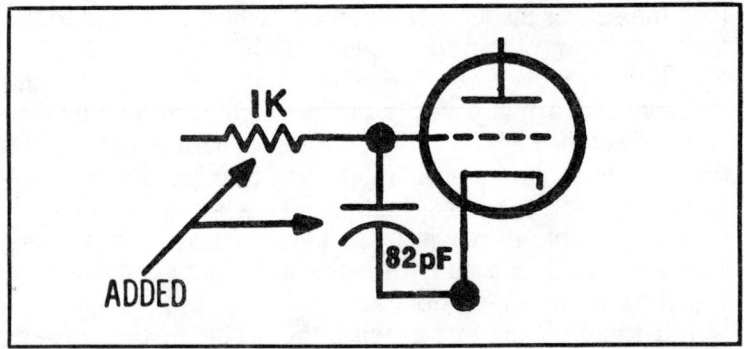

Fig. 13-7. Most cases of audio interference to TV by hams involve detection of the ham signal by the audio circuits of the receiver. This type of interference can also attack hi-fi installations, BC radios, and other entertainment devices. It is almost always due to faulty design of the affected device. Addition of an rf filter composed of a resistor and capacitor as shown here will cure it. Never make the change yourself, though; insist that the work be done by a professional service technician unless you want to be blamed for everything that ever may go wrong with the set after the filter is installed.

diplomacy rather than technical knowledge are the cure to the complaint, but it always helps to be able to show your own TV sitting alongside your station, operating perfectly, to convince the complainer that you aren't deliberately upsetting his reception and that possibly his own set may be contributing to the problem.

IMPROVING THE HAM BAND OPERATION

Interference and congestion are the hallmarks of the amateur bands, which has suffered an overpopulation problem since long before the population explosion became a popular worry.

The cause of the problem is painfully obvious: Several tens of thousands of hams over the world, at any one instant, are attempting to all operate in a segment of the rf spectrum which totals less than 4 MHz, and which furthermore is split up into many smaller sections.

Not all of these small sections are overpopulated. For instance, while the 75-meter phone band may be a mass of interference from one end to the other on a winter weekend night, it's quite likely that the CW portion of 80 meters may be almost vacant at the same time.

Similarly, the low end or "rare DX" region of the 20-meter CW subband may be a howling cacaphony of QRM, while only a few dozen kilohertz away from the pileup, vacant space goes begging.

One cure for congestion would be unpopular, but has been seriously suggested by some hams—that is, to limit the amateur population.

Short of an actual restriction upon numbers of hams, about the only practical solution to the problem is for all operators to adopt procedures which tend to minimize congestion and interference.

Among these procedures are the elimination of unnecessary on-the-air testing, keeping contacts on crowded bands brief, and listening more than you transmit.

Some on-the-air testing is, of course, necessary. You can't tell how a new antenna is going to load without loading it, for instance, and it's difficult to check the efficiency of a TVI-proofing operation unless you put the transmitter on the air and see whether its TVI is eliminated.

But there's no need to perform such tests as these at the times when the band is most crowded. Most on-the-air testing is of a nature which can be done almost any time, and if it's done when the band is relatively unused to obviously won't interfere with as many communications as it would if performed when the band was at its most crowded level.

Along with unnecessary on-the-air tests during crowded operating hours, long-winded CQs can be done away with. Inexperienced operators in particular tend to call CQ for hours on end, without stopping to find out if anyone is answering. While there's no officially recognized record, the fellow who sent nothing but CQ for 15 minutes without a break undoubtedly was a contender for it—not to mention being a rule violator, because the rules require that the transmitting station be identified at least once every 10 minutes.

Normally, a "3 by 3" CQ is adequate for the purpose. A "3 by 3" from W2NSD/1, for instance, would be: "CQ CQ CQ de W2NSD/1, CQ CQ CQ de W2NSD/1, CQ CQ CQ de W2NSD/1." Then pause and tune the band to see if anyone is answering. Since an answer may show up anywhere on the band, it will take much longer to tune for a possible reply than it took to send the call, and so the transmitting/receiving time ratio will go down for each operator—which reduces band congestion.

Keeping contacts brief is a rule of life for DX operators, who are trying to make as many contacts as possible in a limited time. Even for a ragchewer who enjoys conversation for its quality rather than its quantity, though, there's no need to spend all night saying "uhhhh . . . " into the mike. During contests or other periods of high activity, when bands are at their most crowded, the ragchewer might do well to avoid the crowded parts of the bands and wait until later to visit. This will preserve not only the tempers of the other operators, but that of the ragchewer as well, because he won't be continually plagued with interference drowning out part of his contact.

This is not an argument against chewing the rag, because that can be one of the most satisfying aspects of ham radio for those who enjoy conversation. It's merely recognition that there's a time and a place for everything, and a crowded expressway is not place to pull up alongside an old friend, stop, and visit. When expressway conditions prevail on the bands, visiting can move to less crowded regions, and return when the traffic is less dense.

Technically adequate equipment, operated properly, is necessary to reduce congestion. An overmodulated phone rig, whether AM, F, or SSB, can interfere with every contact on an entire band. This, however, is not so much an operating practice as it is a question of meeting the required technical standards (most of which are intended to reduce interference to a minimum).

Proper choice of the type of modulation to suit the purpose of the contact can go a long way toward reducing interference. A CW signal takes up only about 1/600 the bandwidth af an AM phone signal, so for contacts which can be adequately handled by CW, it would be the proper choice from an interference standpoint. RTTY compares favorably to CW in its bandwidth requirement.

Another operating practice which reduces congestion is that of adopting operating standards, then adhering to them. For instance, on a traffic net many operating conventions are established in order to minimize the number of times a message (or part of a message) must be repeated. When the same message is going to several addressees, the text may be sent only once. Conventions of this sort help reduce the amount of time any one station is on the air, which in turn reduces congestion and interference.

Interference can be minimized by "channelizing" operation, and there has been some movement toward doing so on a voluntary basis on some bands. This means that an operator on such a band uses only one of the "channel" frequencies, rather than using a frequency between channels. When the channel frequencies are properly chosen to suit the type of modulation in use, no transmission on any channel will interfere with any signal on any other channel. The only interference then comes from other stations on the same channel, and from those individualists who persist in using the spots between channels. The fellows "in the cracks," however, interfere with not one but two channels, and receive interference from both as well, which tends to discourage them from staying there.

The most important single rule to reduce interference and congestion, though, is one which is not limited to radio. It was enunciated many years before radio was invented, and reads: *Do unto others as you would be done unto.* Any operator who follows this rule, in all cases, should find few problems with interference—and if we all followed it, the interference and congestion problems which have plagued ham bands from the beginning would disappear.

14

Rules and Regulations (Q and A)

The Rules and Regulations of the Federal Communications Commissions (FCC) are what we live by on the amateur bands. Or, perhaps it is better to say what we are *supposed* to live by (there are always some turkeys on the air)! The FCC General and Technician class examinations will have a certain number of questions that deal with the regulations. The idea is to make sure that you understand the legal aspects to operating an amateur radio station. There are a lot of applicants for the amateur license who assume that their knowledge of electronics will allow them to sail past the examination with flying colors. Besides the fact that there are some matters in amateur radio examinations that are just not taught in a general electronics curriculum (engineering or technician level), it is quite possible to flunk the examination on regulations alone. This is why we have placed the Rules and Regulations Question and Answer section in a chapter of its own. If you have a pretty good grasp of electronics, especially practical communications (which is different from "communications" taught in engineering school), then bone up on the regs in this chapter and go try your hand at the FCC license examination. Good luck.

The format to this section will be the traditional Q & A that so many readers have preferred over the years. A question will be given, and then the answer will follow. In some cases, it may be necessary to have a more extensive discussion of the answer in order to improve your understanding. In these cases, the format will be Q, A & D. The Rules and Regulations of the Federal Communications Commission cover more areas than the General/Technician class written examination, so you are advised to read over the R & R (published in their entirety in Appendix E).

QUESTIONS AND ANSWERS 1-16

Q1. What is meant by the "control point" of an amateur radio station?

A. The control point is the operating position of an amateur radio station where the control operator (you) performs his or her function (i.e., the operator operates the station from this point).

Q2. Define "emergency communications" as it applies to the amateur service.

A. Any amateur radio communication directly related to the immediate safety of life of individuals or the immediate protection of property.

Q3. What is a "general state of emergency communications?"

A. The FCC may, at its discretion, declare such a state when an emergency exists. No amateur operation other than those communications directly related to the emergency may take place. You are advised to read Section 97.107 for additional information on this question.

Q4. In general, how much power may an amateur radio operator use in his transmitter?

A. The dc plate/collector power shall not exceed 1000 watts, except for Novice class operators who may only run 250 watts. Any amateur station that runs more than 900 watts must provide an adequate means for accurately measuring the dc power to the plate/collector. There are also some other restrictions on power related to frequency or geography (see 97.67).

Q5. What is the rule regarding the operation of an amateur radio station by a party other than the station licensee?

A. The licensee may allow *any* third party to participate in the operation of the amateur radio station. But, if the third party does not have an amateur license, the control operator must be present to exercise the control functions. Also, the privileges of the control operator shall govern the operation of the station. For example, a Novice class operator may only exercise Novice class privileges when operating a station owned by a higher grade licensee (i.e., General, Advanced or Extra). Regardless of who operates the station, however, the licensee is still responsible for the operation of the amateur station.

Q6. An Advanced class operator is at the controls of a station licensed to a General class operator. The station is operated according to the privileges of the General class operator. Which call sign is used?

A. The call sign of the licensee of the General class station.

Q7. In the situation above, the Advanced class operator desires to use a frequency outside of the General class sub-band. What call sign is used?

A. A combination call sign. For example, if K4IPV desires to operate K3RXK, then the proper call sign will be K3RXK/K4IPV.

Q8. How are you to identify under an interim permit when you wish to use privileges for the higher class for which you just qualified?

A. There will be a special identifier printed on the interim permit. On radiotelephone, say the word "interim," followed by the identifier. On radiotelegraphy, use the slash bar (/) "$\overline{DN}$" followed by the identifier issued by the FCC examiner.

Q9. Under what conditions may a third party, who does not have an amateur radio license, operate an amateur station?

A. The licensee may permit such a third party to use the station provided that there is a licensed control operator present to continuously monitor and supervise the radio communication to insure compliance with the rules and regulations.

Q10. May an amateur operator accept payment or gifts for performing third party communications?

A. Third party traffic involving material compensation, either tangible or intangible, direct or indirect, paid or promised to a control operator or licensee, is prohibited.

Q11. May an amateur operator transmit third party traffic of a business or commercial nature?

A. No, except in a bona fide emergency as defined earlier in this section.

Q12. Is it legal for an amateur operator to perform international third party communications?

A. No, unless the country of origin agrees to allow such traffic.

Q13. May an amateur radio station transmit music?

A. No, the transmission of music is prohibited.

Q14. May an amateur radio station transmit codes of ciphers in order to hide the meaning of the transmission?

A. No, codes and ciphers are prohibited. All communications, regardless of the type of emission employed, must be in plain language except that generally recognized abbreviations established by regulation or custom are permitted where the intent is not to obscure the meaning but to facilitate communication.

Q15. Under what circumstances may an amateur radio station transmit general broadcasts to listeners who are not amateur radio operators?

A. There are no such circumstances. Broadcasting is prohibited in the amateur service.

Q16. What is the regulation regarding the transmission of obscene or profane language.

A. It is prohibited. No licensed radio operator or other person shall transmit communications containing obscene, indecent, or profane words, language, or meaning.

QUESTIONS AND ANSWERS 17-32

Q17. What types of one-way communications are permitted in the amateur service?

A. The following forms of one-way communication are not to be construed as broadcasting, so are permitted in the amateur service: a) emergency communications, including bona fide emergency drill practice transmissions; b) information bulletins consisting solely of subject matter having direct interest to the amateur radio service as such; c) round table and network operations where more than two amateur stations are in communications, each station taking a turn at transmitting to other stations of the group; and d) code practice transmissions intended for persons learning or improving their skill in the International Morse code. The amateur station is also permitted to make one-way communications for the purpose of measurement of emissions, temporary observations of transmission phenomena, radio control of remote objects, etc. (see 97.89 and 97.91).

Q18. What frequencies are available to the Technician class operator?

A. All amateur frequencies above 50.0 MHz, plus the full privileges of the Novice class, which means the following HF bands (CW only): 3700-3750 kHz, 7100-7150 kHz, 21,000-21,200 kHz, and 28,100-28,200 kHz.

Q19. What frequency bands are available to the General class operator?

A. The General class operator may use all authorized amateur frequencies except those specially set aside for Advanced class and Extra class operators. The *forbidden* frequencies are:

Extra Only	Advanced and Extra
3500-3525	3800-3890
3775-3800	7150-7225
7000-7025	14,200-14,275
14,000-14,025	21,270-21,350
21,000-21,025	
21,250-21,270	

D. Be careful of the frequency questions! The FCC, bless their souls (or is that *soles*?), have a habit of nesting an Advanced/Extra or Extra class frequency in a list of frequency bands that are allowed to the General. In the past (like when I took the exam too many years ago to count), they would ask convoluted versions of this question, like "on which frequency of those listed below may the General class *not* use radiotelephone (or A3) emissions?" Then, when you look down the list, all possibilities are General phone frequencies except the correct answer, which is an Advanced/Extra class phone frequency. When you encounter this type of question, read it very, very carefully to make sure that you know the correct meaning. A little word like "not" is a real powerhouse when it comes to knocking you out of the exam!

Q20. All emissions are allowed in which segment of the 20-meter band?

A. The whole nine yards! You can use CW anywhere in the band!

Q21. On which frequencies are A3 emissions allowed in the 20-meter band?

A. 14,200-14,350 (see Part 97.61).

D. Read Part 97.61 of the FCC Rules and Regulations. Memorize the frequency limitations for all of the HF bands and the VHF bands up through 2 meters. When memorizing, make it easier on yourself by recognizing that all emissions are given a letter and a number designation: *A* for *AM* (including on-off keying CW), *F* for *FM*. The emission types are 0 for no modulation, 1 for CW, 3 for radiotelephone, and 5 for television. F5, then, means frequency modulated television emissions (there is such a thing). If you are confused, then let me invite you to curl up with your copy of the FCC R&R, a cold 807, and a good book (for when the Regs bore you 89 milliseconds after the onset of reading). Read Appendix C of the Regs.

Q22. What are the general regulations regarding the selection and use of frequencies within the authorized amateur bands?

A. a) An amateur station may transmit on any frequency within any authorized amateur frequency band; b) sideband frequencies resulting from keying or modulation of a carrier wave must be confined within the amateur band (no splatter on 14356!); and c) the frequencies available for use by a control operator of an amateur station are dependent on the classification of the control operator and are listed in Section 97.7 of the regs.

Q23. How much power may a General class operator run when operating on a frequency of 21,225 kHz?

A. 1000 watts.

Q24. How much power may a General class operator run when operating on a frequency of 21,125 kHz.
A. 250 watts.
D. Trick question! Read Section 97.67 (d)! *All* operator classes are restricted to 250 watts in the *Novice* portions of the amateur bands. Know the Novice frequencies, or this type of question will screw you up. Remember, each question answered incorrectly will reduce the chances of your passing the exam. *Remember, only 250 watts of power is authorized to all classes of operator when using the novice segments of the bands. Know the novice frequencies!*

Q25. Under what circumstances may AØ emission be used?
A. For short tests, and for authorized remote control operation as in radio controlled models.

Q26. Under what conditions may an amateur transmitter be used solely for remote control of model craft?
A. If the power output is 1 watt or less, and a permanent transmitter identification card (FCC Form 452-C), or a label plate made of a *durable material*, is affixed to the transmitter. The licensee's name, address and call sign shall be on the ID plate.

Q27. Is station identification needed for operation of the transmitter in Q26?
A. No, provided that the transmitter is used solely for control of the model craft at that time.

Q28. Is an amateur who transmits codes and control signal modulated onto a carrier wave for the sole purpose of controlling a model craft in violation of the ban on codes and ciphers?
A. No, Part 97.99 (b) specifically exempts such signals.

Q29. Must the operator of a remote control transmitter, as in Q26-Q28, keep a log showing the date and time that operation commenced?
A. No.

Q30. What speeds are available to an amateur operator using radioteletype?
A. The normal transmitting speed shall be as close as possible (+5 WPM) to one of the standard teletypewriter speeds: 60, 67, 75, or 100 WPM.

Q31. What is the deviation permitted between the mark and space signal in F1 (radioteletype) emission?
A. 900 Hertz (see 97.69(c)).
Q32. What is the highest permissable audio frequency when the transmitter is operated in F2 (i.e., audio frequency shift keying, a type of teletype)?
A. 3000 Hz, with the same 900 Hz limitation between mark and space frequencies.
D. Example: If the high tone is 3000 Hz, then the low tone must be not less than (3000-900), or 3100 Hz.

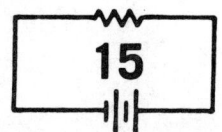

Technical Q and A

The questions and answers in this chapter follow the same format used in the previous chapter. Here, we deal with the technical information you'll need to know.

QUESTIONS AND ANSWERS 1-21

Q1. How much power may a General class operator use on 21,190 kHz?

A. 250 W. This is a Novice class frequency, and *all* operators using this frequency are limited to no more than 250 watts, the Novice "gallon" (see Section 97.67d).

Q2. How much power may a General class operator use on 21,380 kHz?

A. 1000 watts dc input to the stage or stages supplying power to the antenna. This power is measured from the anode-to-cathode or collector-to-emitter.

Q3. What factor does *not* affect the *impedance* of a transmission line?

A. Its length, advice offered by CB "experts" to the contrary notwithstanding.

Q4. What type of vacuum tube most often requires neutralization when used as an rf power amplifier?

A. Triode.

D. The tetrode and pentode power beam tubes have too little interelectrode capacitance to oscillate in most cases. They will, however, occasionally be used in circuits where neutralization is needed. Also, the need for neutralization in triode amplifiers is lessened if the tube is operated in the grounded-grid configuration, a common choice.

Q5. What factors affect the Standing Wave Ratio (SWR) of an antenna system?

A. The impedance matching between the antenna and the transmission line.

Q6. What type of rf power amplifier is most suited to amplifying single sideband (SSB) signals?

A. A *linear* amplifier, such as class AB_1 (first choice) and class B (a close second).

Q7. What type of rf power amplifier is most suited to amplifying FM and CW signals?

A. Class C.

D. The class C amplifier is not linear, so will not properly amplify signals that cannot tolerate distortion of the amplitude. Class C amplifiers are however, a lot more efficient than linear amplifiers (Q6), so are best used when the signal will tolerate amplitude distortion (as in CW and FM).

Q8. What is the function of a diode?

A. It passes current in only *one direction*; i.e., it *rectifies* ac into dc (of sorts).

Q9. Which ratings of a solid-state diode are most important?

A. The *peak inverse voltage* (PIV), also sometimes called the *peak reverse voltage* (PRV), and the *maximum forward current*.

Q10. What is the total length of the wire used to make one element of a quad antenna?

A. One full wavelength.

Q11. What is the length of one side of the quad antenna?

A. Quarter wavelength.

Q12. What is the maximum allowable speed of the CW identifier used with an automatic repeater?

A. 20 WPM.

Q13. Find the total capacitance in Fig. 15-1.

A. $C_t = C1 + C2 + C3$

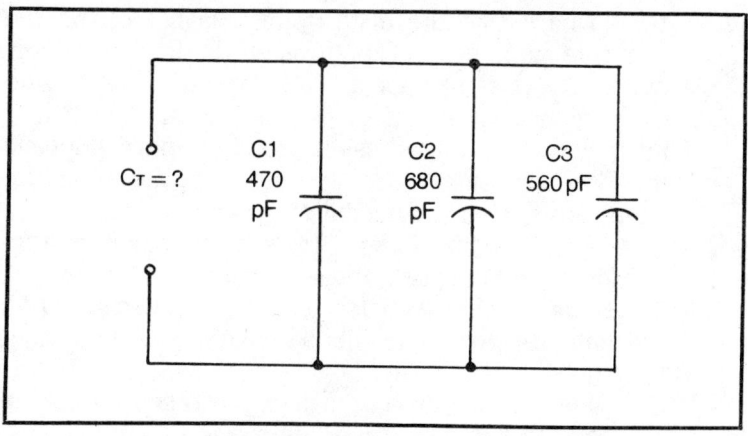

Fig. 15-1. Parallel capacitors.

$$C_t = (470 \text{ pF}) + (680 \text{ pF}) + (560 \text{ pF})$$
$$C_t = 1710 \text{ pF}$$

D. With capacitors in parallel, *add* to find the total capacitance in the circuit.

Q14. Find the capacitance in Fig. 15-2.

A. $C_t = C1 + C2$ (Remember: add capacitances in parallel)
$$C_t = (100 \text{ pF}) + (220 \text{ pF})$$
$$C_t = 320 \text{ pF}$$

Q15. Define *splatter*. How is splatter prevented?

A. Splatter is *spurious emissions* (i.e., emissions from your transmitter that you don't need or want) resulting from overmodulation of the transmitter. In general, splatter is caused when the audio signal drives an AM transmit-

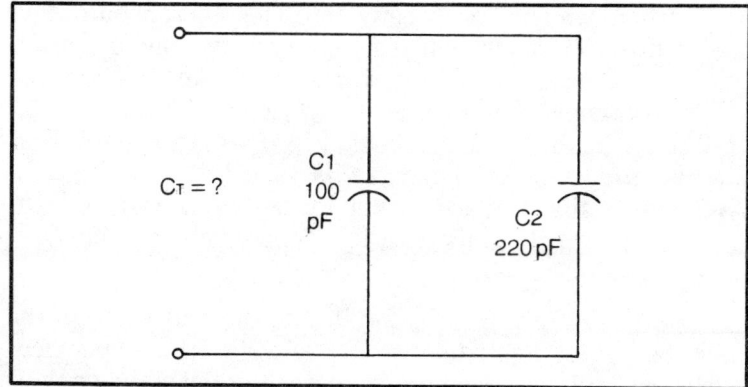

Fig. 15-2. Two capacitors in parallel.

ter so hard that the plate voltage is cut off on the negative audio peak. This turns off the final amplifier, hence the rf signal, for a brief portion of each audio cycle, and this "delinearizes" the signal, which is a fancy and inept way to say it produces extra sideband frequencies. Splatter is prevented by keeping the transmitter from overmodulating.

Q16. What procedure is used to help minimize unintentional interference between amateur stations?

A. Listen on the channel before you call! Don't call CQ on the Halo Missionary net just because *you* can't copy net control!

Q17. What element of a transistor corresponds to the anode in a vacuum tube?

A. Collector.

Q18. What element of a transistor corresponds to the grid of a vacuum tube?

A. Base.

Q19. Which element of a vacuum tube corresponds to the emitter of a transistor?

A. Cathode. (We fooled you! Notice that the statement in Q19 is inverted over the two previous questions! The FCC likes to invert questions to fool test guide authors and trip up would-be amateurs who can't read too good.)

Q20. What is the best procedure for adjusting the *vox* on an amateur transmitter?

A. Adjust the control knob until the transmitter will not turn on from ordinary room noise (which, if you have a 4-year old like me, can be considerable), yet triggers when you speak directly into the microphone. The anti-vox is adjusted to not trip on receiver audio signals. In some cases, the vox, vox delay and Antivox controls are somewhat interactive, so "twiggle" them a little bit.

Q21. What does the symbol in Fig. 15-3 represent? What *unit* is used to express its value?

A. It is a capacitor, and the unit for capacitance is *farad*. Do *not* use *fard*, which actually appeared on an old exam!

TEST TAKERS CAMEO

Test taking is a skill, and it is in the mastering of the skill of test taking that some will do well, others spectacularly, and still others come home with their tail between their ears.

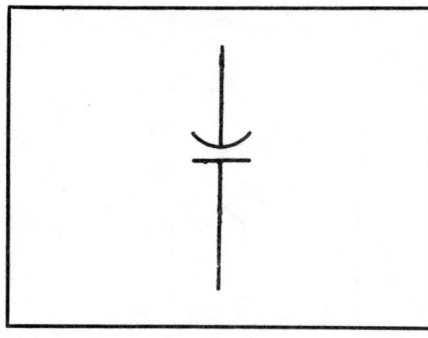

Fig. 15-3. Capacitor symbol.

Here are some rules:
1. Get a boatload of sleep the night before the test, but not too much.
2. Arrive early enough to get a good seat, and calm down.
3. Carry a calculator (or, if you are over 40 and have a technical education, a slip-stick). Make sure that the batteries are charged (or are new). If possible, carry the charger with you, and sit next to an electrical outlet. If your calculator takes ordinary, non-rechargeable batteries, then bring a spare even if the battery in the calculator is new (ever heard of shelf life?). Your calculator should be a minimal scientific model, and include Log_{10}, trig functions, square root, etc. But don't buy one of those super jobs with a million functions unless you know how to run it and need it for something else besides the exam. Having to figure out too much calculator during the exam is a lot like trying to nail Jello to the wall.
4. Use the *multipass* method of taking the test. On the first pass through the test, answer only those questions that you know immediately, and do not require either thought or calculation. On the second pass, get the quick calculation problems, but ignore any that you cannot solve. On the third, fourth and fifth passes (if it takes that many) answer all of the toughies.
5. Keep the answer sheet straight! It is a machine-scored answer sheet, and the numbering sequence on the sheet *must* match the numbering on the exam paper!
6. Don't daydream unless you need a break.

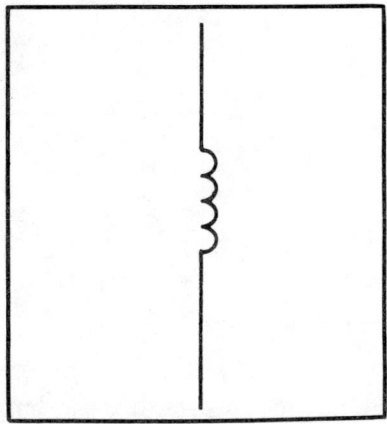

Fig. 15-4. Coil or inductor symbol.

Q22. What is the device represented by the symbol in Fig. 15-4, and what *unit* is used to express its value?

A. It is a *coil*, also called an *inductor* (probably by the FCC, since that is its *real* name!). The *unit* of *inductance* is the *Henry*.

Q23. What is the device represented by the schematic symbol in Fig. 15-5, and what *unit* is used to express its value?

A. It is a resistor, and its value is expressed in Ohms.

Q24. What does the figure in 15-6 represent?

A. It is an oscilloscope (Lissajous pattern) presentation of the mark and space signals in a radioteletype system.

Q25. Identify the *mark* and *space* loops in Fig. 15-6.

A. See Fig. 15-6.

Although it is impossible to predict the mind of the FCC, it is possible to guess something about the exams based on prior experience. The FCC in the past has loved *oscillator identification* questions. When I took the test back in the dark ages, this section was the killer. Most test takers viewed it as a gift from Attila the Hun. But take heart. Over the years I have learned that oscillator identification is just a little matter of learning some salient identifying points. It no longer snaps my mind clean out of its socket! In the questions

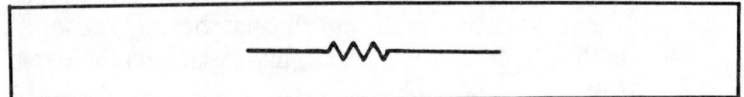

Fig. 15-5. Resistor symbol.

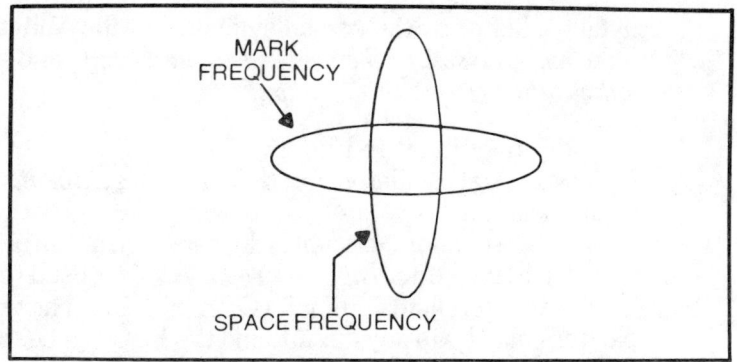

Fig. 15-6. Oscilloscope pattern of mark and space teletype signals.

to follow, we will identify the principal forms of oscillator circuits, and point out the identifying feature in each case. All Hartley oscillators, for example, use a *tapped coil*.

Q26. Identify the circuit in Fig. 15-7.

A. It is a crystal oscillator. More specifically (the FCC sometimes gets specific), it is a *Miller* crystal oscillator, or, simply, Miller oscillator. You can identify any

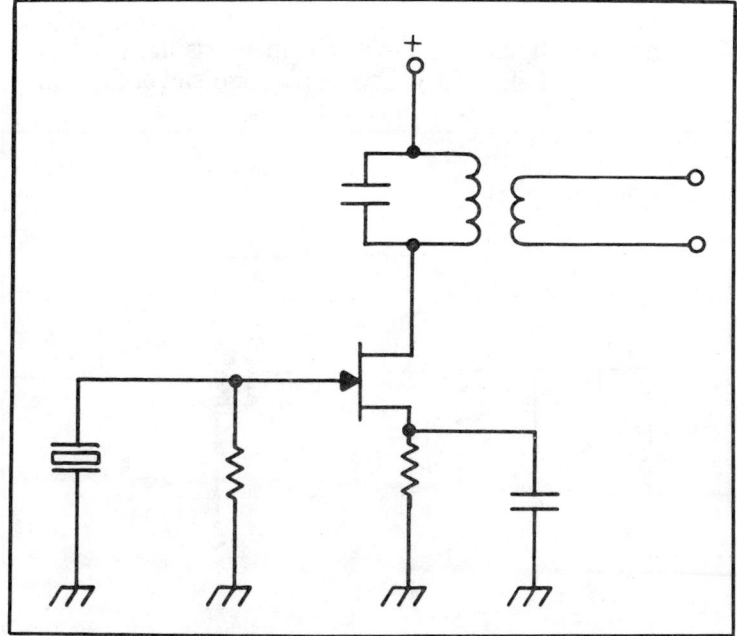

Fig. 15-7. Crystal oscillator circuit.

333

crystal oscillator by the crystal symbol. But the Miller circuit has a *crystal in the grid/base/gate circuit,* and a *parallel tuned circuit in the output.*

Q27. Identify the circuit in Fig. 15-8.
A. It is also a crystal oscillator. But this time, it is a *Colpitts* crystal oscillator.
D. All Colpitts oscillators, crystal or LC tank tuned, can be identified by the series *capacitor voltage divider* used to provide the feedback signal (C1 and C2). These capacitors may be in any circuit, and the FCC has been known in the past to use a circuit with the voltage divider connected with one end to the anode or collector of the oscillator device. It is still a Colpitts circuit if it has the tapped capacitor network.

Q28. Identify the circuit in Fig. 15-9.
A. It is a Hartley oscillator.
D. All Hartley oscillators use a tapped coil in the tuning network. In most cases, the cathode (or emitter or source) is connected to the tap, and the other elements are connected, often via capacitors, to the ends. Again, the FCC likes to screw you up by using a different configuration, such as connecting one end of the coil to

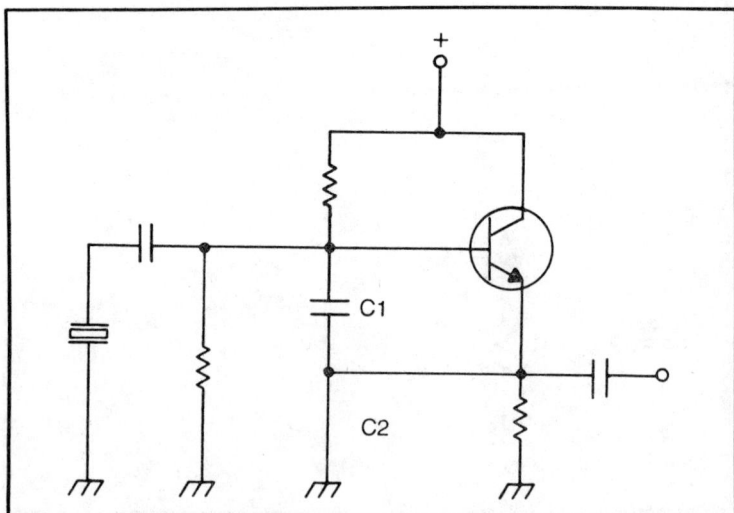

Fig. 15-8. Colpitts crystal oscillator.

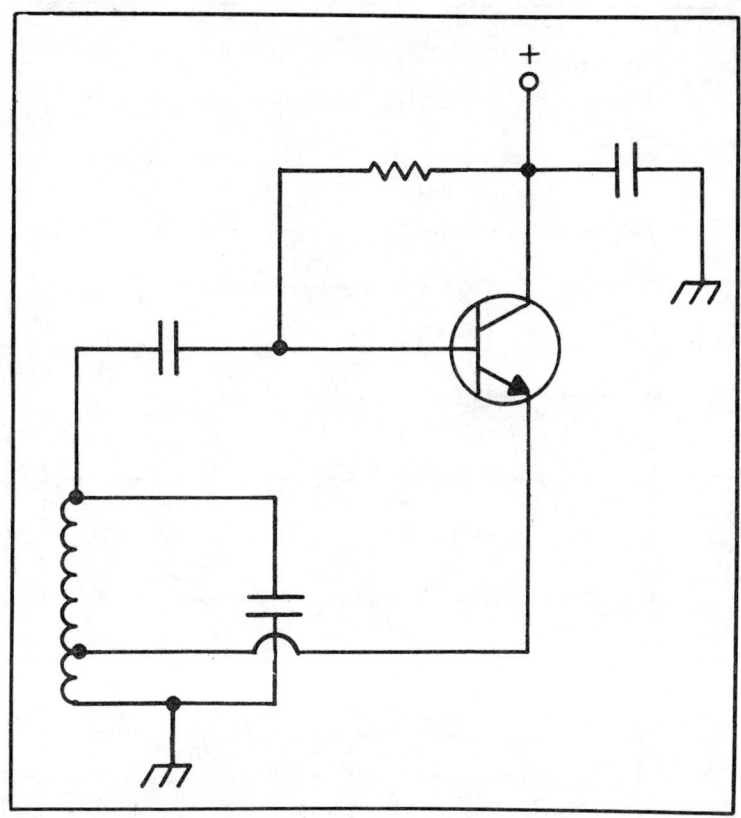

Fig. 15-9. Hartley oscillator circuit.

the plate instead of the ground. It is still a Hartley oscillator if it uses a tapped tuning coil to provide feedback.

Q29. Identify the circuit in Fig. 15-10.
A. It is one of those permuted Hartley oscillators with the coil connections going places that no self-respecting designer would select, even though they officially work (FCC says so). It is still a Hartley oscillator, however.

Q30. Identify the circuit in Fig. 15-11.
A. It is a Colpitts oscillator.
D. Don't be fooled because this is a VFO, and is a permuted circuit. The feedback system is a tapped capacitor voltage divider, so the circuit is the Colpitts.

Q31. Identify the circuit in Fig. 15-12.
A. Colpitts oscillator.

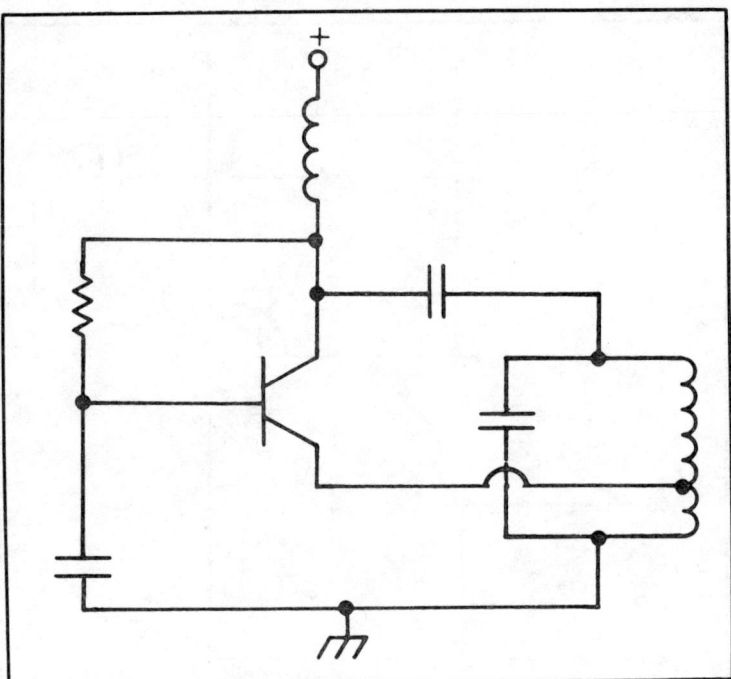

Fig. 15-10. Another Hartley oscillator circuit.

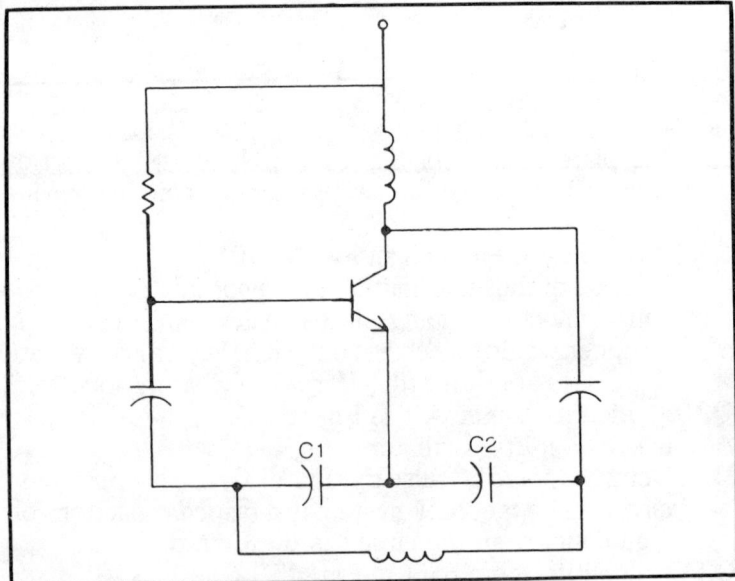

Fig. 15-11. Another Colpitts oscillator configuration.

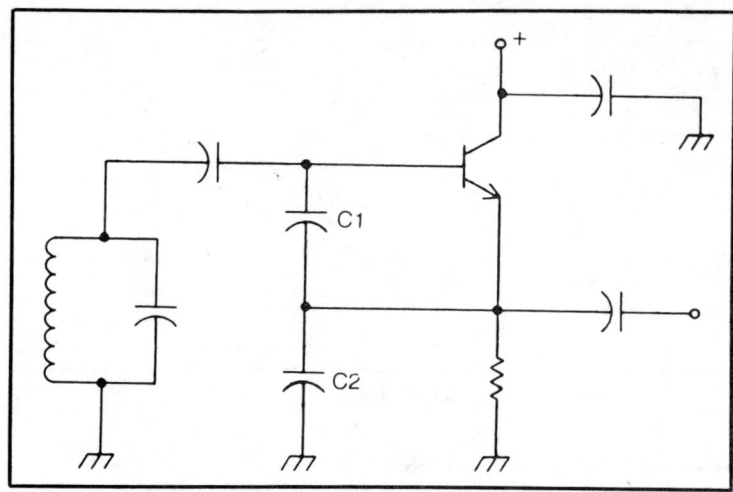

Fig. 15-12. Colpitts oscillator with an LC frequency-determining network.

Q32. Identify the circuit in Fig. 15-13.
A. Half-wave rectifier.
Q33. What is the function of the circuit in Fig. 15-13?
A. It converts bidirectional alternating current into unidirectional direct current, or, more properly, unidirectional *pulsating* dc. It does this neat trick because a diode will conduct current in only *one direction*.
Q34. Identify the circuit in Fig. 15-14.
A. It is a full-wave *bridge* rectifier.
D. Memorize this circuit! The FCC likes to show you several versions, most of which will short out, or simply not work. Only one—the one with all diodes pointed in the same direction—is correct.
Q35. What is the purpose of transistor Q1 in Fig. 15-15?
A. It is a series-pass voltage regulator transistor.

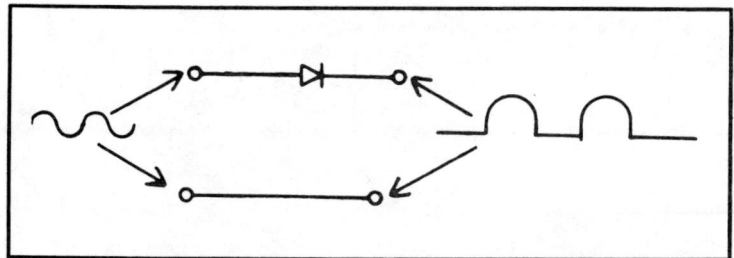

Fig. 15-13. Half-wave rectifier circuit.

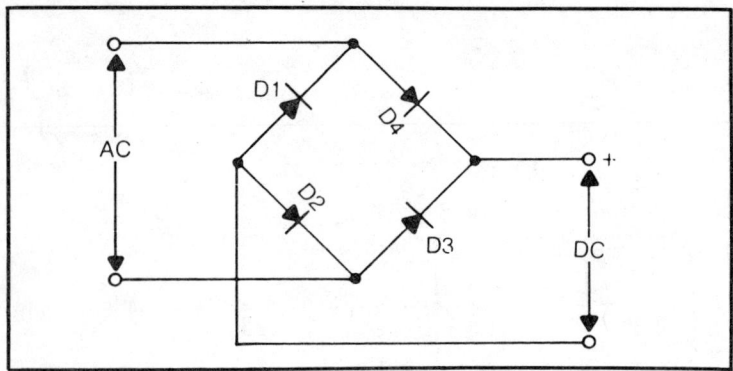

Fig. 15-14. Full-wave bridge rectifier circuit.

Q36. What is the purpose of diode D1 in Fig. 15-15?
A. It provides a *reference voltage* to the base of transistor Q1.
Q37. What type of diode is D1 in Fig. 15-15?
A. Zener diode.
Q38. What is the purpose of capacitor C1 in Fig. 15-15?
A. Filtering the pulsating dc input voltage.
Q39. What is the purpose of capacitor C2 in Fig. 15-15?
A. Improves the transient response, i.e., response to fast changes in current load.
Q40. What type of circuit will be connected to points A and B in Fig. 15-15?
A. The output of the *rectifier* that converts alternating current to pulsating direct current.

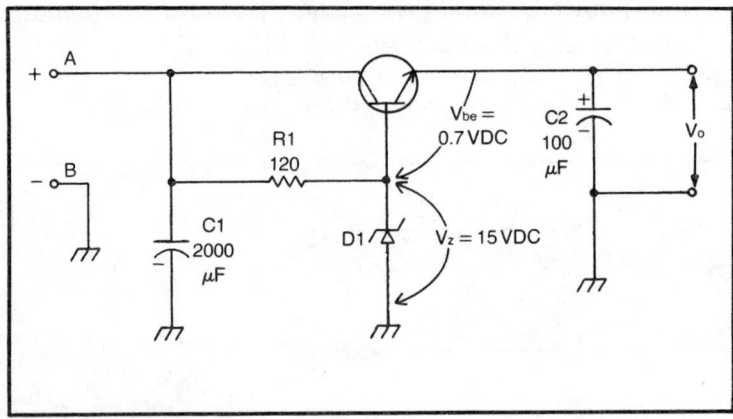

Fig. 15-15. Series-pass voltage regulator circuit.

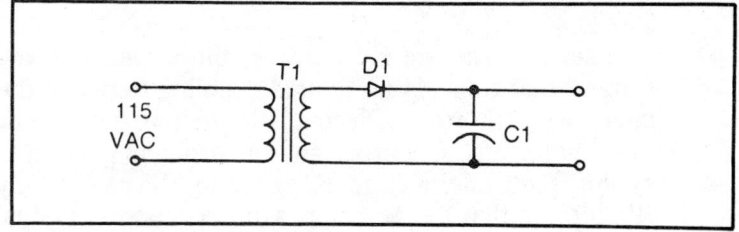

Fig. 15-16. Half-wave rectifier with a capacitor filter.

Q41. What is the approximate output voltage of the circuit in Fig. 15-15?
A. 14.3 volts.
D. The output voltage is the zener potential (V_z) less the voltage drop from emitter-to-base on transistor Q1 (V_{be} = 0.7 volts), so
$$V_o = V_z - V_{be}$$
$$V_o = (15 \text{ VDC}) - (0.7 \text{ VDC})$$
$$V_o = 14.3 \text{ VDC}$$

Q42. Identify the circuit in Fig. 15-16.
A. It is a half-wave rectifier with a capacitor filter.

Q43. What is the purpose of this circuit?
A. The rectifier converts bidirectional ac into unidirectional pulsating dc. The rectifier alone, however, cannot provide the smooth dc needed by most circuits (and required for amateur transmitters below 144 MHz). The filter capacitor provides some of the smoothing needed to make the pulsation dc pretend that it is pure dc.

Q44. What is the total inductance of the circuit in Fig. 15-17?

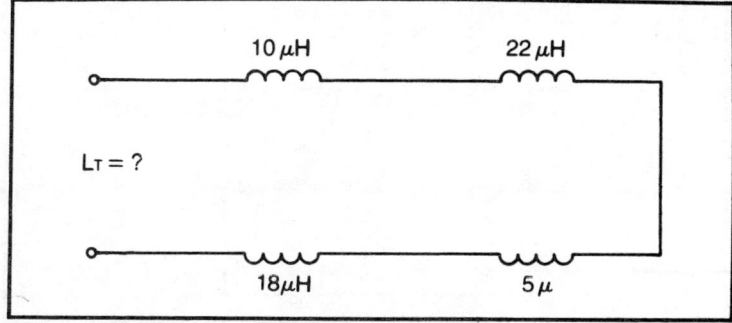

Fig. 15-17. Inductances in series add.

A. 55 μH.
D. In a series circuit of inductances, the values *add*, assuming that there is no mutual coupling between the inductors. If there is such coupling, then we must know the coupling coefficient(s), and the problem gets messy. For the General class examination, the most likely situation is that the question will not mention mutual coupling at all, so the answer will be the simple sum of all inductances in the circuit.

Q45. Find the current I in the circuit of Fig. 15-18.
A. 0.36 amperes
D.
$$I = E/R$$
$$I = (12)/(33)$$
$$I = 0.36 \text{ amperes}$$

(Note: The FCC might try to fool you by giving the answer in milliamperes. Know how to convert from one to the other. One ampere is 1000 mA, so 0.36 amperes must be 360 mA).

Q46. Find the power dissipated in the restrictor of Fig. 15-18.
A. 4.3 watts.
D. There are two ways to solve the problem: $P = E^2/R$ or I^2R.

$P = E^2/R$	$P = I^2R$
$P = (12)^2/(33)$	$P = (0.36)^2(33)$
$P = 144/33$	$P = (0.130)(33)$
$P = 4.3$ watts	$P = 4.3$ watts

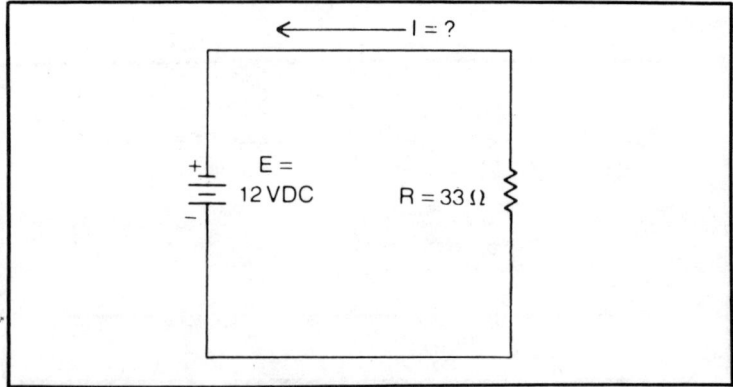

Fig. 15-18. Simple series circuit with current unknown.

Q47. Calculate the applied voltage in Fig. 15-19.
A. 10 volts.
D.
$$E = IR$$
$$E = (0.01 \text{ A})(1000 \text{ ohms})$$
$$E = 10 \text{ volts}$$

Q48. Calculate the power dissipated in the resistor in Fig. 15-19.
A. 0.1 watt (also written as 100 mW).
D. Again, there are two ways, and one of them does not require that we first calculate the applied voltage:

$$P = I^2R \qquad\qquad P = E^2/R$$
$$P = (0.01)(1000) \qquad P = (10)^2/(1000)$$
$$P = 0.1 \text{ watt} \qquad\qquad P = 0.1 \text{ watt}$$

Q49. Calculate the voltage in Fig. 15-20.
A. Aha! A *hard* one! Not really. Notice that the resistance across which the voltage is measured is approximately 1/100 of the total resistance. So, we could reasonably expect the voltage to be 1/100, or, 1 volt. If we want to solve it the *correct* way, or, if the FCC is not clever enough to select a resistance ratio that is solvable by inspection, then we must use the *voltage divider formula*:

$$E = \frac{E_1 R_2}{R_1 + R_2}$$
$$E = \frac{(100V)(100)}{(9900 + 100)}$$

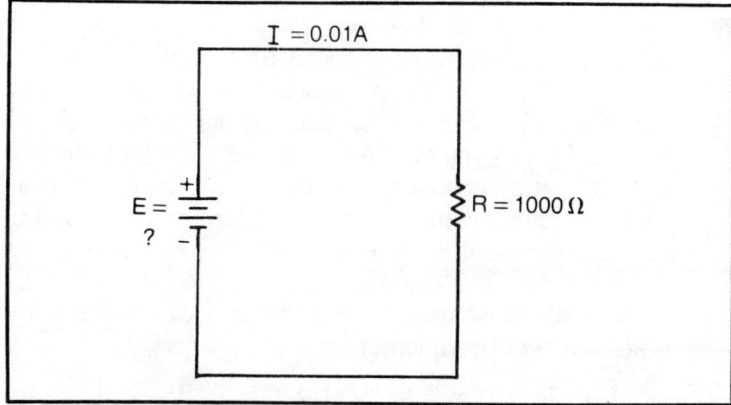

Fig. 15-19. Simple series circuit with unknown source voltage.

$$E = \frac{10{,}000}{10{,}000}$$

$$E = 1 \text{ volt}$$

The "trick" is to note that the voltage and the resistor across which the output voltage is taken are in the numerator ("upstairs"), while the sum of all resistances in the circuit is in the denominator ("downstairs").

Q50. How much power is dissipated in resistor R2 in Fig. 15-20?
A. 0.01 watt, or, 10 mW.
D. The easiest way to solve this problem is by E^2/R:

$$P = E^2/R$$
$$P = (1)^2/(100)$$
$$P = 1/100$$
$$P = 0.010$$

Q51. What is the voltage drop across resistor R1 in Fig. 15-20?
A. 99 volts.
D. If we had solved the previous problems first, then we could note that, if $E = 1$ volt, then the drop across R1 had to be (100V - 1V), or 99V. We could have also used the voltage divider formula with R1 in the numerator instead of R2.

Q52. Which amateur bands are *most* susceptible to ionospheric absorption during daylight hours?
A. 160, 80, 40 meters.
D. The FCC may not give you a choice of three bands. In general it is the low bands that are affected. This is why 80 meters seems terribly "local" during the day, but becomes darn near transcontinental during the night. If the FCC gives you only one of these three as a choice, then it is the correct one. If two or more are offered, then one choice will probably be of the nature "A and C above." Fortunately, the new crew at FCC seems to have eliminated some of the "all of above" answers, at least that's what they claimed when I called them for guidance over this update!

Q53. What is a result of ionospheric absorption during daylight?

A. The most practical result is that communications distances become very short compared with the distances possible at night. On 40 meters, for example, it is usual to be able to work only a few hundred miles during the day. From K4IPV in Virginia, for example, daytime DX on 40 meters seems to be all east of the Mississippi River. But, at night, when I can crush through a hole in the QRM from Radio Moscow, it is possible to work New Zealand and Australia. California stations are commonplace.

Q54. What is the length of the sunspot cycle?
A. 11 years.
D. Do not get confused over the sunspot period of 27 days.
Q55. What is the effect of the sunspot cycle?
A. During periods of high sunspot activity, it takes little effort to make long distance communications on the higher amateur bands. Little more than a transistor QRP oscillator loaded into a wet string can run up high DX countries total during those periods. But, when the sunspot count is low, it is a lot more difficult to make contacts on the higher bands (20-10 meters), and a lot of activity goes to the lower bands. But this does not mean that the upper HF bands are dead, just seem lousy much of the time. But never fear, the sunspot count on Ol' Sol regroups every 11 years and all of that juicy DX returns.

Q56. Not counting repeaters, what are two methods by which a two-meter signal can be propagated beyond the line-of-sight radio horizon?
A. Tropospheric propagation by *refraction* and *ducting*.

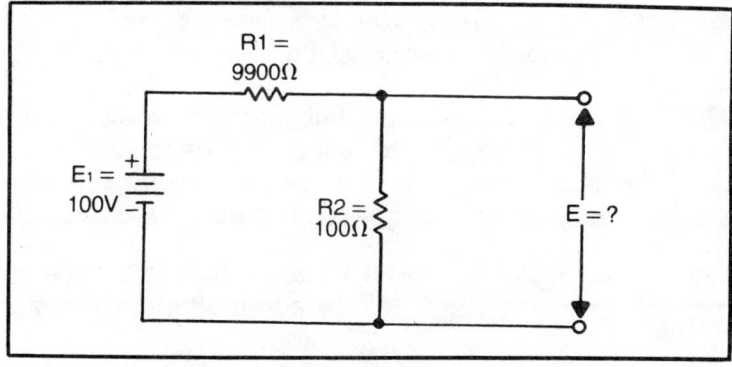

Fig. 15-20. Simple series circuit with unknown output voltage.

Q57. What will the sound be like when an AM amateur transmitter interferes with the operation of a phonograph or electronic organ?
A. Clear and distinct voice from the loudspeaker that is all too easy to identify as *you*.
Q58. What will the sound be like when an SSB transmitter interferes with the operation of a phonograph or electronic organ?
A. The sound is garbled and almost indistinguishable (which allows you some anonymity, maybe).
Q59. On which HF bands does the Technician class licensee have no privileges?
A. 160 meters and 20 meters.
D. The Technicial class licensee has Novice class HF privileges.
Q60. What is audio rectification?
A. Modulated RF signal applies to the input of an audio amplifier (as in a TV receiver, radio receiver, phonograph or electronic organ) will sometimes drive the amplifier into clipping, which serves to rectify (i.e., *detect*) the signal and extract the audio modulation. This signal is then amplified and produces a loud output signal. It is the type of interference mentioned in Q57 and Q58.
Q61. What is the highest percentage of modulation allowed on an AM transmitter operated on a frequency of 14.357 MHz?
A. 100 percent.
D. It is always 100 percent. The frequency is irrelevant and was thrown in to make you wonder a little.
Q62. A _____ test is used to check for distortion of an SSB transmitter on an oscilloscope.
A. Two-tone.
Q63. What is the FCC regulation regarding serious emissions radiated from an amateur transmitter?
A. The spurious radiation must be 40 dB below the main unmodulated carrier signal, and in no event exceed 50 mW.
Q64. The FCC requires all amateur transmitters operating below _____ to use a well filtered DC power supply.
A. 144 MHz.

Q67. Name three types of maps that may be used to determine the beam heading relative to true north from your operating location.
A. Great circle, azimuthal and Lambert conformal.
Q66. What type of maps may not be used to determine the beam antenna heading?
A. *Mercator* projections.
Q67. A circuit contains the following capacitors in parallel: 0.01 μF, 0.01 μF and 0.022 μF. What is the total capacitance?
A. 0.01 μF + 0.01 μF + 0.022 μF = 0.042 μF
Q68. How can you correct audio rectification interference to an electronic organ or phonograph?
A. Filter the power line, signal lines and speaker lines to and from the device. The addition of a low-pass filter designed for audio equipment is sometimes helpful, as is reducing your power output. Shielding may help, too. *But never undertake the modification or service of your neighbor's equipment yourself. Refer them to a qualified repairer. Once you touch that rig you're married to it, and the neighbor will hold you responsible. More than one ham has been hauled into small claims court to answer charges of busting the thing even though the new trouble was totally unrelated to the interference problem. Note:* Not all interference problems are your fault. The problem is not because you are doing something to their equipment, but that their equipment responds inappropriately to your signal! A low-pass filter on your rig is a good idea, but will *not* solve most audio rectification problems. This problem is caused by rectification of your *fundamental* by the equipment experiencing interference.
Q69. What steps can be taken to eliminate or reduce the effects of television interference by your amateur transmitter?
A. At the transmitter end, which is the only end that you can affect without the cooperation of your neighbor, you should: a) ensure a good ground for the transmitter; b) use a low-pass filter between the transmitter and the antenna; and c) use an antenna coupler
D. The FCC publication on TVI is one of the best sources of information on this topic, and is reproduced in the

Appendix. Note that the FCC recommendation for users of linear amplifiers is to use a low-pass filter between the exciter and the linear amplifier, as well as another low-pass filter between the output of the amplifier and the antenna. It is also a good idea to use an antenna tuner between the output of the low-pass filter and the antenna. The purpose of a low-pass filter is to pass only those frequencies below the cutoff frequency. Most amateur radio LPFs have a cutoff frequency between 35 and 45 MHz. Since TV stations operate on frequencies *above* 54 MHz, the LPF will take out any amateur band harmonics that might interfere with operation of the TV receiver.

But some TVI is *caused* not by your transmitter doing something to the other guy's TV receiver, but results from the TV receiver responding inappropriately to your signal. In those cases, we find that a high-pass filter sometimes helps, although sometimes a repair is needed. *See the warning about repairing the set above.*

Q70. What is the difference between a sudden ionospheric disturbance and an ionospheric storm?

A. The sudden ionospheric disturbance is caused by an electromagnetic wave originating from eruptions on the surface of the sun. Since it is an electromagnetic wave, it travels at the speed of light, so will arrive at the earth approximately 8 seconds after the disturbance. The ionospheric storm, on the other hand, is caused by charged particles from the disturbance on the sun. Since these particles travel at a lot slower speed, they will require 18 hours to several days to reach the earth's atmosphere. Both disrupt radio communications.

Q71. The nominal radiation resistance of a horizontal half wavelength dipole antenna is____.

A. 73 ohms.

Q72. The nominal radiation resistance of a quarter wavelength vertical antenna is____.

A. 52 ohms.

Q73. The nominal radiation resistance of a half wavelength inverted vee antenna is____.

A. 50 to 75 ohms, depending upon actual length and bend angle.

Q74. What is the typical range of radiation resistances observed on a horizontal half wavelength dipole near the ground?
A. 40 to 125 ohms.
D. The 73-ohm valve in Q71 is for an antenna in free space. We define this location as being many wavelengths from any conducting surface, which just about eliminates all amateur radio station locations. The FCC, however, still likes this question.

Q75. What is the range of radiation resistances found on practical quarter wavelength vertical antennas?
A. 10 to 70 ohms.

Q76. What type of capacitors are suitable for use in the filter of a high voltage DC power supply used in a transmitter?
A. Oil-filled or electrolytics.
D. The capacitor must be a value four to several hundred microfarads, at voltages to several kilovolts, depending upon the transmitter. Only oil-filled and electrolytics will give these values in reasonably sized packages. (The FCC likes electrolytics.)

Q77. Define MUF.
A. MUF stands for *maximum usable frequency*, the highest frequency that you can use to transmit from one point on the earth's surface to another point using the sky wave (i.e., ionospheric bounce).

Q78. Which type of common amateur radio antenna might help reducing interference to your receiver from man-made noise?
A. Horizontally polarized antennas.
D. There are two reasons why a horizontally polarized antenna will work best in this case: 1) most man-made noise is vertically polarized, so will not be optimally received by a horizontal antenna, and 2) the horizontal antenna is *directive*. If the noise source is off the ends of the antenna, then it will be attenuated.

Q79. How must you deal with spurious radiation?
A. (See FCC R&R Part 97.73): You must "take steps to eliminate the interference in accordance with good engineering practice."

Q80. What is a bandpass filter?
A. A bandpass filter passes only those frequencies that are above a low-frequency cutoff point, and below a high-frequency cutoff point. The difference between the two frequencies is the bandwidth of the filter.

Q81. How are bandpass filters used in amateur SSB transmitters?
A. The bandpass filter is used to remove the unwanted sideband from double sideband, suppressed carrier signal produced by the balanced modulator.

Q82. What is the typical frequency split between transmitter and receiver sections of a 2-meter repeater?
A. 600 kHz.

Q83. What is the typical frequency split between transmitter and receiver sections of 220 MHz repeater?
A. 1600 kHz.
D. The splits given in Q82 and Q83 are not dictated by the FCC, but are used by common practice and consensus among amateurs who build repeaters.

Q84. What are the best bands for communications during periods of minimum sunspot activity?
A. 80 and 40 meters.

Q85. What is a high-pass filter?
A. A high-pass filter passes only those frequencies that are above a specified cutoff frequency. A principal use of HP filters is to prevent harmonics from a nearby amateur transmitter from getting into the tuner of a TV receiver or FM receiver. The HP filter, which has a cutoff frequency between 30 and 54 MHz, is placed as close as possible to the terminals of the tuner.

Q86. What is a speech processor?
A. A speech processor is a device that is used to boost the effectiveness of an SSB transmitter by reducing the peak excursions of the signal and increasing the average power.

Q87. How much power is in the sidebands of an AM transmitter that is modulated 100 percent with a 1000 Hz sine wave?
A. Approximately 33 ⅓ percent of the power, with 66 ⅔ being in the carrier.

Q88. What is the radiation resistance of an antenna?

A. It is defined as the resistance portion of the antenna impedance, which is the voltage at the feedpoint divided by the current at the feedpoint:

$$R_r = \frac{E_{feedpoint}}{I_{feedpoint}}$$

Q89. Why are repeaters used on VHF and UHF?
A. Because a repeater will increase the distance that mobiles, hand-held portables, and low-power fixed stations can communicate. Because most repeaters are built at high locations, it is impossible to greatly extend the effective line-of-sight distance between two communicating stations.

Q90. Define power factor.
A. There are at least three ways to express power factor, and it will be wise to memorize all three. Note that these are not three different methods, but three expressions of the same phenomena:

a). Power factor is the ratio of the resistance to the reactance in an ac circuit:

$$P.F. = \frac{R}{X}$$

b). Power factor is the ratio of the *true power* to the *apparent power* (i.e., power read on ammeters and voltmeters):

$$P.F. = \frac{True\ Power}{Apparent\ Power}$$

c). Power factor is the *cosine* of the *phase angle* between *current* and *voltage*:

$$P.F. = \cos \Theta_{ie}$$

Q91. The mode of communications used by most EME ("moonbounce") operators is:
A. Simplex.

Q92. What factors affect the impedance of parallel transmission line?
A. The radius of the conductors and the center-to-center spacing of the conductors.

Q93. What factor does not affect the characteristic impedance of open-wire transmission line?
A. Length.
Q94. What is "unity" SWR?
A. It is a standing wave ratio of 1:1.
D. A unity SWR means that the antenna is perfectly matched to the transmission line.
Q95. What is one advantage of having a unity SWR in an antenna system?
A. A unity SWR indicates that the maximum possible power transfer is taking place between the transmitter and antenna, via the transmission line.
Q96. Does SWR affect the signal attenuation on a coaxial transmission line?
A. Yes, but the effect becomes noticeable only when the SWR is 2:1 or more.
Q97. What is the principal factor that affects the total attenuation of a coaxial transmission line?
A. If the frequency is constant, then the length is the principal factor.
D. Both length and operating frequency affect the attenuation factor of coaxial cable. The attenuation increases with frequency. The answer selected was due to the fact the frequency-caused changes in attenuation are negligible on any given band. Also, below about 30 MHz the attenuation per unit length is constant. The attenuation factor is specified in terms of dB attenuation per hundred feet of cable.
Q98. Does the characteristic impedance of coaxial cable affect the attenuation factor?
A. No.
Q99. What is a synonym for *characteristic impedance?*
A. Surge impedance.
Q100. What is 100 percent modulation for FM transmitters.
A. There is no physical meaning to 100 percent modulation for FM transmitters. The definition of 100 percent modulation is set by either convention (in the amateur service), or by regulation (all other services). In the FM broadcasting service, 100 percent modulation is a deviation of 75 kHz, while a television sound transmitter is 100 percent modulated at 25 kHz of deviation. In

the landmobile service, on the other hand, a transmitter is considered to be 100 percent modulated with a deviation of 5 kHz. The amateur service convention approximates the practice in the landmobile service, or makes 100 percent modulation deviation of 5 kHz. In the bands under 30 MHz, however, however, one is cautioned that the bandwidth of an FM transmission must be no more than that of an AM transmission, or 6 kHz. This means that the maximum deviation is 6 kHz/2, or 3 kHz. The definition of 100 percent modulation for these bands is, therefore, 3 kHz.

Q101. When is an AM transmitter modulated 100 percent by a sine wave audio signal?

A. The AM transmitter is 100 percent modulated by a sine wave when the rf signal amplitude doubles on positive audio peaks and drops to exactly zero on negative audio peaks.

Q102. How much audio power is required to 100 percent modulate at AM transmitter that has a dc input power of 150 watts?

A. 75 watts.

D. The audio power required to 100 percent modulate an AM signal is approximately one-half of the dc input power to the modulated stage.

Q103. What is the legal form of station identification when using RTTY?

A. Either CW (A1 emission), or radiotelephone (either A3 or F3 emission).

Q104. What is the maximum permissable speed of a CW identification when using an RTTY transmitter?

A. 20 WPM, or less.

QUESTIONS AND ANSWERS 105-150

Q105. How much power is dissipated when a potential of 25 volts causes a current of 1.5 amperes?

A. $P = EI$
$P = (25V)(1.5A)$
$P = 37.5$ watts

Q106. How much power is dissipated when a potential of 12 volts causes a current of 0.6 amperes?

A. $P = EI$
$P = (12V)(0.6A)$
$P = 7.2$ watts

Q107. Find the power dissipated when a current of 50 mA is developed by a potential of 15 volts.

A.
$$P = EI$$
$$P = (15V)(0.050A)$$
$$P = 0.75 \text{ watts, or } 750 \text{ milliwatts.}$$

D. Note that the question asked for a power when the current was in milliamperes. But the formula requires the current in amperes. The means that we must first convert the current to its equivalent: 50 mA is the same as 0.050 watts. Know your units conversions.

Q108. Find the total capacitance if a 0.01 µF capacitor is in series with a 0.022 µF capacitor.

A.
$$C_t = \frac{C1 C2}{C1 + C2}$$

$$C_t = \frac{(0.01)(0.022)}{0.01 + 0.022}$$

$$C_t = \frac{0.00022}{0.032}$$

$$C_t = 0.0069 \ \mu F$$

D. In all problems involving *either capacitors in series* or *resistors in parallel*, the total value of the result will be less than the lowest value in the problem. For example, in this problem the lowest value was 0.01 µF, so we knew that the answer could be less than this! If the FCC puts just one possible value on the multiple-guess question that is lower than the lowest value on the test, then *that* is the answer. For example:

Find the total capacitance when a 0.01 µF capacitor is in series with a 0.022 µF capacitor:
  a) 0.01 µF
  b) 0.0069 µF
  c) 0.032 µF
  d) 0.1 µF

Obviously, only b) is smaller than the smallest capacitor in the circuit, so it must be the answer! Also, note answer c): it is the addition of the two capacitance.

Q109. Find the total capacitance of a series circuit consisting of 470 pF, 270 pF, and 390 pF capacitors.

A.
$$C_t = \frac{1}{\frac{1}{470} + \frac{1}{270} + \frac{1}{390}}$$

$$C_t = \frac{1}{0.0021 + 0.0037 + 0.0026}$$

$$C_t = \frac{1}{0.0084}$$

$$C_t = 119 \text{ pF}$$

D. Again, if the FCC gives only one value that is less than the lowest value capacitor in the *series* circuit, then use that. In this case, the lowest value is 270 pF.

Q110. Find the reactance of 2 Henry inductor at 1000 Hz.

A.
$X_L = 2\pi FL$
$X_L = (2)(3.14)(1000 \text{ Hz})(2 \text{ Hy})$
$X_L = 12,560$ ohms

D. "pi" (π) is always 3.14 for radio problems. Whenever you see 2π, therefore, think 6.28.

Q111. Find the reactance of a 10 μH choke at a frequency of 7.3 MHz.

A.
$X_L = 2\pi FL$
$X_L = (2)(3.14)(7,300,000)(0.000010\text{H})$
$X_L = 458$ ohms

Q112. Find the total capacitance of a series network containing 100 pF, 200 pF and 75 pF capacitors.

A. On the exam, if there is only one choice less than 75 pF, the smallest capacitance in the network, then select it without bothering to calculate the capacitance. Otherwise:

$$C_t = \frac{1}{\frac{1}{100} + \frac{1}{200} + \frac{1}{75}}$$

$$C_t = \frac{1}{0.01 + 0.005 + 0.013}$$

$$C_t = \frac{1}{0.028}$$

$$C_t \approx 36 \text{ pF}$$

Q113. Calculate the total capacitance of a network consisting of a 100 pF capacitor in series with a 220 pF capacitor.

A.
$$C_t = \frac{C1\ C2}{C1 + C2}$$

$$C_t = \frac{(100)\ (220)\ \text{pF}}{(100 + 220)\ \text{pF}}$$

$$C_t = \frac{22,000}{320}$$

$$C_t = 69\ \text{pF}$$

Q114. Calculate the capacitive reactance of a 100 pF capacitor at 3.5 MHz

A.
$$X_c = \frac{1}{2\pi FC}$$

$$X_c = \frac{1}{(2)\ (3.14)\ (3,500,000)\ (100 \times 10^{-12})}$$

$$X_c = 455\ \text{ohms}$$

Or, we can make the equation easier when the capacitance is in picofarads and the frequency is in megahertz

$$X_c = \frac{1,000,000}{(2)\ (3.14)\ (3.5)\ (100)}$$

$$X_c = 455\ \text{ohms}$$

Q115. Find the beta of a transistor in which a base current of 0.5 mA causes a collector of 50 mA.

A.
$B = I_c/I_b$
$B = (50\ \text{mA})/(0.5\ \text{mA})$
$B = 100$

Q116. A transistor with a beta of 60 has a collector current of 2.5 amperes. Find the base current of this transistor.

A. Solve the equation in Q117 for I_b:
$I_b = I_c/B$
$I_b = 2.5\ \text{A}/60$
$I_b = 0.041\ \text{ampere}\ (41\ \text{mA})$

Q117. What is the VSWR on a transmission line when the voltage is constant all along the line?
A. 1:1.
D. The line is "flat," i.e., has equal voltage at all points, only when the VSWR is unity.

Q118. The RF current in a transmission line is measured over a distance of 3/4 wavelength. It is found that the maximum current is 1.5 amperes and the minimum current is 600 mA. Find the SWR.
A.
$$SWR = I_{max}/I_{min}$$
$$SWR = 1.5A/0.6A$$
$$SWR = 2.5:1$$

Q119. Find the power applied to a dipole antenna if the feedpoint impedance is 73 ohms and the feedpoint current is 800 mA.
A.
$$P = I^2R$$
$$P = (0.8)^2(73)$$
$$P = 47 \text{ watts}$$

Q120. The voltage along a transmission line is measured for a distance of more than a half wavelength. It is found that the maximum voltage is 170 volts RF, while the minimum potential is 125 volts RF. Find the SWR.
A.
$$SWR = E_{max}/E_{min}$$
$$SWR = (170)/(125)$$
$$SWR = SWR = 1.36:1$$

Q121. An rf wattmeter at the feedpoint of an antenna measures a forward power of 150 watts and a reflected power of 25 watts. Find the SWR.
A.
$$SWR = \frac{1 + \sqrt{P_r/P_f}}{1 - \sqrt{P_r/P_f}} \qquad SWR = \frac{1 + (6)^{1/2}}{1 - (6)^{1/2}}$$
$$SWR = \frac{1 + (150/25)^{1/2}}{1 - (150/25)^{1/2}} \qquad SWR = \frac{1 + 2.45}{1 - 2.45}$$
$$SWR = 2.76:1$$

Q122. Find the length of a quarter wavelength vertical radiator for operation on a frequency of 146.31 MHz.
A. $L_{feet} = \dfrac{234}{F} = \dfrac{234}{146.31 \text{ MHz}} = 1.6 \text{ feet}$

Q123. Convert the answer found in Q121 to inches.
A. There are 12 inches to the foot, so:
1.6 feet x 12 in/foot = (1.6)(12) inches = 19.2 inches.

Q124. What is the turns ratio required of an audio output transformer to match an 8-ohm loudspeaker to a 2500-ohm plate resistance?

A.
$$n = \sqrt{\frac{Z_p}{Z_s}}$$

$$n = \sqrt{\frac{2500}{8}}$$

$n = \sqrt{313}$
$n = 18$, so the turns ratio is 18:1.

This problem used to be a real bear unless the FCC was nice enough to make the ratio something like 25, so everybody would *know* the square root! Of course, even a cheap calculator will work "SQRTs," so there is no excuse.

Q125. What is the turns ratio of a transformer if 220 volts AC applied to the primary produces 6.3 volts AC across the secondary?

A.
$$\frac{N_p}{N_s} = \frac{E_p}{E_s} = n$$

so,
$$n = 220/6.3$$
$$n = 35:1$$

Q126. Find the total capacitance if a 40 μF capacitor is connected in series with a 60 μF capacitor.

A. If the only selection given by the FCC is less than 40 μF, then it has to be the correct value. If there is more than one selection less than 40 μF, then proceed as follows:

$$C_t = \frac{C1\,C2}{(C1 + C2)} \qquad C_t = \frac{2400}{100}$$

$$C_t = \frac{(40)(60)}{(40 + 60)} \qquad C_t = 24\ \mu F$$

Q127. What is the total capacitance if two 40 μF capacitors are connected in series.

A. 20 μF.

D. If all capacitors in a series network have the same value, then the total capacitance is simply the capacitance of one unit, divided by the total number of capacitors. For example, in the problem as stated, there are two capacitors of 40 μF each, so the total capacitance of the series network is 40 μF/2, or 20 μF. Similarly, if there had been three capacitors in series, then the total would have been 40/3, or 13.3 μF. If there had been four 40 μF units, then the total would have been 10 μF. Alternatively, you could work the problem out long-hand, as in the previous problem:

$$C_t = \frac{C1\,C2}{C1+C2}$$
$$C_t = \frac{(40)(40)}{(40+40)}$$
$$C_t = \frac{1600}{80}$$
$$C_t = 20\,\mu F$$

Q128. Find the total capacitance of a 20 μF and a 40 μF capacitor in parallel.

A.
$$C_t = C1 + C2$$
$$C_t = 20\,\mu F + 40\,\mu F$$
$$C_t = 60\,\mu F$$

Q129. Find the total capacitance of a series circuit containing a 100 pF capacitor and a 470 pF capacitor.

A.
$$C_t = \frac{(C1)(C2)}{C1+C2}$$

$$C_t = \frac{(100)(470)}{(100+470)}$$

$$C_t = \frac{47{,}000}{570}$$

$$C_t = 82\,pF.$$

D. Again, note that the correct answer is less than the smallest capacitance in the circuit. If there is only one selection on the FCC exam "shopping list," then it has to be the correct choice.

Q130. Find the gain of a power amplifier, expressed in decibels, if 100 watts of excitation produces 600 watts of RF output power.

A. $dB = 10 \; Log_{10}(P_o/P_i)$
$dB = 10 \; Log_{10}(600/100)$
$dB = 10 \; Log_{10}(6)$
$dB = (10)(0.78)$
$dB = 7.8$

D. Decibel notation invariably confuses the newcomer. It is merely a number used to express a gain or a loss. That's all we really have to know about it. There are two kinds of dB, one for voltages and currents, and another for power. The constant for power (as in this example is 10, while that for I and E is 20. We also know that, for power dB, the following are approximately true:

1 dB = 25 percent change
2 dB = 50 percent change
6 dB = 200 percent change
10 dB = 100 percent (i.e., 10X) change

Q131. Find the loss in decibels of a coaxial transmission line if 100 watts applied at the transmitter end results in 56 watts at the antenna.

A. $dB = 10 \; Log_{10}(P_o/P_i)$
$dB = 10 \; Log_{10} \; (56/100)$
$dB = 10 \; Log_{10} \; (0.56)$
$dB = (10) \; (-0.25)$
$dB = -2.5$

Q132. Find the loss in dB if 1000 μV applied to the input of a resistive attenuator produces an output potential of 400 μV.

A. $dB = 20 \; Log_{10} \; (E_o/E_i)$
$dB = 20 \; Log_{10} \; (400/1000)$
$dB = 20 \; Log_{10} \; (0.4)$
$dB = (20) \; (-0.39)$
$dB = 7.8$

Q133. Find the physical length needed to make an electrical half wavelength transmission line on 21.39

MHz if the coaxial cable has a velocity factor of 0.66.

A.
$$L_{ft} = \frac{492 \times V}{F}$$

$$L_{ft} = \frac{(492)(0.66)}{21.39}$$

$$L_{ft} = \frac{324.7}{21.39}$$

$$L_{ft} = 15.2 \text{ feet}$$

D. What is this thing called *velocity factor?* The velocity factor of a transmission line is the fraction of the speed of light that radio waves travel in the transmission line. Normally, in free space, a radio wave travels at a velocity equal to the speed in light (indeed, both light and radio waves are examples of electromagnetic waves), which is approximately 300,000,000 meters per second, or 186,200 miles per second. But, in transmission lines the velocity of a radio wave is less than this. If the velocity factor is 0.66, then we are saying that the velocity of a radio wave in that transmission line is 66 percent of the speed of light, or 0.66 × 300,000,000 m/s, which is 198,000,000 m/s. Common values of velocity factor for coaxial cable are:

Regular dielectric 0.66
Foam dielectric 0.80
Teflon dielectric 0.70
Twin-lead 0.82

All transmission lines will be shorter in physical length (i.e., the actual number of feet) than their electrical length. This is a direct result of the velocity factor.

Q134. Find the physical length of a piece of coaxial cable (VF = 0.8) that is to be a half wavelength on 14.250 MHz.

A. L = 492V/F
L = (492)(0.8)/(14.25)
L = 394/14.25
L = 27.6 feet

Q135. What is the physical length of a piece of coaxial cable for a half wavelength on 3,900 kHz if the velocity factor is 0.7?

A.
$L = 492V/F$
$L = (492)(0.7)/(3.9)$
$L = 344/3.9$
$L = 88$ feet

Q136. What is the physical length of a piece of coaxial cable for a quarter wavelength on 21.250 MHz if the velocity factor is 0.66.

A.
$L = 492V/F$
$L = (492)(0.66)/(21.25)$
$L = 162.4/21.25$
$L = 7.6$ feet

Q137. Convert the answer to 0138 meters.

$$7.6 \text{ feet} \times \frac{1 \text{ meter}}{32.8 \text{ ft}} = 7.6/3.28 = 2.3 \text{ meters}$$

Q138. Find I_p in Fig. 15-21.

A. In a triode vacuum tube, the plate current is equal to the cathode current, so $I_p = I_k = 2.5$ mA. In a tetrode or pentode vacuum tube, we would have to subtract the screen grid current from the cathode current in order to find the plate current.

Q139. Find the potential on the cathode of the tube in Fig. 15-21.

A. Use Ohm's law:
$E = I_k R_2$
$E = (0.0025A)(1000 \text{ ohms})$
$E = 2.5$ volts dc

Q140. What is the grid bias voltage on the tube in Fig. 15-21?

A. -2.5 volts.

D. This circuit uses cathode bias, which means that we do not actually place the grid at a minus voltage, but rather, place the cathode at a positive voltage. Since the grid is effectively grounded through resistor R1, the cathode will be 2.5 volts more positive than the grid, which is effectively the same as making the grid 2.5 volts more negative than the cathode.

Q141. Find voltage drop E3 in Fig. 15-21.

A. Use Ohm's law. We known that $I_p = 2.5$ mA, so:
$$E3 = I_p R3$$
$$E3 = (0.0025)(20,000)$$
$$E3 = 50 \text{ volts}$$

Q142. Find voltage E_B in Fig. 15-21.

A. Voltage E_B is the voltage between the cathode and the plate of vacuum tube V1. We know that E_B, E_k and E3 must be equal to E_{BB}, and so:
$$E_B = E_{BB} - E_k$$
$$E_B = (100V) - (50V) - (2.5V)$$
$$E_B = 100V = 52.5V$$
$$E_B = 47.5 \text{ volts}$$

Q143. What is the plate resistance of the tube in Fig. 15-21.

A. Use Ohm's law:
$$R = E_B / I_p$$
$$R = (47.5V)/(0.0025A)$$
$$R = 19,000 \text{ ohms}$$

Q144. What is the voltage at point A in Fig. 15-22?

A. Point A in Fig. 15-22 is the screen grid terminal for the vacuum tube. The current flowing in the resistor determines the voltage drop, and is the screen current, so:
$$E_A = E_B - (I_s R_1)$$
$$E_A = 750 - (0.012 \times 25,000)$$
$$E_A = 750 - 300$$
$$E_A = 450 \text{ volts}$$

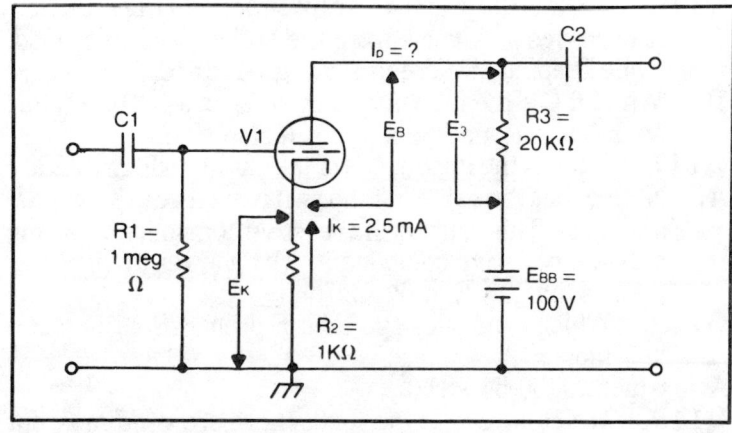

Fig. 15-21. Typical triode vacuum tube amplifier circuit.

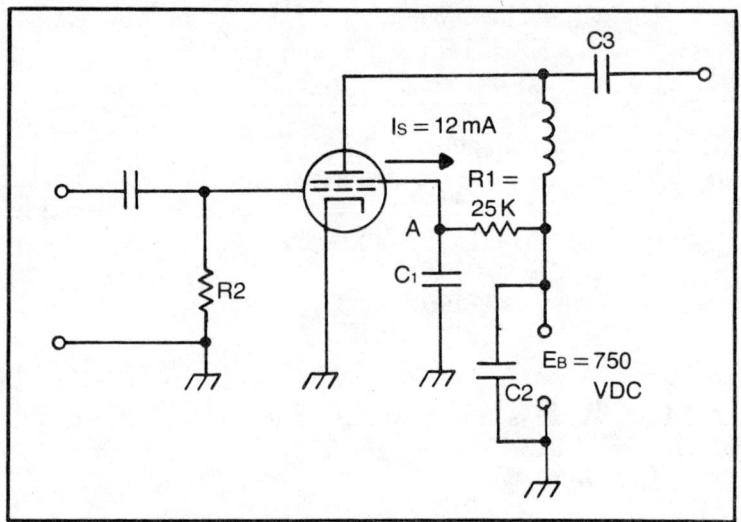

Fig. 15-22. Tetrode vacuum tube amplifier circuit.

Q145. What power rating should be used for R1?
A. The power dissipated is:
$$P = I^2 R$$
$$P = (0.012)^2(25,000) = 3.6 \text{ watts}$$
It is generally considered to be good practice to use the next higher standard power resistor, unless the actual power is very close to that value (i.e., 4.9 watts for a 5-watt resistor). Always try to give yourself at least 10 percent. So, use a 5-watt resistor for this application.

Q146. What is the purpose of capacitor C1 in Fig. 15-22?
A. It bypasses ac signals from the screen grid to ground, while keeping the screen above ground at dc.
D. Without C1, the tube may very well act as an oscillator, so the capacitor serves to keep it stable.

Q147. What is the impedance of the circuit in Fig. 15-23?
A. Notice that the values of the two reactances (X_c and X_L) are equal. This cancels the reactive component, leaving only the resistive component. The answer in this case, then, is that $Z = R$, so $Z = 50$ ohms.

Q148. Which band does sporadic-E propagation affect the most?
A. 6 meters (50-54 MHz).

Q149. Which layers of the ionosphere are used most for DX communications using sky wave?

A. The F layer (the FCC selection might use F, F_1 or F_2).

Q150. What procedure should be followed when using VOX on a transmitter?

A. It is good manners, and good operating procedure, to let the VOX drop now and again, by pausing for a few seconds, to see if anyone else wants to interrupt.

QUESTIONS AND ANSWERS 151-232

Q151. What does neutralization of a final power amplifier prevent?

A. Self-oscillation and parasitic oscillation. The former is oscillation on the frequency that the amplifier is tuned to, and the latter is oscillation on another frequency that is not harmonically related to the operating frequency, usually in the VHF region. Parasitic oscillation is caused by stray inductances and capacitances forming resonant circuits.

Q152. What are "Q signals," and when are they used? When are they not used?

A. Q signals are standard 3- and 4-letter symbols, beginning with the letter Q, that are used as a kind of shorthand between radio operators. They are used on radiotelegraph transmission, but are not appropriate for radiotelephone. Of course, there are a lot of amateurs on the air who really dig Q signals because it makes them sound like they know what's going on . . . to outsiders.

Q153. What are the principle advantages of Quad and Yagi antennas?

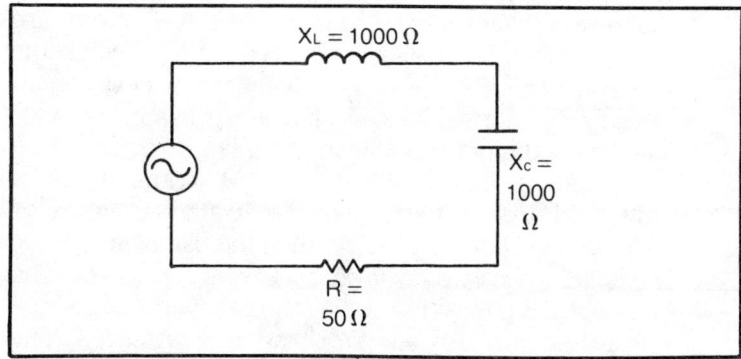

Fig. 15-23. Series resonant circuit with equal reactances.

A. Apparent gain and directivity on both receive and transmit. The directivity focuses all of your power in one direction on transmit, making it sound like a lot more than it is. On receive, the directivity of the beam antenna (quad or yagi) reduces some of the interference arriving from directions outside of the beam pattern. If you live in Chicago, for example, and want to work Europe, then you will receive less interference from stations in the south and west because of the directivity of the beam. Most people have the erroneous notion that the beam is used to make them "get out better." But this is not true for most amateurs most of the time. If you run anything over 100 watts, especially if you run a kilowatt, then the gain of the beam doesn't help much, except under the worst of conditions, when it is too much trouble to be on the air in the first place. But the real glory of the beam is on receive. It lets you cut out a lot of the noise being received, which is something a dipole does a little bit and a vertical not at all.

Q154. What is the relationship between the modulated and unmodulated power in an AM signal when the modulating signal is a sine wave?

A. The modulated power is 1.5 times the unmodulated power.

$$P_{mod} = 1.5\, P_{unmod}$$

Q155. Which form of emission has the narrowest bandwidth?

A. This question is often tricky because some people have disagreed with the FCC over the correct answer. Very few would argue that the CW (i.e., A1 emission) is the narrowest bandwidth. But if all the selections are radiotelephone emissions (AM, SSB, FM), then some disputes may arise. AM is usually the widest signal, because the normal bandwidth of an AM signal is twice the bandwidth of the modulating signal. For example, if the speech amplifier in an AM transmitter is bandwidth-limited to pass only those audio frequencies below 3000 Hz (i.e., 3 kHz), then the bandwidth of the modulated rf carrier will be 2 x 3 kHz, or 6 kHz. The SSB signal has exactly one-half the bandwidth of the AM signal because only one sideband is transmitted. The SSB transmitter has the bandwidth of the audio system.

For example, in the case of a 3000-Hz audio signal, the bandwidth of the AM transmitter is 6 kHz and that of the SSB transmitter is 3 kHz.

The rub comes when you consider the FM transmitter. The bandwidth of the FM transmitter may be no wider than that of an equivalent AM transmitter (6 kHz), but this rule applies in the HF bands. In the VHF bands, however, the bandwidth can be wider than an AM transmitter. The disagreement stems from reports that the FCC will use FM as the correct answer, and the engineering community may not be in universal agreement with this choice.

Q156. What is a low-pass filter? What are its characteristics and uses?

A. A low-pass filter is an electronic circuit designed to pass frequencies from dc up to some specified cutoff frequency. All signals with frequencies greater than cutoff are greatly attenuated.

The principal use of the low-pass filter in the amateur service is to attenuate harmonics from the transmitter that may interfere with other services. For example, a popular series of amateur low-pass filters have cutoff frequencies in the 30-45 MHz range (which is between 10 meters (the highest HF band) and 6 meters (the lowest VHF band and the first TV channel, channel 2, 54-60 MHz). One model offers a cutoff frequency of 35 MHz with 80 dB of attenuation at 54 MHz, TV channel 2. If the amateur transmitter emits any signals with harmonics that fall into a TV channel, then the low-pass filter will reduce their strength to a point where they will not cause any harm—most of the time.

The low-pass filter must be placed between the antenna and the output of the transmitter or linear amplifier. Although any amateur should use a low-pass filter, the use of a high-power linear amplifier makes it all but mandatory!

The FCC publication on TVI (see Appendix of this book) suggests that the users of linear amplifiers place a low-pass filter between the exciter and the input of the linear amplifier and another one between the output of the linear amplifier and the antenna. It is also recommended that users of high-power linear amplifiers use an antenna coupler to further reduce the strength of harmonics. The low-pass filter should be placed between the output of the transmitter and the input of the antenna coupler.

Q157. What can be done to prevent self-oscillation?
A. Neutralize the final amplifier of the transmitter.

Q158. What is the principal cause of self-oscillation in rf amplifiers?
A. The interelectrode capacitance causes a feedback path that results in an in-phase signal being applied to the input.

Q159. What is FM deviation?
A. Deviation in an FM system is the amount of frequency change the signal undergoes from the unmodulated frequency. The unmodulated frequency is typically changed in the upward direction for positive audio signals and in the downward direction for negative signals. The deviation is the change from *either* the lowest frequency or the highest frequency to the unmodulated carrier frequency. The total frequency swing, or "bandwidth," when a sine wave modulating signal is used, will be twice the deviation.

Q160. May an amateur operator use special coded tones to activate a remote controlled teleprinter?
A. Yes, provided, of course, that the station is properly identified.

Q161. Will a built-in S-meter yield precision measurement of signal strength?
A. No.

Q162. What is meant by "full break-in operation?"
A. Full break-in is a mode of CW operation in which the receiver is turned back on between *dits* and *dahs* so that the operator can hear the other operator when they wish to interrupt.

Q163. When do the F_1 and F_2 layers exist?
A. During daylight hours. During the night hours, the F_1 layer seems to disappear.

Q164. At what altitudes does one find the F_1 and F_2 layers of the ionosphere?
A. 100 to 200 miles up.

Q165. What procedure should be followed in case of electrical shock?

A. Turn off the power immediately, and ground any exposed filter capacitors in order to dump their charge. *Do not touch the victim* until the two previous actions have been taken, or *you* may also become a victim. If the victim is unconscious, then administer cardiopulmonary resuscitation (CPR), as needed. Call for an ambulance. The best "therapy" for electrical shock is prevention. Observe the safety rules when working on electrical devices.

Q166. At what intervals do sunspot problems typically reappear in HF communications?

A. If a problem is due to a specific sunspot formation, then the problem will reappear every 27 days, which is the period of rotation of the sun.

Q167. Identify the type of antenna shown in Fig. 15-24.

A. Folded-dipole.

Q168. What is the device at point A in Fig. 15-24?

A. Balun coil.

Q169. What is the purpose of the balun coil at point A in Fig. 15-24?

A. It converts the unbalanced circuit of the transmission line to a balanced circuit of the antenna. In addition, it also transforms the impedance of the transmission line to the feedpoint impedance of the antenna. Balun transformers are generally available in 1:1 and 4:1 impedance ratios.

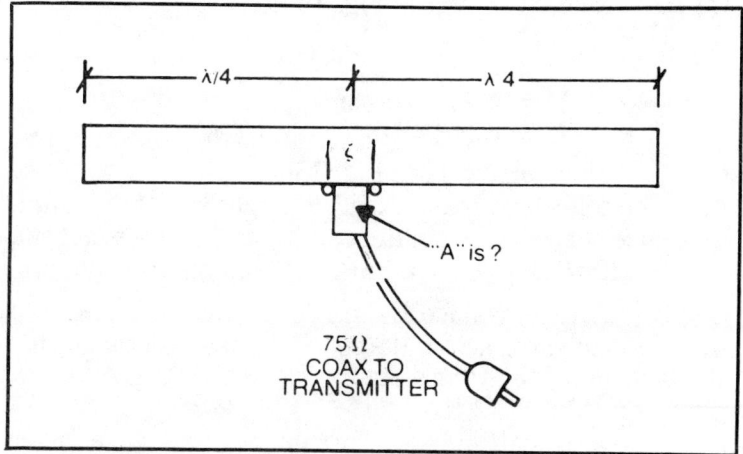

Fig. 15-24. Half-wave antenna.

Q170. What is the typical feedpoint impedance of a folded dipole in freespace?
A. 300 ohms.
D. The actual impedance is 250 to 300 ohms, but the answer usually sought is 300 ohms, probably because amateurs routinely build folded dipoles with 300 ohm twin-lead transmission line.

Q171. What is an advantage of the folded dipole antenna?
A. Broader bandwidth, which is translated to mean a lower SWR over a larger range of frequencies.

Q172. What is the purpose of resistor R1 in Fig. 15-25?
A. Collector load.

Q173. What is the purpose of resistor R2 in Fig. 15-25?
A. Emitter load, which leads to better thermal stability in the operation of the transistor.

Q174. What is the purpose of resistor R3 in Fig. 15-25?
A. Supplies bias current to the base-emitter junction.

Q175. What is voltage E_B in Fig. 15-25?
A. 1.9 volts dc.
D. This is a *silicon* NPN transistor, according to the diagram. The b-e voltage for silicon transistors is 0.6 to 0.7 volts, and in an NPN transistor the base potential is more positive than the emitter potential. Therefore, the base potential will be 1.2 volts + 0.7 volts, or 1.9 volts. The 1.2 volts is the emitter potential, and is taken from the diagram.

Q176. Find current I1 in Fig. 15-25.
A. Use Ohm's law:
$$I1 = E_e/R2$$
$$I1 = 1.2V/250 \text{ ohms}$$
$$I1 = 0.0048 \text{ Amperes, or } 4.8 \text{ mA}$$

Q177. Find current I2 in Fig. 15-25.
A. The collector current in an NPN transistor is the difference between the emitter current and the base current. So, $I2 = I1 - I_B = 4.8 \text{ mA} - 0.240 \text{ mA} = 4.56 \text{ mA}$

Q178. What is voltage E1 in Fig. 15-25?
A. E1 is the drop across the collector load resistor and has a value of (by Ohm's law):
$$E1 = I2 \, R1$$
$$E1 = (0.00456)(1,500)$$
$$E1 = 6.84 \text{ volts}$$

D. *Note:* FCC calculators often have only one or two digits. This means that the answer selection that is correct for this type of question on the exam may be 6 volts, 6.8 volt or 7 volts. Be aware of rounding in the possible answer selections of all numerical problems!

Q179. What is wrong in the circuit of Fig. 15-26?

A. The battery polarity is backwards. In a PNP transistor, as shown in Fig. 15-26, the collector is supposed to be negative with respect to the emitter. In this case, the positive terminal of the collector-emitter power supply is connected to the collector.

Q180. What bands are affected most by sudden ionospheric disturbances?

A. The lower HF bands, primarily. This means 160, 80 and 40 meters. Later in the event, and sometimes not at all, the upper HF bands are affected (20, 15 and 10 meters).

Q181. What geographical areas of the world are most affected by a sudden ionospheric disturbance?

A. Areas near the equator.

Q182. What times of the year do SID events occur most frequently?

A. There are no specific seasons. The SID can occur anytime.

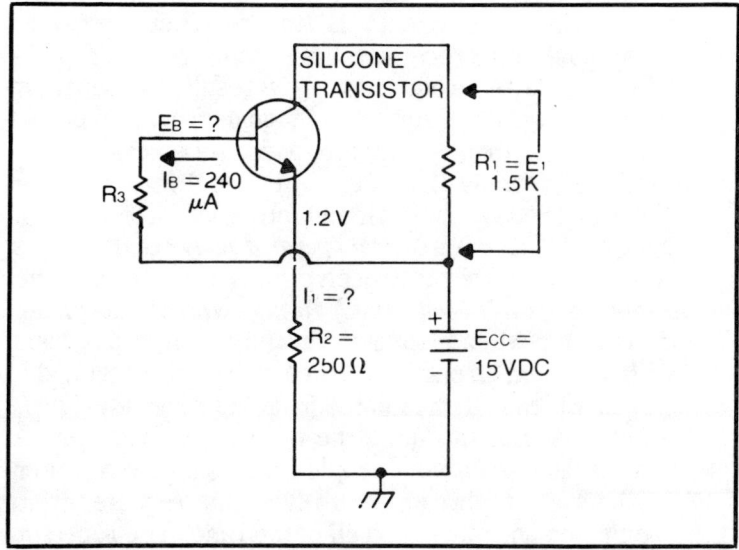

Fig. 15-25. Transistor amplifier circuit.

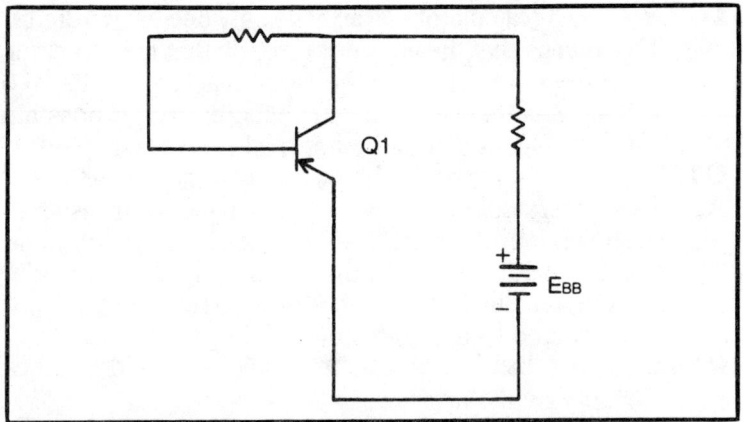

Fig. 15-26. There's a defect in this circuit. It will not operate as shown.

Q183. What seasons of the year are likely for ionospheric storms?

A. Fall, winter and spring. There is less liklihood of an ionospheric storm during the summer months.

Q184. Generally describe the opeation of an amateur 2-meter repeater.

A. A repeater consists of a transmitter on frequency A and a receiver tuned to frequency B. The users of the repeater have exactly the opposite combination: i.e., a transmitter on frequency B (the repeater's receiver frequency) and a receiver on frequency A (the repeater's transmitter frequency). As a result, the repeater can pick up and retransmit the signals from low-power portable and mobile stations. Most repeaters are automatic, meaning that they will respond to an input signal and retransmit it without intervention of a control operator. There is a *carrier operated relay* (COR) that is connected to either the receiver squelch system or the *automatic gain control* (AGC) voltage which turns on the transmitter when a signal is in the receiver passband with sufficient strength to be heard. The receiver audio signal, which is demodulated from the received signal, is then used to modulate the repeater transmitter. A *timer* is used to prevent people from using the repeater too long at one breath, or to keep the repeater from hanging up and staying on all of the time. The repeater is identified either by radiotelephone (FM), when being operated manually, or by an automatic CW identifier that operates at a speed less than 20 WPM.

Q185. K4IPV has a repeater. What is the proper form of CW ID?

A. Either "K4IPV/R", or "K4IPV/RPT"

Q186. What constitutes good repeater usage practice?

A. Stop yakking and listen every now and then to see if anyone else wants to use the facility; it is not a private line! Especially listen for stations wishing to transmit emergency traffic or phone patch (autopatch) traffic.

Q187. Why would an amateur station use a multi-element Yagi antenna on 2 meters?

A. To take advantage of the *gain* and *directivity* of the Yagi antenna.

Q188. What polarity Yagi antennas are generally used on 2 meters?

A. Either vertical or horizontal. Vertical polarity is a little more popular because: a) it is easy to implement on 2 meters because of the short element lengths; and b) it is consistent with the polarity of most mobile and repeater (omnidirectional) antennas.

Q189. Why is speech clipping sometimes used in SSB transmitters?

A. To reduce the ratio between the peak and average energy in the audio signal. This has the effect of raising the average power of the SSB signal, so allegedly produces more "oommphh" in the transmitted signal. But it also tends to cause distortion and overmodulation if not applied correctly.

Q190. What is the name of the network consisting of L2-C3-C4 in Fig. 15-27?

A. Pi-network.

Q191. What is the purpose of the network in the previous question?

A. To match the high plate resistance of tube V1 to the low impedance (50-100 ohms) of the antenna.

Q192. What is the purpose of C2 in Fig. 15-27?

A. It is a high-voltage DC blocking capacitor and is used to prevent high voltage from the anode of V1 from appearing on the antenna output connection—a very dangerous situation!

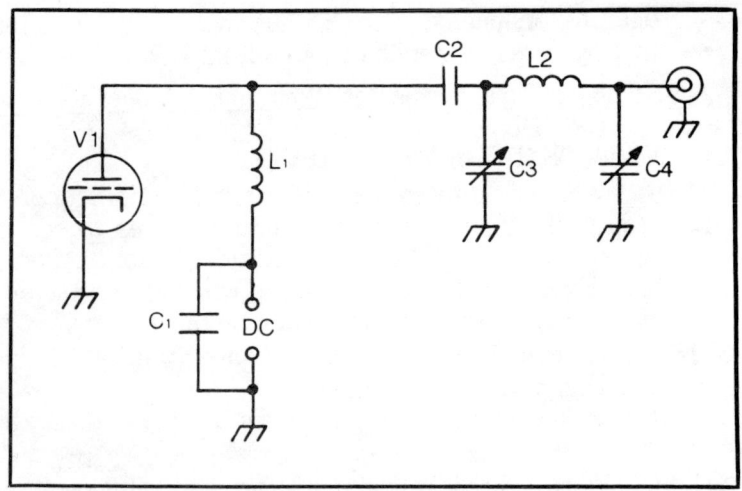

Fig. 15-27. Typical transmitter output stage circuit.

Q193. Coil L1 in Fig. 15-27 is an:
A. RF choke
Q194. What is the purpose of L1 in Fig. 15-27?
A. It prevents RF from the anode of the vacuum tube amplifier from getting into the power supply, and thereby being coupled to previous stages.
Q195. What is the purpose of C1 in Fig. 15-27?
A. It is a power supply decoupling, or "bypass" capacitor. It is used to place the power supply at a low impedance to ground for RF, yet at a high voltage for DC. The capacitor bypasses any RF that gets past the RF choke, so prevents coupling to previous stages.
Q196. What is the purpose of capacitor C1 in Fig. 15-28?
A. Neutralization of the final amplifier.
Q197. Figure 15-29 shows the circuit for 220-volt AC wiring in your house. The transformer shown is mounted on a pole outside of the house somewhere. Which lead is grounded in normal practice?
A. Lead B, the transformer center tap. You can also tell by the fact that lead B goes to the odd terminal on the 220-volt outlet.
Q198. Where would the fuses be placed in Fig. 15-29?
A. In leads A and C.
Q199. Describe the differences between AM and SSB signals.

A. Both AM and SSB are amplitude modulated signals. In the process of amplitude modulation, the amplitude of a carrier (RF) signal is varied in step with a modulating audio signal. This process creates three signals to be transmitted: carrier, upper sideband and lower sideband. The *upper* sideband consists of the carrier frequncy *plus* the audio frequency, while the *lower* sideband frequency is the carrier *minus* the audio frequency. For example, if a 1 kHz tone is used to modulate a 1000 kHz RF carrier, then the carrier frequency is 1000 kHz, the USB is 1001 kHz, and the LSB is 999 kHz. In an AM transmitter, all three signals, carrier, LSB, and USB, are transmitted. In an SSB transmitter, however, the carrier and one of the sidebands is eliminated. This has the practical effort of reducing system bandwidth requirements by *two* (i.e., only one-half the bandwidth is needed, so theoretically twice as many stations can use any given band), and all of the transmitter power goes to transmission of speech information. In an AM transmitter, only the power of one sideband—33-1/3 percent of the total—is actually useful.

Q200. What is an oscilloscope?
A. An oscilloscope (Fig. 15-30) is an electronic instrument that displays a signal waveform on the viewing screen of a cathode ray tube.

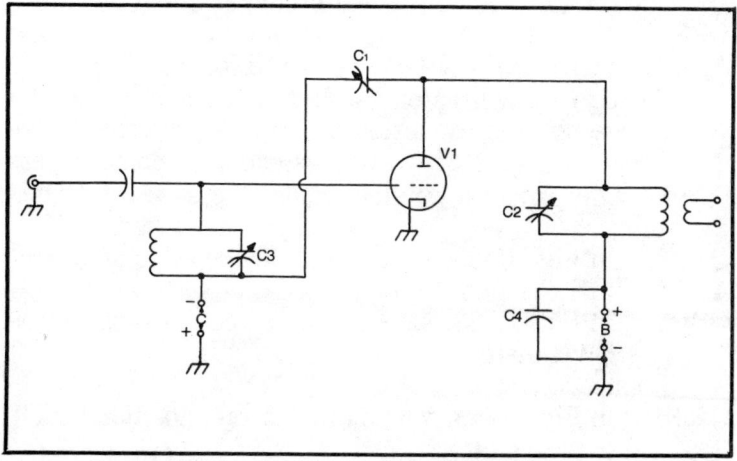

Fig. 15-28. Transmitter output with neutralizing adjustment (C1).

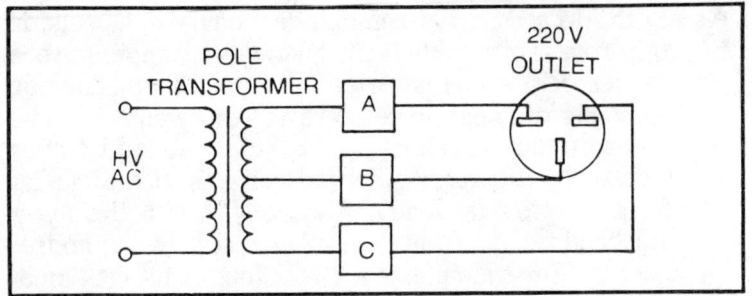

Fig. 15-29. Typical electrical power transformer and outlet.

Q201. What is a dummy load?

A. A dummy load (example shown in Fig. 15-31) is a non-radiating load for a transmitter that substitutes for the antenna for purposes of testing and adjusting transmitters. The dummy load must have the same resistance as the radiation resistance of the antenna that it replaces, and must be non-reactive (no inductive or capacitive reactance). In addition, it must be adequately shielded to prevent radiation. The purpose of using the dummy load is to allow you to test and adjust a transmitter without going on the air. Presumably, a defective transmitter will have impure emissions, so should not be connected to the antenna until it is repaired. In addition, it is plain poor manners to test a transmitter on a heavily used band when a few lousy bucks spent on a dummy load would prevent problems altogether.

Q202. What is a keyer, and how is it used?

A. An electronic keyer (see Fig. 15-31 for an example) is a device that will automatically form *dits* and *dahs* for Morse code transmission in response to contact closures on a special two-way paddle key. Some keyers have the paddle key built it, while others use a detached key.

Q203. In Fig. 15-33, what is the device at point A?

A. Antenna changeover relay.

Q205. In Fig. 15-33, what is the correct location for the SWR meter?

A. Point C.

Q205. In Fig. 15-33, what is the correct position for the low-pass filter?

A. Point B.

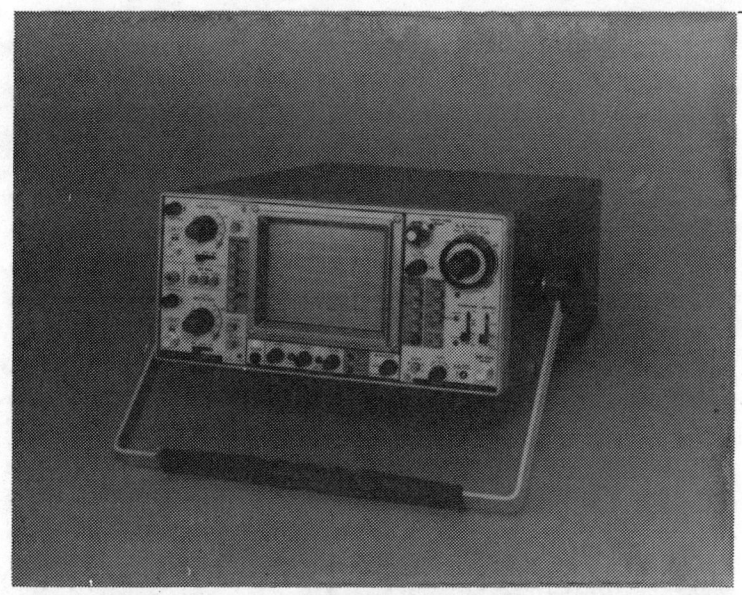

Fig. 15-30. A typical oscilloscope (courtesy of Kikusui International).

Q206. In an AM signal, what is the relationship between the audio modulating frequency and the instantaneous value of carrier amplitude?

A. The instantaneous change of the RF carrier is proportional to the audio modulating frequency.

Q207. How is a Sudden Ionospheric Disturbance (SID) recognized?

A. By being able to work DX stations when the MUF is too low for the frequency that you are using, and the distance is too great for ground wave. An example of this might be working Europe on 40 meters during daylight hours.

Q208. What are AØ emissions?

A. Unmodulated RF carrier, i.e., when you hold the key down on your transmitter, you are sending out an AØ emission.

Q209. Under what circumstances are AØ emissions allowed in the HF bands?

A. AØ may not be used, except for brief tests below 51 MHz.

Q210. F3 emissions are _____.

A. Frequency modulated radiotelephony.

Fig. 15-31. Typical transmitter dummy load.

B. Know all of the principal emission type designations: AØ, A1, A2, A3, F1, F2, plus A5/F5.

Q211. What must you do when permanently moving the control point of your amateur radio station?

A. Use Form 610 to notify the FCC. Send it to the Gettysburg address on the form, not to your local FCC office.

Q212. The FCC cannot impose quiet hours on an amateur radio station.

A. They *can* and *will* if you are uncooperative about resolving a TVI complaint, or fail to use "good engineering practice" to limit TVI complaints.

Q213. What is the length of time between ID in repeater operation? In net operation?

A. In *all*, repeat *all*, forms of operation (except radio model control where no ID is required) the period between station ID may not exceed 10 minutes. Beware of trick questions that try to make you think that the type of operation designated in the question is somehow special.

Q214. What is the period between ID when in auxiliary operation?

A. If you forgot the answer to Q213, reread it.

Q215. What is the maximum code speed allowable for repeater ID?

A. 20 WPM.

Q216. What is the maximum code speed allowable for RTTY ID?

A. Again, 20 WPM.

Q217. May you operate another amateur radio station if you left your license at home hanging in a frame above your transmitter?

A. No. The license must be with you whenever you operate an amateur radio station. It is good practice to make a photocopy of the license for the frame in the shack, and then carry the original with you.

Q218. If you fail the General class test, how long must you wait before trying it again?

A. 30 days.

Q219. May an interim permit be renewed?

A. Technically, no; practically, maybe—depending upon the mood of the local FCC. The correct answer, though, is *No*.

Fig. 15-32. Typical keyer.

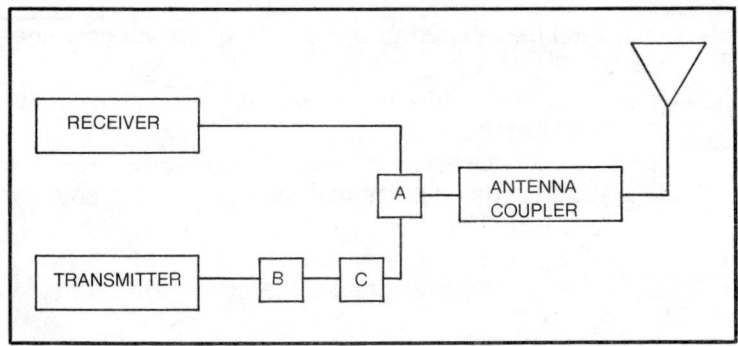

Fig. 15-33. Block diagram of antenna connection to a transmitter and receiver.

Q220. What is portable operation?
A. Operation when you are away from your primary station location.

Q221. What are the two components of your amateur radio license?
A. 1) *station license* (providing the call sign), and 2) an operator license.

Q222. What is a declaration of emergency?
A. See Part 97.107. When amateurs request a declared emergency, the local FCC office can set aside certain requested frequencies for exclusive use in emergency operations.

Q223. On which of the following frequencies may a General class operator use A3 emission?
A. This is a favorite FCC question. Listed below are all of the General class A3 (radiotelephone) frequencies in the HF bands:
1800-2000 kHz
3890-4000 kHz
7225-7300 kHz
14,275-14,350 kHz
21,350-21,450 kHz
28,500-29,700 kHz
MEMORIZE THEM!

Q224. What are emergency communications?
A. Communications regarding an imminent threat to life or property.

Q225. What is the control point of an amateur radio station?
A. The place where you control the station.

378

Q226. Under what circumstances may a third party who is unlicensed use your rig?

A. When you are present to supervise the operation to ensure that none of the regs are violated.

Q227. What is proper ID procedure on the amateur bands.

A. The call sign of the station that you are calling, the words, "this is" or the Morse code equivalent DE, followed by your call sign. For example, if K4IPV is working WA2GBC, the procedure is, "WA2GBC this is K4IPV," on radiotelephone, and "WA2GBC DE K4IPV," on CW.

Q228. Suppose that you are an Advanced class operator and are operating the station of a Novice class operator (WN4AAA). What is the proper call sign procedure?

A. Use the call sign of the station that you are operating, in the manner given in Q227.

Q229. Suppose that you are operating the station of a Novice class operator (WN4XYZ) and want to use General class frequencies. What is the proper call sign procedure?

A. The call sign of the station being operated, followed by a slash bar, and then your call sign. For example: WN4XYZ/K4IPV.

Q230. Define "amateur radio license."

A. It is the instrument that is issued by the FCC to allow you to operate an amateur radio station.

Q231. May you make up your own call sign?

A. No!

Q232. Under what circumstances do you not have to ID?

A. When using a transmitter of less than 1 watt output power to remotely control a model craft.

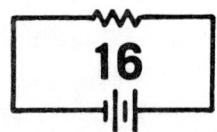

16

Learning the Code

Presumedly, anyone who is trying to learn the material for the General class examination has already passed the Novice hurdle, and is at least acquainted with the International Morse Code (the version of Samuel F.B. Morse's code that the FCC and most other Governments recognize for "wireless"). If you have not, then don't sweat it! The Morse code is a lot less of a hurdle than most people think, and it can be made a lot easier than it was when some of us were first learning.

MEMORIZING THE CODE

The first job is to memorize the code, including all alphabet characters, numerals, and common punctuation or special symbols. Do not memorize the code in the form "dot-dash!" Instead, learn early in the game to make sound pictures of the characters so that it is easier for you to make the jump to listening to a code practice machine later on. This means that we use the phoneme *di* for dot and *dah* for dash. The Morse code letter A (.-), then, becomes *not* dot-dash but *didah*. Generations of soldiers, sailors and ARRL code practice classes cannot be wrong, can they?

LEARNING AIDS

Once you start memorizing the code, supplement your training with a code practice tape or record that is specifically labeled for beginners. The kind that you want will have an announcer that says the letter (i.e., "this is an A"), followed by the Morse code character. This will reinforce your sound picture of the character. You must learn to recognize the timing of the characters, and this method is the best!

The timing of the properly transmitted Morse code character is as follows: If the *di* is taken to be *one* unit of time, then the *dah* is *three* units. Furthermore, we place a single unit of time between characters of the same word and three units of time between words in the same sentence.

STUDY HABITS

Morse code is best learned through *practice*. But don't overdo it. Some instructors recommend that the new student spend only 15 minutes a day at actual practice, 30 minutes at most (unless you have a hard callous on your buttocks, in which case 45-60 minutes might be your maximum). For most learners, however, a daily time limit of 15 to 30 minutes should be imposed.

Of course, you could improve somewhat, after you have the code memorized, if you try to spell out in Morse code all the signs and names of people in your day. If you want people to think that you are a real, gold-plated weirdo, then whistle the code spelling! But, if you are a little more concerned with your reputation as a clear-thinking, level-headed dude, then try it in your head.

Code tapes and records are reasonable for many people who are just starting out. But there is a built-in problem with these training aids: *memorization*! It seems that, after a few passes through the same tape, that you memorize its contents! Even though you may not be consciously aware of the process, the tape is memorized none the less. The apparent speed you have gained is not true code speed, but is due partially to the fact that you can anticipate what's coming next.

Code tapes are too costly to own enough to prevent memorization. So why not get some buddies together and each buy different code tapes and then swap back and forth!

Also, consider buying (maybe as a group or club purchase) one of those new electronic Morse code trainers such as the AEA MT-1 and KT-1, the MFJ *Professor Morse,* and the Curtis Keyer. The AEA, which is what one of your authors (JJC) uses to train new operators, contains a series loop of 24,000 characters of all kinds. The instructor (or student)selects from one of ten possible starting locations, or a random sequence. It is almost impossible to memorize that many Morse code characters.

Another source of code practice that doesn't grow" on you is the ARRL Morse code practice transmissions from their station W1AW in Newington, Ct. (and a West Coast station as well). These code practice sessions are scheduled in advance, and each one gives the QST text in which the answers can be found (instant feedback). The W1AW code practice transmissions are on several HF bands, and at several times during the day. The schedule is periodically printed in the *Operating News* section of *QST* magazine. The sechedule changes often enough to make it unreasonable for us to include here, so if you cannot find it in any recent *QST* (it is not in every month), then write to the ARRL at 225 Main Street, Newington, Ct., 06111, and ask them for the W1AW code transmission schedule.

I personally prefer the type of Morse training in which the code characters are sent individually at 12 to 17 WPM, and the code speed is adjusted by the quiet period between characters. This makes it less of the new experience when you start listening to the faster characters than you normally hear in "slow code" training. The AEA Morse Trainer (MT-1) adds another little catch to the job: they imperceptibly increase the speed on you as you practice. Before you realize it, you are copying 15 WPM!

THE FCC TEST

The FCC code test has changed somewhat over the years. When most of your authors took the test, you had to copy (5 × 13) or 65 characters in a row without an error somewhere in the five minutes of the test. That was somewhat more rugged than the current method. Now, they have you copy one side of a mock QSO (contact) and then ask you questions about the material transmitted. Such questions as the call signs of the stations, names of the operators, locations, etc. As of this writing, there is a little ambiguity over whether the test is fill in the blanks or multiple guess. In any event, if you can copy 10 WPM dead solid, and can get 85 percent of the 13 WPM, then it is a good bet (a probability, not a guarantee!) that you can receive enough to pass the test.

At one time, the Morse code test was the hurdle to pass! It is no longer such a fearsome mess. With the changes wrought by the FCC's new look in Federal Regulation, and the training aids that are available, the code test is a lot easier, No longer is learning the Morse code a lot like trying to nail Jello to the wall!

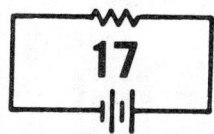

Applying for your License

For your Novice license you will have to write to the nearest FCC office, or to Room B 10, FCC, Washington, D.C. 20554, for a copy of FCC Form 610. Everything is done by mail for this class of license. If you are upgrading your license to General, Advanced, or Extra Class you can obtain, and fill out, a copy of Form 610 when you appear at the FCC office to take the examination. Like most government forms, the 610 is not difficult if you read the instructions carefully. (See Fig. 17-1).

The information that you have been given, regarding the amateur radio licenses, is correct as of the date of publication of this book. However, the FCC is going through the period of change for amateur licenses as discussed in chapter one, and some of the regulatory practices are being challenged in court. At the moment, for example, there are no fees charged for amateur licenses, but this may be changed by Congressional action in the near future.

You will find that the amateur radio magazines keep up-to-date on the changes in the regulations and will keep you posted by special columns and articles on the subject. The local radio club will be able to help you, too, and the ARRL will be glad to answer any questions that you might have.

If you are a Novice and have taken the General test, you may find that you will get commercial literature from dealers and manufacturers while you are still waiting for the mailman to deliver the final copy of your license. When your new license is printed at the FCC a computerized list is generated which is available to the

United States of America
Federal Communications Commission
Washington, D.C. 20554

GAO Approved
B-180227(R0036)

APPLICATION FOR INDIVIDUAL AMATEUR RADIO STATION AND/OR OPERATOR LICENSE

ATTACH HERE the original or photocopy of your primary station license and an original or photocopy of any additional station license(s) as required by instruction F. If a license has been lost, mutilated, or destroyed, the circumstances must be explained in this block.

FCC USE ONLY (Examiner's Report)

ELEMENT	1(A) SPEED	1(B) SPEED	1(C) SPEED	2	3	4(A)	4(B)
PASSED							
FAILED							
DATE OF EXAM			FCC EXAMINER			DISTRICT NO.	

FCC USE ONLY

OPERATOR CLASS FOR WHICH QUALIFIED

CODE	NO	YES	EXAM	NO	YES	SCREEN	NO	YES	SIGN	NO	YES

I. READ ACCOMPANYING INSTRUCTIONS BEFORE FILLING OUT FORM. USE TYPEWRITER OR PRINT CLEARLY.

1. Licensed Amateur Radio Operators give the following information

PRIMARY STATION CALL SIGN	OPERATOR CLASS	EXPIRATION DATE

2. LAST NAME	FIRST NAME	MIDDLE INITIAL	3. DATE OF BIRTH (Month, Day, Year)

4. MAILING ADDRESS (all applicants must complete both items 4 and 5)

5. STATION LOCATION FOR FIXED OPERATION (You cannot use a Post Office Box No. or General Delivery Address)

NUMBER AND STREET	NUMBER AND STREET (or other indication of location)

CITY	STATE	CITY	STATE

COUNTY	ZIP CODE	COUNTY	

Fig. 17-1. FCC Form 610.

public under the Freedom of Information Act. A commercial mailing house gets the list and prints labels that are sold to the trade for commercial use. This all happens while your license is on its way to you through all of the paperwork. So, don't think that your license is lost in the mail if some junk mail lists your call letters before you know them yourself.

INSTRUCTIONS FOR FORM

A. *For Item 1* —This item is to be completed by all applicants. For applicants applying for a club station license, or for a station license for recreation under military auspices, the following applies:

(1) The applicant for a club station license must be the trustee of a *bona fide* amateur radio organization or society. The trustee must be an FCC licensed amateur radio operator of other than the Novice Class and must have been designated by the amateur radio organization involved to hold a club station license for the organization.

NOTE: a club station license always expires on the same date as the trustee's basic operator license.

(2) An amateur station license may be issued to an individual (other than an alien or a representative of any alien or of any foreign government) who is in charge of a proposed amateur station for recreation under military auspices (only of the Armed Forces of the United States) which is to be located in approved public quarters, but not operated by the United States Government.

B. *For Item 4* —Criminal Conviction information. If you answered "Yes" regarding criminal convictions, furnish the following in a separate attached statement.

(1) For each such conviction, give these details:

(a) The nature of the offense, the date of conviction, and the name and address of the court.

(b) The sentence imposed, including the fine and imprisonment; if execution of the sentence was suspended in whole or in part, state to what extent it was suspended.

(c) The dates of commencement and termination of actual imprisonment, if any.

(d) If released on parole or placed on probation, give the dates of commencement and termination of parole or probation, and state whether or not parole or probation was successfully completed without incident; if

presently on parole or probation, give name and address of parole officer, or give name and address of any probation officer or authority to whom you must report.
- (e) State whether or not radio facilities were used in committing the offense; if such facilities were so used, describe how they were used.
- (2) A statement as to the nature and duration of your present employment or business activity, and the name and address of your employer.
- (3) A description of the proposed use of Amateur Radio if the requested license were to be issued.

C. *For Item 5A* —
- (1) Applicants for the Novice or Technician operator licenses must pass an examination which is normally conducted by mail under the supervision of a volunteer examiner. (See Section 97.28 of the Commission's Rules for the necessary procedure.) Applications involving examination by mail must include a written request from the volunteer examiner for appropriate examination papers and must be submitted to the Federal Communications Commission, 334 York St., Gettysburg, Pennsylvania 17325.

  NOTE: when applicant desires to take Novice and Technician exams at the same time, only one application Form 610 should be used and only one written request by a volunteer examiner is required.

- (2) The volunteer examiner's written request must include the following:
  - (a) Name and permanent addresses of both the examiner and the applicant.
  - (b) Qualifications of volunteer examiner to administer the examination. (Examiner must indicate that he is at least 21 years of age and that he is the holder of an Amateur Extra, Advanced, or General Class Amateur Radio operator license, or the holder of a commercial radiotelegraph operator license issued by the Commission, or presently employed in the service of the United States as the operator of a manually operated radiotelegraph station.)
  - (c) A statement that "The applicant successfully passed the required code test within ten (10) days prior to making application."

17A. Do you have any other amateur radio application on file with the Commission that has not been acted upon? If yes, answer items 17B and 17C. YES ☐ NO ☐	17B. Purpose of other application	17C. Date Submitted
18A. Have you failed an amateur examination within the last 30 days? If yes, answer items 18B and 18C. YES ☐ NO ☐	18B. Class of Examination	18C. Date of Exam

19. City and State where you will take the amateur examination (See instruction 19)

20. Have you been convicted in a Federal, State, or local court of any crime for which the penalty imposed was a fine of $500 or more or an imprisonment of six months or more within 10 years previous to the date of this application? (If "yes", see instruction 20). ☐ YES ☐ NO

CERTIFICATION

I CERTIFY that all statements herein and attachments herewith are true, complete and correct to the best of my knowledge and belief and are made in good faith; that I am not a foreign government or a representative thereof; that I waive any claim to the use of any particular frequency or of the ether as against the regulatory power of the United States because of the previous use of the same whether by license or otherwise and that the station to be licensed, if any, will be inaccessible to unauthorized persons.

WILLFUL FALSE STATEMENTS MADE ON THIS FORM OR ATTACHMENTS ARE PUNISHABLE BY FINE AND IMPRISONMENT. U.S. CODE TITLE 18, SECTION 1001	21. SIGNATURE OF APPLICANT (Do not print)	22. DATE SIGNED

II. REQUEST FOR EXAMINATION PAPERS BY VOLUNTEER EXAMINER (Applies only to Novice Class examinations by mail)

1. NAME OF VOLUNTEER EXAMINER	2. NAME OF APPLICANT (must be same as item 2 on front side)

3. MAILING ADDRESS OF VOLUNTEER EXAMINER (Number, Street, City, State and Zip Code)

4. Check the operator license you have which allows you to administer the examination.

☐ GENERAL CLASS ☐ ADVANCED CLASS ☐ AMATEUR EXTRA CLASS ☐ CALL SIGN

	5. Expiration Date of License
6A. Do you have a renewal application on file with us? If yes, answer item 6B. YES ☐ NO ☐	6B. Date You Filed Renewal

CERTIFICATION OF VOLUNTEER EXAMINER

I certify that I am a licensed amateur radio operator, General Class or above; that I am at least 18 years of age; and that: *(Check appropriate box)*

☐ I have examined the above named applicant within the last ten days and he has successfully completed the prescribed telegraphy examination element set forth in Section 97.21 of the Commission's Rules.

☐ I have not examined the applicant in the International Morse Code since the applicant claims code test credit. Applicant's statement supporting code test credit is attached.

7. SIGNATURE OF VOLUNTEER EXAMINER *(Do not print)*	8. DATE SIGNED

III. PHYSICIAN'S CERTIFICATION
(Required only if applying for an examination based on a protracted disability)

☐ I certify that I have examined the applicant within the last 30 days and because of the protracted disability described below found him/her unable to travel to the nearest Commission examining point located at:

City and State

1. BRIEFLY DESCRIBE WHY APPLICANT CANNOT TRAVEL TO THE SPECIFIED NEAREST EXAMINING POINT.

2. NAME, ADDRESS, AND TELEPHONE NUMBER OF PHYSICIAN

3. SIGNATURE OF PHYSICIAN *(Do not print)*	4. DATE SIGNED

FCC Form 610 (back)
September 1977

Fig. 17-1. FCC Form 610 (continued from page 387).

(d) Written signature of the volunteer examiner.
(3) If application is for what used to be the Conditional Class license, applicant must provide a statement to show which of the following is the basis for his eligibility:
   (a) Unable to appear for supervised examination because of protracted physical disability (certification from physician must be enclosed) or because of active service in the Armed Forces (certification of Commanding Officer must be enclosed).
   (b) Applicant submits evidence of future temporary residence outside of the United States, its territories or possessions, for a continuous period of one year or more.
(4) When applying for a new General or Amateur Extra Class license, the applicant should, by mail or in person, submit his application to the Commission Field Office which will conduct the examination.

D. *For Item 5F* — Unless it is also renewed, a license which is modified will bear its present expiration date. To obtain a new license term (normally, five years) both renewal and modification of license can be requested at the same time. This may be done on a single application by checking the boxes in items 5F and 5G and by showing in item 5G that the operating time and code speed requirements for renewal of license have been met.

E. *For Item 5G* —
(1) When a trustee for renewal of a club station license or when an individual applies for renewal of his additional station license, he should file a separate application at the same time for renewal of his basic operator and station license and should complete item 7 on both applications. This should be done because the expiration date of the club or additional station license may be no later than that of the basic license which is held by the trustee or individual.
(2) The operating time requirement for renewal of license is that the applicant has lawfully accumulated, at an amateur station licensed by the Commission, a minimum total of either two hours operating time during the last three months of the license term or five hours operating time during the last 12 months of the license term. Such operating time, for the purpose of renewal, shall be counted as the total of all that time between the entries in

the station log showing the beginning and end of transmission both during single transmissions and during a sequence of transmissions.

(3) The code speed requirement for renewal of license is that the applicant can send by hand key, i.e., straight key or any other type of hand-operated key such as a semi-automatic or electronic key, and receive by ear, in plain language, messages in the International Morse Code at a speed of not less than that which is required in qualifying for an original license of the class being renewed. (See Rule Section 97.21.)

F. *For Item 8*—Normally, new call signs for amateur stations will be assigned systematically by the Commission, and requests for special call signs will not be considered. However, Section 97.51(a) of the Rules sets forth certain exceptions to systematic assignment. If applicant satisfies one of these exceptions and desires a special call sign, he should indicate the special call sign requested in the space provided in item 8. (Changes have been proposed for this special call sign.)

G. *For Item 9*—This item must be completed as follows:

(1) Furnish the address of the station location if it is different than the mailing address shown in item 2; or

(2) Write "see item 2" if the station location is the same as the mailing address. EXCEPTION: When you have furnished a Post Office box number, or the like, or just the name of a city or town as your mailing address, a more complete description must be given for your station location. If a street or road address cannot be furnished, give a brief geographical description of the station location, such as its exact distance and direction from nearby identifiable locations, such as major road intersection or city.

NOTICE: Do not telephone the Commission's office in Gettysburg concerning the status of pending applications or other information relating to the Amateur Radio Service. The Gettysburg office is not equipped to answer telephone inquiries. Any inquiries concerning applications pending more than about eight weeks should be directed, preferably in writing, to the Commission's office in Washington, D.C. 20554.

NOVICE

A volunteer examiner shall be selected by the applicant to administer the code and written parts of the examination. The examiner must be at least 21 years of age, hold a valid Amateur license of the General, Advanced, or Extra Class Grade, or be the holder of a valid Radio Telegraph Commercial Radio Operator License, or be employed in the service of the United States as operator of a manually-operated radiotelegraph station. He must not be related to the applicant.

1. Applicant obtains and fills in Form 610 which may be obtained by writing a post card or letter to one of the FCC field Offices.
2. Examiner administers appropriate code test, sending and receiving.
3. If code test is passed, applicant—within 10 days—submits the following to the Federal Communications Commission, Gettysburg, Pa. 17325.
    a. Form 610.
    b. Letter from examiner requesting written part of examination, also including:

1. Names and addresses of examiner and applicant.
2. Age of examiner.
3. Class of license held.
4. Examiner's statement that applicant has passed the code test within 10 days prior to submission of the letter.
5. Examiner's written signature.

Examination paper will then be sent to the examiner from Gettysburg.

GENERAL/ADVANCED/EXTRA CLASS

The applicant obtains FCC Form 610, fills it out, and brings present ham license to nearest FCC examination point.

WHERE TO TAKE YOUR EXAM

If you're going for your General, Advanced, or Extra Class license, you'll want to take your written and code examination at the rearest location to where you live. If you are not sure where to appear, you can telephone the FCC Engineer-In-Charge of a district office and tell him what county you are in. He'll advise you where to appear and when.

EXAMINATION POINTS

FCC Field Office	Examination Schedule
ALABAMA 439 Federal Bldg. 113 St. Joseph St. Mobile 36602 (205) 690-2808	By appointment only. Call on Monday the week of the examination.
ALASKA U.S. Post Office Bldg. 4th & G Sts. Anchorage 99510 (907) 265-5021	Code-Mon. thru Fri. by appointment No code-Mon. thru Fri. 8 A.M. thru 12 noon.
CALIFORNIA 3711 Long Beach Blvd. Suite 501 Long Beach 90807 (213) 426-4451	Code-8:30 A.M. and 1:00 P.M. No code-8:30 A.M..–2:00 P.M.
1245 Seventh Ave. San Diego 92101 (714) 293-5460	Make appointment one week in advance.
323A Custom House 555 Battery St. San Francisco 94111 (415) 556-7700	Extra Class Friday 8:30 A.M. Code-Friday 10:00 A.M. No code-Friday 8:30 A.M.
COLORADO Suite 2925 1405 Curtis St. Denver 80202 (303) 837-4053	First and Second Wednesday of each month 8:30 A.M.
DISTRICT OF COLUMBIA 1919 M St., N.W. Rm 411 Washington, D.C. 20554 (202) 655-4000	Friday 9:00 A.M. and 10:30 A.M.
FLORIDA 919 Federal Bldg. 51 S.W. First Ave. Miami 33130 (305) 350-5541	Code-Thursday 9:00 A.M. No code-Tuesday and Wednesday 8: A.M. - 1 P.M.

738 Federal Bldg
500 Zack St.
Tampa 33602
(813) 228-2605

Make appointment one week in advance.

GEORGIA
Rm 440
1365 Peachtree St., N.E.
Atlanta 30303
(404) 881-7381

Code-Friday 8:30 A.M.
No code-Tuesday and Friday
8:30 A.M.-12:00 noon.

238 Federal Office Bldg.
125 Bull St.
Savannah 31402
(912) 232-4321 ext. 320

Make appointment one week advance.

HAWAII
335 Merchant St.
Honolulu 96808
(808) 546-5640

Wednesday 8 A.M. Other times by appointment.

ILLINOIS
3935 Federal Bldg.
230 S. Dearborn St.
Chicago 60604
(312) 353-0195

Friday 8:45 A.M.

LOUISIANA
829 E. Edward Herbert
Federal Bldg.
600 South St.
New Orleans 70130
(504) 589-2094

Code-Tuesday 8 A.M.
No code-Tuesday and Wednesday 8:30 A.M.-12 noon.

MARYLAND
Rm. 823 Fallon Federal Bldg.
31 Hopkins Plaza
Baltimore 21201
(301) 962-2727

Code-Monday 8:30 A.M.
No code-Monday and Friday
8:30 A.M.-12 noon.

MASSACHUSETTS
1600 Customhouse
165 State St.
Boston 02109
(617) 223-6608

Wednesday 9 A.M.

MICHIGAN
1054 Federal Bldg.
231 W. Lafayette St.
Detroit 48226
(313) 226-6078

Friday 9 A.M.

MINNESOTA
691 Federal Bldg.
316 N. Robert St.
St. Paul 55101
(612) 725-7819

Friday 8:45 A.M.

MISSOURI
1703 Federal Bldg.
601 E. 12st St.
Kansas City 64106
(816) 374-5526

Tuesday 9 A.M.

NEW YORK
1307 Federal Bldg.
111 W. Huron St.
Buffalo 14202
(716) 842-3216

Code-Friday 9 A.M.
No code-Friday 10 A.M.

201 Varick St.
New York 10014
(212) 620-3436

Wednesday 9 A.M.

OREGON
1782 Federal Office Bldg.
1220 S.W. 3rd Ave.
Portland 97204
(503) 221-3097

Friay 8:45 A.M.

PENNSYLVANIA
11425 Byrne Federal
Court House
601 Market St.
Philadelphia 19106
(215) 597-4410

Code-Tuesday and Wednesday
8:30 A.M.
No code-Mon.-Wed.
10 A.M.-12 noon.

PUERTO RICO
U.S. Post Office
Rm. 323
San Juan 00903
(809) 722-4562

Code-Friday 10 A.M.
No code-Thursday and Friday
8:30 A.M.
(or 1:00 P.M. by appointment).

TEXAS
13E7 Federal Bldg.
1100 Commerce St.
Dallas 75242
(214) 749-3243

Tuesday 9 A.M.

Rm. 323 Federal Bldg.
300 Willow St.
Beaumont 77704
(713) 838-0271 ext. 317

Make appointment one week in advance.

5636 Federal Bldg.
515 Rusk Ave.
Houston 77002
(713) 226-4306

Code-Wednesday 8 A.M.
No code-Wedensday 9 A.M.-12 noon.

VIRGINIA
Military Circle
870 North Military Highway
Norfolk 23502
(804) 461-4000

Thursday 9 A.M.

WASHINGTON
3526 Federal Bldg
915 Second Ave.
Seattle 98174
(206) 442-7610

Friday 8:30 A.M.

The FCC conducts examinations at other cities throughout the country where there are no field offices. The location and schedule for these tests follow.

If you wish to take the tests at one of these cities you should submit your application, Form 610, as far in advance as possible to the FCC field office responsible for administering the examination. Tell them the city in which you want to take the test and the month. The FCC will then notify you when and where to appear for the examination.

STATE	CITY	MONTH OF EXAM	RESPONSIBLE FCC OFFICE
ALABAMA	Birmingham	Mar., Sept.	Atlanta
	Montgomery	June, Dec.	
ALASKA	Fairbanks	Jan., Apr., July, Oct.	Anchorage
	Juneau	May, Nov.	
	Ketchikan	May, Nov.	
ARIZONA	Phoenix	Jan., Apr., July., Oct.	Los Angles
	Tuscon	Apr., Oct.	

STATE	CITY	MONTH OF EXAM	RESPONSIBLE FCC OFFICE
ARKANSAS	Little Rock	Feb., May, Aug., Nov.	New Orleans
CALIFORNIA	Bakersfield	May, Nov.	Los Angeles
	Fresno	Mar., June, Sept., Dec.	San Francisco
CONNECTICUT	Hartford	Jan., Apr., July, Oct.	Boston
FLORIDA	Jacksonville	Apr., Oct.	Miami
GEORGIA	Albany	Feb., Aug.	Atlanta
GUAM	Agana	July, Sept., Nov., Jan., Mar., May	Honolulu
HAWAII	Hilo	Aug., Nov., Feb., May	Honolulu
	Lihue	Sept., Dec., Mar., Jun.	
	Wailuku	Aug., Nov., Feb., May	
IDAHO	Boise	Apr., Oct.	Portland
	Pocatello	Nov., June	
ILLINOIS	Rock Island	Feb., May, Aug., Nov.	Chicago
INDIANA	Fort Wayne	Feb., May, Aug., Nov.	Chicago
	Indianapolis	Jan., Apr., July, Oct.	
IOWA	Des Moines	Mar., June, Sept., Dec.	Kansas City
KANSAS	Wichita	Mar., Sept.	Kansas City
KENTUCKY	Louisville	Mar., June, Sep., Dec.	Chicago
LOUISIANA	Shreveport	Apr., Oct.	New Orleans
MAINE	Bangor	Feb., Aug.	Boston
	Portland	May, Nov.	
MICHIGAN	Grand Rapids	Jan., Apr., July, Oct.	Detroit
	Marquette	May, Nov.	St. Paul
MINNESOTA	Duluth	Mar., Sep.	St. Paul
MISSISSIPPI	Jackson	June, Dec.	New Orleans
MISSOURI	St. Louis	Feb., May, Aug., Nov.	Kansas City
MONTANA	Billings	Apr., Oct.	Seattle
	Helena	Apr., Oct.	
NEBRASKA	Omaha	Jan., Apr., July, Oct.	Kansas City
NEVADA	Las Vegas	Jan., July	Los Angeles
	Reno	Apr., Oct.	San Francisco
NEW MEXICO	Albuquerque	Apr., Oct.	Denver
NEW YORK	Albany	Mar., June, SepT., Dec.	New York
	Syracuse	Jan., Apr., July, Oct.	Buffalo
NORTH CAROLINA	Wilmington	June, Dec.	Norfolk
	Winston-Salem	Feb., May, Aug., Nov.	
	Charlotte	Jan., July	
NORTH DAKOTA	Bismark	Feb., Aug.	St. Paul
	Fargo	June, Dec.	
OHIO	Cincinnati	Feb., May, Aug., Nov.	Detroit
	Cleveland	Mar., June, Sept., Dec.	
	Columbus	Jan., Apr., July, Oct.	
OKLAHOMA	Oklahoma City	Jan., Apr., July, Oct.	Dallas
	Tulsa	Feb., May, Aug., Nov.	
OREGON	Medford	Sept., May	Portland
PENNSYLVANIA	Pittsburgh	Feb., May, Aug., Nov.	Buffalo
	Wilkes-Barre	Mar., Sept.	
SOUTH CAROLINA	Columbia	May, Nov.	Atlanta
SOUTH DAKOTA	Rapid City	May, Nov.	Denver
	Sioux Falls	Apr., Nov.	St. Paul
TEXAS	Corpus Christi	June, Dec.	Houston
	El Paso	June, Dec.	Dallas
	Lubbock	Mar., Sept.	
	San Antonio	Feb., May, Aug., Nov.	Houston
TENNESSEE	Knoxville	Mar., June, Sept., Dec.	Atlanta
	Memphis	Jan., Apr., July, Oct.	
	Nashville	Feb., May, Aug. Nov.	
UTAH	Salt Lake City	Mar., June, Sept., Dec.	Denver

STATE	CITY	MONTH OF EXAM	RESPONSIBLE FCC OFFICE
VERMONT	Burlington	Apr.,Oct.	Boston
VIRGINIA	Salem	Apr.,Oct.	Norfolk
WASHINGTON	Sokane	Feb.,May,Aug.,Nov.	Seattle
WEST VIRGINIA	Charleston	Mar.,June,Sept.,Dec.	Detroit
WISCONSIN	Milwaukee	Mar.,June,Sept.,Dec.	Chicago
WYOMING	Casper	May,Nov.	Denver

Appendix

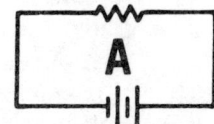

Element 3 Syllabus

Following is a listing of study topics for the Technician/General class license examination.

RULES AND REGULATIONS

(1) Control point 97.3(p)
(2) Emergency communications 97.3(w); 97.107
(3) Amateur radio transmitter power limitations 97.67
(4) Station identification requirements 97.84(b), (f), (g); 97.79(c)
(5) Third party participation in amateur radio communications 97.79(d)
(6) Domestic and international third party traffic 97.114; Appendix B, Art. 41, Sec 2
(7) Permissible one-way transmissions 97.91
(8) Frequency bands available to the Technician class 97.7(d)
(9) Frequency bands available to the General class 97.7(b)
(10) Limitations on use of amateur radio fequencies 97.61
(11) Selection and use of frequencies 97.63
(12) Radio controlled model crafts and vehicles 97.65(a); 97.99
(13) Radioteleprinter emissions 97.69

Prohibited Practices:

(14) Broadcasting   97.113
(15) Music   97.115
(16) Codes and ciphers   97.117
(17) Obscenity, indecency, profanity   97.119

The numbers following subject listings above refer to sections of the FCC's Part 97 rules.

OPERATING PROCEDURES

(1) Radiotelephony
(2) Radio teleprinting
(3) Use of repeaters
(4) Vox transmitter control
(5) Full break-in telegraphy
(6) Operating courtesy
(7) Antenna orientation
(8) International communication
(9) Emergency preparedness drills

RADIO WAVE PROPAGATION

(1) Ionospheric layers; D, E, F1, F2
(2) Absorption
(3) Maximum usable frequency
(4) Regular daily variations
(5) Sudden ionospheric disturbance
(6) Scatter
(7) Sunspot cycle
(8) Line-of-sight
(9) Ducting, tropospheric bending

AMATEUR RADIO PRACTICE

Safety Precautions:

(1) Household ac supply and electrical wiring safety
(2) Dangerous voltages in equipment made inaccessible to accidental contact

Transmitter Performance:

(3) Two-tone test
(4) Neutralizing final amplifier
(5) Power measurement

Use of Test Equipment:
- (6) Oscilloscope
- (7) Multimeter
- (8) Signal generators
- (9) Signal tracer

Electromagnetic Compatibility; Identify and Suggest Cure:
(10) Disturbance in consumer electronic products caused by audio rectification

Proper Use of the Following Station Components and Accessories:
- (11) Reflectometer (VSWR meter)
- (12) Speech processor - rf and af
- (13) Electronic T-R switch
- (14) Antenna tuning unit; matching network
- (15) Monitoring oscilloscope
- (16) Non-radiating load; dummy antenna
- (17) Field strength meter; S-meter
- (18) Wattmeter

ELECTRICAL PRINCIPLES

Concepts:
- (1) Impedance
- (2) Resistance
- (3) Reactance
- (4) Inductance
- (5) Capacitance
- (6) Impedance matching

Electrical Units:
- (7) Ohm
- (8) Microfarad, picofarad
- (9) Henry, millihenry, microhenry
- (10) Decibel

Mathematical Relationships:
- (11) Ohm's law
- (12) Current and voltage dividers
- (13) Electrical power calculations

(14) Series and parallel combinations; of resistors, of capacitors, of inductors
(15) Turns ratio; voltage, current, and impedance transformation
(16) Root mean square value of a sine wave alternating current

CIRCUIT COMPONENTS

Physical Appearance, Types, Characteristics, Applications, and Schematic Symbols for:

(1) Resistors
(2) Capacitors
(3) Inductors
(4) Transformers
(5) Power supply type diode rectifiers

PRACTICAL CIRCUITS

(1) Power supplies
(2) High-pass, low-pass, and band-pass filters
(3) Block diagrams showing the stages in complete AM, SSB, and FM transmitters and receivers

SIGNALS AND EMISSIONS

(1) Emission types A0, A3, F1, F2, F3
(2) Signal; information
(3) Amplitude modulation
(4) Double sideband
(5) Single sideband
(6) Frequency modulation
(7) Phase modulation
(8) Carrier
(9) Sidebands
(10) Bandwidth
(11) Envelope
(12) Deviation
(13) Overmodulation
(14) Splatter
(15) Frequency translation; mixing, multiplication
(16) Radioteleprinting; audio frequency shift keying, mark, space, shift

ANTENNAS AND FEEDLINES

Popular Amateur Radio Antennas and Their Characteristics:
(1) Yagi antenna
(2) Quad antenna
(3) Physical dimensions
(4) Vertical and horizontal polarization
(5) Feedpoint impedance of half-wave dipole, quarter wave vertical
(6) Radiation patterns; directivity, major lobes

Characteristics of Popular Amateur Radio Antenna Feedlines; Related Concepts:
(7) Characteristic impedance
(8) Standing waves
(9) Standing wave ratio; significance of
(10) Balanced, unbalanced
(11) Attenuation
(12) Antenna-feedline mismatch

Appendix

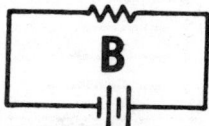

International Morse Code

A	.—
B	—...
C	—.—.
D	—..
E	.
F	..—.
G	——.
H	....
I	..
J	.———
K	—.—
L	.—..
M	——
N	—.
O	———
P	.——.
Q	——.—
R	.—.
S	...
T	—
U	..—
V	...—
W	.——
X	—..—
Y	—.——
Z	——..

Punctuation	Symbol	Code
Period	.	.—.—.—
Comma	,	——..——
Colon	:	———...
Question mark, or request for repetiton of a transmission not understood	?	..——..
Apostrophe	'	.————.
Dash or hyphen	—	—....—
Fraction bar	/	—..—.
Parenthesis (before and after words)	( )	—.——.—
Underscore (before and after words or part of sentence)		..——.—.
Equal sign	=	—...—
Understood		...—.

Ä (German) ·—·—	
Á or Å (Spanish-Scandinavian) ·———·—·—	
CH German-Spanish ————	
É (French) ··—··	
Ñ (Spanish) ——·——	
Ö (German) ———·	
Ü (German) ··——	

1 ·————	
2 ··———	
3 ···——	
4 ····—	
5 ·····	
6 —····	
7 ——···	
8 ———··	
9 ————·	
0 —————	

Error	········
Cross or end-of-telegram or end-of-transmission signal	·—·—·
Invitation to transmit	—·—
Wait	·—···
End of work	···—·—
Starting signal (beginning every transmission)	—·—·—
Separation signal for transmission of fractional numbers (between the ordinary fraction and the whole number to be transmitted) and for groups consisting of figures and letters (between the figure-groups and the letter-groups)	·—··—

Appendix

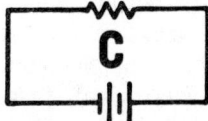

TVI/BCI/Hi-Fi-I

One Saturday morning in 1959, I was busily enjoying a well earned sleep-in when the telephone rang. An irate neighbor *demanded* that I get off the air and stop interfering with his television reception. He was really hot! The week before, I had put up a simple 40/20/15-meter dipole, and he had seen it. Trouble was I didn't even own a transmitter. I had just passed my novice license examination and was busily waiting out the 6 to 10 weeks required to grade the exam and process the license by getting ready to go on the air. My brand new Heathkit DX-20 transmitter kit hadn't even been received in the mail yet!

Sooner or later, if you operate an amateur radio station, you will be faced with the irate neighbor problem. The interference may be real or imagined, but the result is the same: *you* get the blame. From the moment that your antenna goes up, you will automatically spring to mind whenever any neighbor's TV reception goes "bloop." You might, like one amateur was, be blamed for damage to the owner's TV set.

The best advice is to stay cool and be informed of the problem and possible solutions. Do not, under any circumstances, become irate, even if you are only matching your neighbor tit-for tat.

There are several causes of interference. It is possible that your equipment is at fault. This must be taken care of before you can be on really firm ground in any dispute. Some of the material in this chapter will help you determine whether there is a high probability

409

that your rig is producing the problem. These are not absolute techniques, but are often good enough to resolve problems.

One word of warning: *Do not attempt to modify any equipment owned by a neighbor*! There are methods, including both addition of filtering and modification of circuits, that often help a receiver that does already not reject unwanted signals properly. Even if you are an experienced, qualified, certified, bottle-in-bond, 100-proof consumer electronics service technician, make no attempt to resolve the neighbor's problem yourself. If you do, then you will be permanently married to that set. From then on, every needed repair will automatically—in your neighbor's mind—become your responsibility. If you are *not* qualified *professionally* as a consumer electronics technician (and I don't mean to exclude electronics engineers of any type, or electronics technicians from fields other than consumer electronics), then you are just plain stupid to attempt to make modifications to the neighbor's equipment. Let the neighbor know the types of things that might be needed and then tell him to call a TV/hi-fi repairman to actually do the job.

One frequency problem with consumer electronics devices is that they are unable to reject your signal—and they are supposed to! Even some equipment by the best manufacturers is designed poorly in the interference rejection department. The owner of that $1000 color TV or $1500 hi-fi, though, sees *you doing something* that fouls up previously good operation of their expensive gear. They do not appreciate the fact that the equipment is unable to do its total job. A TV or hi-fi is not only required to respond to the desired signal, but must also be capable of rejecting the undesired signals.

TVI: TYPES OF INTERFERENCE

It is possible to diagnose the type of interference experienced by a TV receiver by noting how it affects the picture. Not all forms of video interference are caused by radio transmitters. Much interference can be traced to defective devices in the house of complainer. Furnace controllers, blower motors, defective thermostats, and elements in electrical resistance heaters, SCR lamp dimmers, vacuum cleaner or pump motors, and a host of other devices will create a snow-like interference in the picture. Such signal sources also sometimes produce a hash-like, raw sound in the audio of the TV receiver. If the interference is noted at times when you were not on the air, then it may well be that the problem is right in the home of the person who raised the complaint.

Fig. C-1. TVI from a CW/AM/SSB transmitter.

An example of genuine radio transmitter interference is shown in Fig. C-1. This interference is characterized by horizontal or slanted bars in the picture that *might* move in step with the modulation of the transmitter. Note that a similar form of disruption occurs if the sound carrier of the TV is getting into the picture, and will manifest itself when the fine tuning of the TV receiver is incorrectly set. Also, be sure to not confuse this form of interference with horizontal sync problems in the TV. Of course, both of these problems become a mute issue if they exist at times when your transmitter is turned off, but try to convince your neighbor of that fact, especially if the person is someone trying to get some ham/CB operator for a free TV repair job.

Figure C-2 shows the herringbone pattern that is typical of interference from FM transmitters. This will normally be limited to channel 6, and possibly one additional channel in the VHF band, if the source is a local FM broadcasting station. If there is a local broadcasting station, or some other station that uses FM, such as the commerical land-mobile two-way, service then the source of interference might be that transmitter. Make sure that your own FM operations are not at fault, though.

An example of co-channel interference is shown in Fig. C-3. If your neighbor complains of seeing your picture on the TV, you might want to chuckle and investigate the possibility of co-channel interference. The channel assignments for TV stations are usually

Fig. C-2. FM interference to a TV receiver.

made at geographical distances that limit this possibility, but atmospherics that excite your VHF DXer buddy will also cause this type of interference to VHF TV receivers. The problem is rarely chronic, however, except at locations approximately equidistant between two transmitters operating on the same channel. If the two signal strengths are approximately the same, the receiver will respond equally. The solution, however, is simple: use a high-gain directional antenna on the receiver.

There is one way to gain at least a psychological advantage over your irate neighbor—take a shotgun. Oh, not really, but the way to drive home your point that his or her equipment *might* be at fault is to keep your own equipment free of interference. If the neighbor sees similar equipment operating right in the same house with the transmitter, then he sees things your way and blames the cheap manufacturer who failed to shield and bypass his TV or hi-fi properly.

STEPS TO TAKE

The material in this section was adopted directly from an FCC publication titled *How to Identify & Resolve Radio-TV Interference*

Problems (stock No. 004-000-00345-4, sold for $1.50 by the Superintendent of Documents, Government Printing Office, Washington, D.C., 20402). It is also available from the Consumer Information Center, Pueblo, CO, under the same number at the same price. We make no apologies for using this material, as it is quite well written and presented. Have several copies of this booklet available to give out to your neighbor, especially if you plan a new installation or the purchase of a high-power linear amplifier.

Installing A High-Pass Filter

There are not set procedures for eliminating television interference—it is a matter of eliminating the most likely sources of interference at step at a time. The first step is to install an inexpensive high-pass filter on the back of your TV set. In making this installation, follow these procedures:

1. Determine the type of antenna wire that is connected to your TV set. There are two possibilities:

Coaxial Cable—a round lead-in wire which requires a filter impedance of 75 ohms (see Fig. C-4A).

Twin Lead Wire—a flat wire which requires a filter impedance of 300 ohms (see Fig. C-2B).

Fig. C-3. Co-channel interference to a TV receiver.

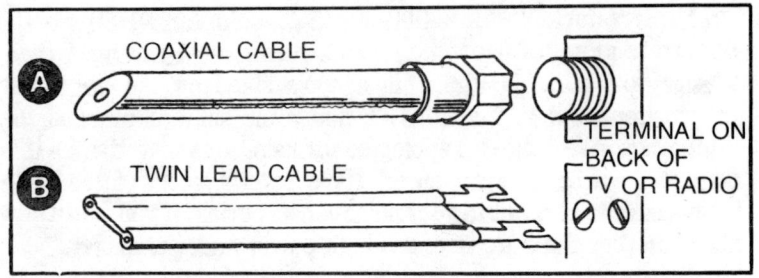

Fig. C-4. Installing a high-pass filter with coaxial cable at A and twin-lead at B.

2. Purchase the filter which matches the type of antenna wire coming from your set. The impedance information mentioned above will be on the filter label. Do not use a combination of twin-lead and coaxial cable without proper matching transformers (often called baluns). Filters are available in most stores that sell or repair television sets.

3. Carefully read the instructions that are provided with the filter. You will be installing the filter on the back of your TV set, as near to the antenna terminal as possible. The antenna terminal and the filter terminal will look like either Fig. C-4A or C-4B depending upon the type of wire you are using—coaxial or twin-lead.

4. If you are on a cable system, you may still install the filter at the antenna terminal. However, if the interference continues, contact the cable company repair service for assistance. Do not attempt to modify the cable system yourself.

5. The following information or installing the filter should answer any additional questions you may have.

Disconnect the antenna wire (twin-lead or coaxial) from the television set antenna terminals. Connect the wire from the antenna to the input terminals of the filter. For twin-lead wire, connect a very short (1" to 2") "jumper" wire from the antenna input terminals of the set to the filter (see Fig. C-5). For coaxial cable, it will be necessary to obtain a jumper cable that has the proper connectors already installed. This can be purchased at the time you buy the coaxial filter.

Be sure that in the case of twin-lead wire, the actual wires are making contact with the terminals. For coaxial cable, be sure the connector plugs are properly installed on the coaxial cable.

If you have an amplifier in your antenna system, you should have a filter installed ahead of the amplifier and another filter ahead of the TV receiver input terminals (see Fig. C-6). If the amplifier is located close to the receiver, then install the filter before the amplifier only.

Booster amplifiers usually are located near the back of the TV set; mast mounted (outdoor) amplifiers are usually located on the antenna; and distribution amplifiers are usually located somewhere in the distribution system. If a distribution amplifier is in your antenna system, then be sure to trace the entire length of the antenna system, because amplifiers are usually in out-of-the-way places; for example, clothes closets, basements, etc.

The connecting wires between the filter and amplifier, and between the amplifier and antenna terminal, should be as short as possible.

The instructions provided with the filter you bought may call for a ground connection. The wire should be as short as possible and connected between the high-pass filter ground terminal and a metallic cold water pipe or a ground rod. Use bell wire for this connection (see Fig. C-5). Bell wire can be obtained from most variety stores.

If installation of the filter at the TV antenna terminals does not entirely eliminate the interference, you should then contact your service represetive to install a high-pass filter inside the TV set at the tuner input terminals. Internal modification to your set should be done *only* by a service representative.

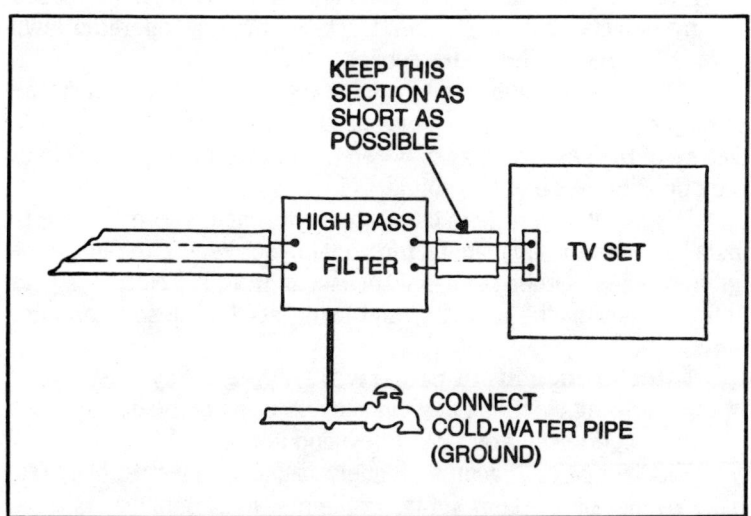

Fig. C-5. Connecting a high-pass filter to twin-lead.

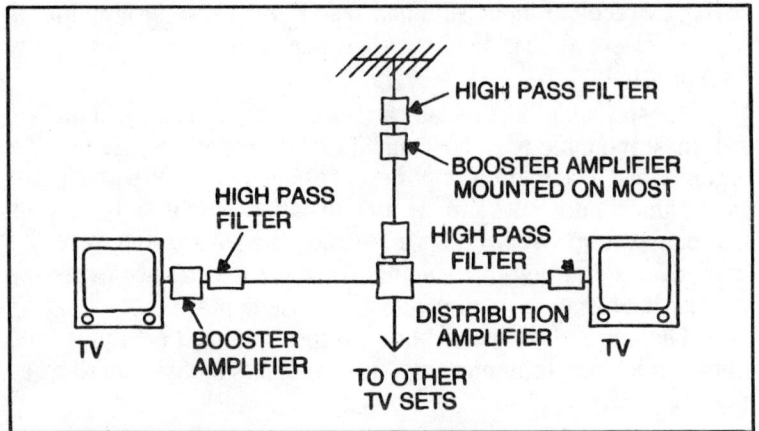

Fig. C-6. High-pass filter installation when a filter is used.

Home Remedies For Resolving Electrical Interference

Electrical interference is caused by two sources: vehicle ignition systems and electrical devices. The first step in attempting to resolve electrical interference problems is to locate the source of interference.

Interference From Vehicle Ignition System. Ignition interference sounds like a "popping" noise in the sound system of your TV that rises in intensity; the "pops" occur closer and closer together as the speed of the engine speeds up. This can be caused by any vehicle ignition system, such as gasoline operated lawn mowers, snowmobiles, automobiles, etc.

If the interference is to television receivers, you may hear the same popping noise in the sound and also see "dancing dots" in the picture of the set. You may only see the interference, and not hear the "popping noise in the sound."

If your own vehicle is causing interference, you may wish to install a commerically manufactured kit in your vehicle to reduce the ignition noise. Other remedial measures include relocating your antenna, raising the antenna, and using shielded lead-in antenna wire.

Interference From Electrical Devices. Any one or more of the following electrical devices may be causing the interference you are experiencing on your television set or AM/FM radio:

Electric razor, vacuum cleaner, fan, drill, electric blankets, bake ovens, fluorescent lights, arc lights, light dimmer controls, relays, static from machinery, lightning arrestors, adding machine, cash register, circuit breakers, ultraviolet lamps, germicidal lamps, defective wiring, loose fuse, arc welder, switch contacts (such as on dishwashers and other home appliances), refrigerator, water pump, sewing machine, light blinkers (including Christmas tree light blinkers), electric heating pads, aquarium warmers, neon signs, door bell circuits/transformers, toys (such as electric trains), sign flashers, antifriction bearings, printing press static eliminators, calculators, insulators, incandescent lamp (new or old), sun lamps, electrical pole (ground wire cut or poor contact), loose electrical connection, electric fence unit, furnace controls, power company transformers, and smoke precipitators.

In attempting to locate the specific device causing the interference, consider the following suggestions. If you have a portable radio that is affected by the interference, use the radio as a detection device to assist in locating the source of interference. With the portable radio, move from room to room and determine in which room the interference appears to be the loudest. Then look for one of the devices listed above and unplug it to see if the interference disappears. If several devices listed above are in the room, unplug them, one at a time, until the interference disappears.

If a portable radio is not affected, you can go to the main fuse or circuit breaker box in your home, remove one fuse at a time, or shut off one breaker at a time, and see if the interference goes away.

If it does not go away when the first fuse or circuit breaker is off, replace the fuse or turn the circuit breaker back on and continue on until the interference does disappear. When the circuit that supplies the power to the TV or radio is turned off, it will be necessary to plug that device into some other circuit to determine if the interference is being generated by a device in the same room as your TV or radio.

When the interference disappears with a fuse removed or circuit breaker off, you should go to the room supplied by that circuit and look for any of the devices listed above. If any of the listed devices are found in the room, replace the fuse of turn the circuit breaker back on. Then unplug the device suspected of causing the interference. If several devices are in the room, unplug them, one at a time.

If you are unable to locate within your own home, the device that is causing the problem, the interference may be coming from a device located in your neighbor's home. With the cooperation of your neighbor, follow the same procedures described above.

If your investigation leads you to suspect that a power line or power company equipment is the source of interference, you should contact the power company to assist you in resolving the problem.

Short duration interference, such as that from electric drills, and saws, may be very costly to attempt to eliminate; you may just want to "live with it."

To *resolve* electrical interference, modifications must be made to the interfering device. This should only be done by a qualified service representative. Information for your service representative is contained in the Technical Information for Service Representative section.

Home Remedies For Resolving FM Interference

The installation of an inexpensive FM band rejection filter is the first step to take in resolving FM interference. In making this installation, follow these procedures:

1. Determine the type of antenna wire you have connected to your TV set. There are two possibilities:

Coaxial Cable—a round lead-in wire which requires a filter impedance of 75 ohms (see Fig. C-4A).

Twin Lead Wire—a flat wire which requires a filter impedance of 300 ohms (see Fig. C-4B).

2. Purchase the appropriate filter, according to the type of antenna wire you have. The impedance information mentioned above will be on the filter lable. Do not use a combination of twin-lead and coaxial cable without proper matching transformers (often called baluns). Filters are available in most stores that sell or repair television sets.

3. Carefully read the instructions that are provided with the filter. You will be installing the filter on the back of your TV set, as near to the antenna terminal as possible. The antenna terminal and the filter terminal will look like either Fig. C-4A or C-4B depending upon the type of wire you are using—coaxial cable or twin-lead wire.

4. If you are on a cable system, you may still install the same FM band rejection filter at the antenna terminal. However, if the interference continues, contact the cable company repair service for assistance. Do not attempt to modify the cable system yourself.

5. The following information on installing the filter should answer any additional questions you may have.

Disconnect the antenna wire (twin-lead or coaxial) from the television set antenna terminals.

Connect the wire from the antenna to the input terminals of the filter.

For twin-lead wire, connect a very short (1" to 2") "jumper" wire from the antenna input terminals of the set to the filter (see Fig. C-5). For coaxial cable, it will be necessary to obtain a jumper cable that has the proper connectors already installed.

Be sure that in the case of twin-lead wire, the actual wires are making contact with the terminals. For coaxial cable, be sure the connector plugs are properly installed on the coaxial cable.

If you have an amplifier in your antenna system, you should have a filter installed before the amplifier and another filter ahead of the TV receiver input terminals (see Fig. C-6). If the amplifier is located close to the receiver, then install the filter before the amplifier only.

The connecting wires between the filter and amplifier, and between the amplifier and antenna terminal, should be as short as possible.

The instructions provided with the filter you bought may call for a ground connection. The wire should be as short as possible and connected between the FM band rejection filter ground terminal and a metallic cold water pipe or a ground rod. Use bell wire for this connection (see Fig. C-5). Bell wire can be obtained from most variety stores.

If the filter does not entirely eliminate the interference, you should call your service representative.

Identification Of And Resolving Audio Interference

Interference to audio devices, such as tape recorder, record players, electronic organs, telephones, hi-fi amplifiers, etc., is caused when the equipment responds to the transmission of a nearby radio transmitter. Audio interference (often called audio rectification) may also affect the sound (audio) portion of your TV and AM/FM radio.

When this type of interference is occuring, you will hear the voice transmissions of the radio transmitter and/or the volume level of the audio device you are using may decrease. If you have determined that this is the type of interference you are receiving, refer to the following suggested methods for eliminating audio interference.

Audio interference is a condition that usually requires internal modification of your equipment. For safety reasons, it is recommended that any modifications be made by a qualified service representative.

Due to the complexity of resolving interference to an electronic organ, again, servicing should be done only by an experienced service representative. More detailed information should be obtained from the equipment manufacturer.

For telephone interference, contact your local telephone company. They can install a 1542A or similar inductor in the telephone instrument to resolve the problem. The information provided in this bulletin applies primarily to privately-owned equipment and should not be applied to equipment owned by the telephone company. Bell system personnel can obtain additional data in Section 500-150-100 of the *Bell System Practices—Plant Series* manual.

For all other audio devices, you may wish to take the following steps before calling your service representative. Replace unshielded wire between the amplifier and speakers with shielded wire. Ground the affected equipment to a metallic cold water pipe or ground rod. A ground connection can be made with a short piece of bell wire which can be obtained at most variety stores. Do not ground AC/DC-type devices. Normally device which may safely be grounded will provide a grounding terminal. If no terminal is provided, then you should consult a qualified service representative for advice.

If the interference is not eliminated after taking these steps you must call a qualified service representative. You may also wish to discuss the matter with the operator of the radio transmitter.

Resolving Radio Transmitter Interference

There are no set procedures for eliminating television interference—it is a matter of eliminating the most likely sources of interference a step at a time. You may be required to take several steps before the interference problem is resolved. Once you have installed the filter called for, or made the adjustment that you were instructed to do, leave the modifications in place and proceed to the next step.

To begin, check to see if a high-pass filter has been installed on the TV set at the antenna terminals. If the interference is still present after the installation of a high-pass filter proceed with the following steps.

Check Radio Transmitter. Contact the operator of the radio transmitter identified as the source and, with his/her cooperation, determine if the transmitter is operating properly.

Is the transmitter properly grounded? This means a good radio frequency (**rf**) ground. A single piece of wire to a ground rod may be an open circuit to **rf**.

Are harmonics and/or spurious emissions present? Is the transmitter cabinet radiating energy?

If the transmitter is not grounded, connect the chassis to a good earth ground with large diameter wire or copper strap. This should assist in eliminating radiation of energy from the cabinet.

Next, install a low-pass filter on the transmitter antenna circuit to see if any difference occurs in the interference pattern. If a change occurs, the interference is probably caused by harmonics and/or spurious emissions from the transmitter. If no change occurs in the interference pattern, it is probably being generated at some point in the TV reception.

Check TV Reception System. Conduct a visual inspection of the TV antenna, lead-in wire, and lightning arrestors. This may reveal a source of trouble. Corroded connections or deteriorated lead-in wire could be at fault and should be repaired.

Assuming no faulty conditions are found, or if found, they are corrected, and the interference is still present, look for an amplifier in the line. Amplifiers are highly susceptible to radio frequency (**rf**) energy.

If an amplifier is in the system, remove it from the circuit. If you find that this eliminates the interference, reconnect the amplifier, but protect the amplifier by grounding, enclosing it in a metallic rf-proof housing and grounding the housing, or installing a high-pass filter at the input to the amplifier. If one filter improves the condition, but does not entirely eliminate the interference, install two filters in series. If no amplifier is utilized, or the interference still persists after following one or all of the above steps, check the TV receiver system.

Check TV Receiver System. An AC power line **rf** filter should be installed to determined if the **rf** from the transmitter is entering the TV via the power cord. (A line filter may be either purchased or one may be constructed by following the schematic shown in Fig. C-7.

If no change is found with the power line filter installed, and the antenna disconnected, then the set itself is responding to the **rf** energy. The most likely internal circuit in the set to be affected by a radio transmitter is the tuner. Disconnect the antenna input lead inside the set directly at the tuner. If the interference is eliminated, then install a high-pass filter at the tuner.

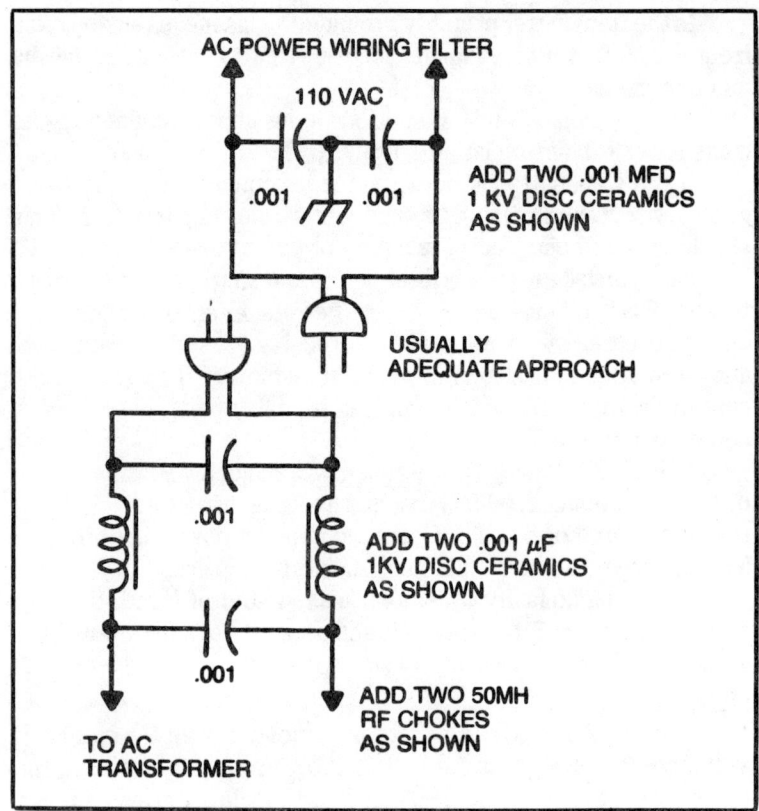

Fig. C-7. Two methods of installing an ac power line rf filter.

If the interference is still present after installing the filter at the tuner, it will be necessary to refer to service data for the set and check each stage of the set for undesired response.

CB Interference To TV Channel 2. Second harmonic interference from a CB transmitter to channel 2 television may exist even though the transmitter meets FCC specifications for harmonic radiation. In such cases, a tuned filter across the antenna terminals of the television should help. The filter may be an inductor and capacitor in series as in Fig. C-8. The filter should be tuned for minimum interference.

A second method is to put an open circuit, quarter-wave, tuned stub across the antenna terminals. The stub should be made of the same type of wire as the antenna input terminals of the television. The initial stub length should be 37" for RG-59/U coax; and 48" for 300-ohm twin-lead.

After connecting the stub, cut the unterminated end of the stub off in ⅛" to ¼" sections until the interference is eliminated. Refer to Fig. C-9. For harmonics falling on other TV channels, such as channel 5, 6, or 9, length of the stub may be appropriately shortened according to the following formula.

$$\text{Length in inches} = \frac{2952V}{f}$$

where **v** is the velocity factor of the line and f is the frequency in megahertz.

Amateur Interference To TV Channel 2. One additional type of interference from a nearby transmitter is unique to the amateur 6-meter band—50-54 MHz. Since 6 meters is immediately adjacent to channel 2 television (54-60 MHz), interference to channel 2 may occur.

In most cases, installation of an open circuit, quarter-wave, tuned stub at the antenna terminals of the television set should be effective. It should be connected as shown in Fig. C-9.

If RG-59/U is used as the TV lead-in wire, the initial length of the stub should be 42". If 300-ohm twin-lead is used, the initial length should be 53".

After the stub is attached to the television, begin cutting off the unterminated end of the stub ⅛" to ¼" at a time until the interference is eliminated. If the interference is reduced, but not eliminated by this method, add a second stub directly to the input terminals of the tuner. The theoretical final length of the stub should be:

$$\text{Length in inches} = \frac{2952V}{f}$$

where **V** is the Velocity factor of line and f is the frequency in megahertz.

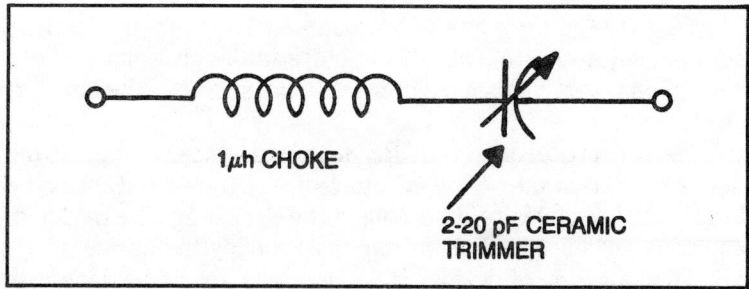

Fig. C-8. Installation of a tuned filter across the antenna terminals of a TV set.

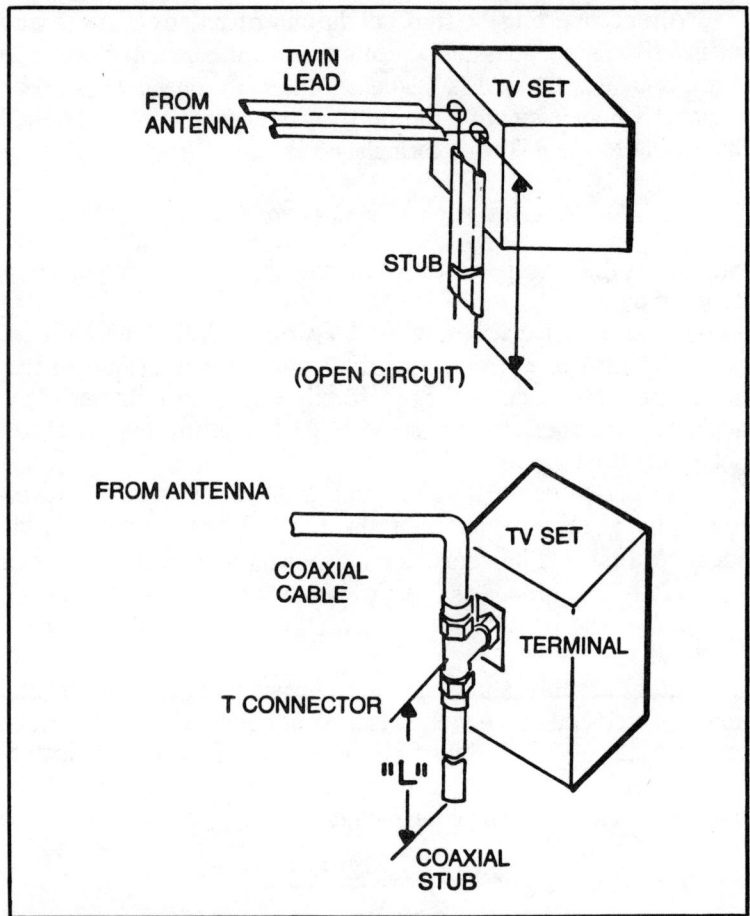

Fig. C-9. Installation of an open circuit, quarter-wave, tuned stub for both twin-lead and coaxial cable systems.

Resolving Electrical Interference

Please read through the procedures outlined in the previous sections before proceeding. If the steps in the previous sections have been taken, you should now know the source of the interference.

Before proceeding with the following steps to modify the device located as the source of interference, you should check the local electrical codes to determine if the device may be modified, and whether a licensed electrician must modify the device.

All bypassing of devices with capacitors should be done with extreme care to insure that the capacitors do not short out the AC line. Dangerous voltages exist which can cause electrocution if mishandled. Also, avoid power wiring which can cause the full AC line voltage to appear on the case of the device.

Since interference from an electric drill or saw may be of short duration, we suggest no modifications be made to the device—mainly because it may be very difficult and time-consuming to modify the device. If, however, interference is of long duration, and you wish to take on this task, proceed as follows.

Interference from a drill or saw is actually caused by arcing between the brushes and commutator. The interference then is transmitted through the power core. Bypassing each side of the line to ground with a capacitor, and each side to the other may be helpful. Also bypass the switch. Figure C-10 shows the schematic involved. The bypassing should be internal to the device in question.

Electric blankets, fish tank heaters, and other thermostatically controlled appliances with worn and pitted contacts cause interference because of contact arcing of the breaker points. This can be eliminated by bypassing the contacts with a 0.001 μF capacitor or replacing the worn or pitted contacts (see Fig. C-11).

Defective devices such as doorbell transformers should be replaced. Dimmer switches that utilize an SCR or triac can produce tremendous interference, and it is very difficult to eliminate. This is

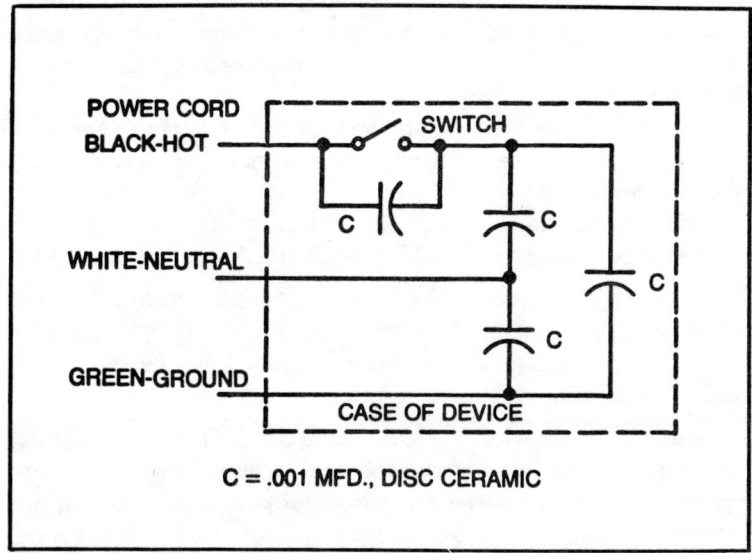

Fig. C-10. Capacitor bypassing to remove interference from arcing.

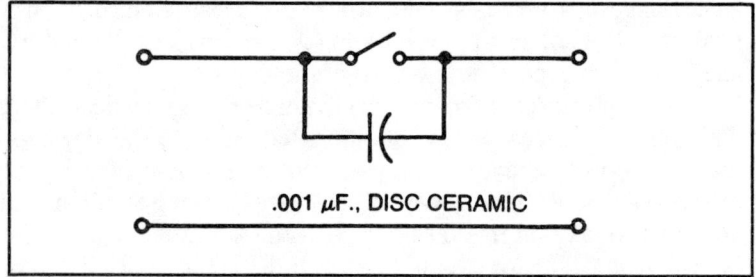

Fig. C-11. Capacitor bypassing to remove interference from thermostatically controlled appliances.

due to the approximate square wave output that is produced by the switching at the SCR or triac. However, bypassing in a manner shown in Fib. C-12 may be helpful.

Since resolving electrical interference has to proceed on a case-by-case basis, you should always consider adequately bypassing any component of the circuit that arcs or distorts the AC sine wave with ceramic condensers.

Resolving FM Interference

There are no set procedures for eliminating FM interference—it is a matter of eliminating the most likely sources of interference a step at a time. You may be required to take several steps before the interference problem is resolved. Once you have installed the filter called for, or made the adjustment that you were instructed to do, leave the modifications in place and proceed to the next step.

To begin, check to see if an FM band rejection filter has been installed on the TV set at the antenna terminals. If not, read the previous section.

If the installation of an FM band rejection filter is not effective, then a tuned stub trap should be constructed (see the example in Fig. C-13). The trap should be placed on and parallel to the lead-in and tuned for minimum interference. Then slide the trap along the line to further reduce interference. Finally, tape the trap to the lead-in in the most effective position.

Another type of stub, called an open circuit quarter-wave type, can be made from the same type of wire as the antenna lead-in wire (see Fig. C-9). The initial length of the stub should be 24" for RG-59/U coaxial cable or 29" for 300-ohm twin-lead wire. For other cables, the initial length can be determined by the general formula:

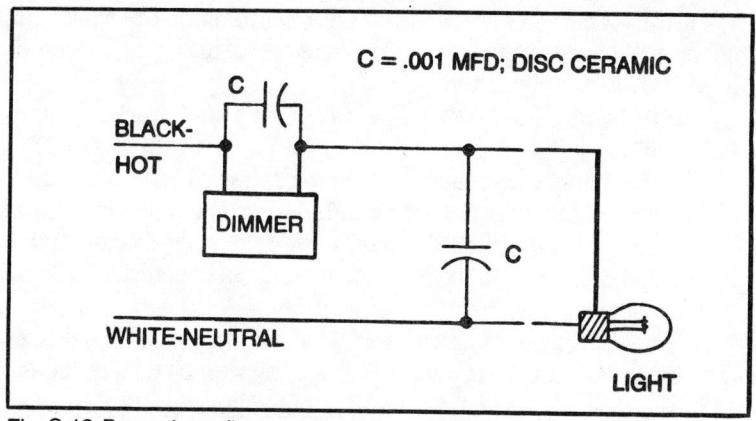

Fig. C-12. Bypassing a dimmer switch.

Length in inches = (35) (Velocity factor of line)

If F-type tee connectors are not available, use BNC-type connectors.

If connecting the stub to the antenna terminals is not completely effective, connect a second stub of the same length directly to the input terminals of the tuner, inside the television set. This should eliminate the interference.

Resolving Audio Interference

Audio interference is defined as reception of radio frequency (**rf**) energy by an audio amplifier. The **rf** energy is then rectified, or more properly detected, by an electron tube, transistor, diode, poor solder joint or ground or integrated circuit, The detected signal is then treated identically as a normal audio signal appearing at the amplifier input terminals. The effects of audio interference vary with the type of modulation employed by the transmitter. The following shows expected effects:

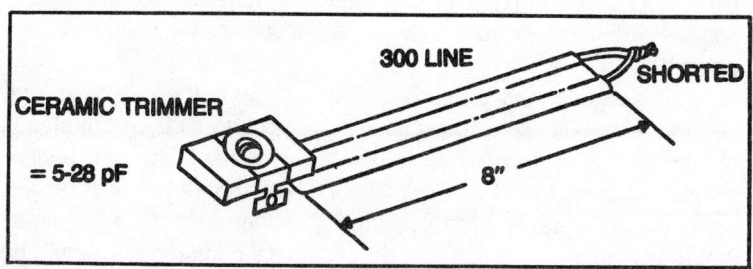

Fig. C-13. A tuned stub trap.

AM—The voice or music will be heard as any normal audio signal applied to the amplifier. The voice or music may be extremely loud and slightly distorted.

SSB (single sideband)—The voice will sound practically unintelligible and garbled.

FM—Usually no sound will be heard; however, a decrease in the volume of the amplifier will be noted when the radio transmitter is on. Clicks may be heard when a two-way radio transmitter is keyed and unkeyed. A "frying" noise (such as bacon sizzling) may also be heard.

TV—Audio rectification of a TV signal will sound like a buzz. The buzz will change its sound as the television picture changes.

In attempting to isolate where in the audio chain the rectification is taking place, check to determine if the volume control has any effect on the interference. If the volume of the interfering signal changes with a change in the volume control, then the rectification is occurring *before* the volume control. If the volume control has minimal or no effect, the rectification is occurring *after* the volume control. You should next proceed to the appropriate set of solutions. If the solutions described below do not resolve the audio interference problem, contact the manufacturer of the audio device for further assistance.

Rectification Before The Volume Control. A multiple input audio amplifier may be susceptible to audio interference on only one or some of the available inputs. Generally, low-level, high-impedance inputs, such as those in turntables, cartridges, tape heads, or microphones, are the most susceptible. If, for example, the only input affected is from a turntable, then disconnect the turntable cartridge from the amplifier at the input terminals of the amplifier.

If the interference is eliminated, then the cartridge, or wire between the cartridge and amplifier, is sensing the **rf**. Proper grounding, connections, shielding, and **rf** bypassing are the keys to solving audio rectification. Often, a "process of elimination" approach must be used.

Grounding. All grounding should be to a good earth ground such as a metallic cold water pipe or 8' ground rod. Ground leads should be as short as possible. Remember, a DC ground may appear as an open circuit to **rf** energy. Ground leads should be of as large a diameter wire as practicable. Finally, grounding of the chassis, shields of speaker leads, and other external connections should be made to a common point to avoid ground loops. (Ground loops are circuits that form a DC ground, but contain rf circulating currents.) Figure C-14 shows the correct and incorrect methods of grounding components.

Caution: Some equipment chassis are at line voltage potential and cannot be connected directly to ground. In these circumstances, a ceramic capacitor of 0.001 μF at 1kV should be placed in the ground lead. This capacitor appears as a short to rf, but an open circuit to AC.

Shielding. All speaker leads from audio equipment should be made of two conductor shielded wires. The shield should be grounded only at the amplifier end, and should not be used as an audio conductor. The two internal wires should be connected to the speaker.

Power Line Filter. RF may be entering the audio device through the AC power line. Several power line filters are commerically available. If necessary, a power line filter like the one shown in Fig. C-7 may be constructed, placing the filter as close as possible to the point where the AC cored enters the amplifier.

Poor Electrical Connections. Occasionally, poor solder connections or old electrolytic capacitors might be the cause of the audio rectification problem. If tests to this point have failed, try resoldering all connections in the amplifier and replacing electrolytic capacitors. Before actually replacing the electrolytic capacitor, try paralleling the capacitor with another one of like value. This should reveal the presence of a band capacitor.

Rectification After The Volume Control. When the volume control is in its minimum position, and the interference is still heard, then an rf filter is required in the audio amplifier. It is extremely important that the filter does not affect the audio response of the amplifier.

Tube Type Equipment. Interference in tube type equipment can be avoided by connecting an rf choke (ranging in value from 2 milihenry to 5 milihenry) in the upper end of the cathode circuit as shown in Fig. C-15.

The choke coil must *not* be bypassed by a capacitor because the DC resistance of such coil is generally quite low and the bias voltage is not greatly affected. However, if the DC resistance does affect the bias voltage, the value of the bias resistor may be decreased to compensate for the DC resistance of the choke.

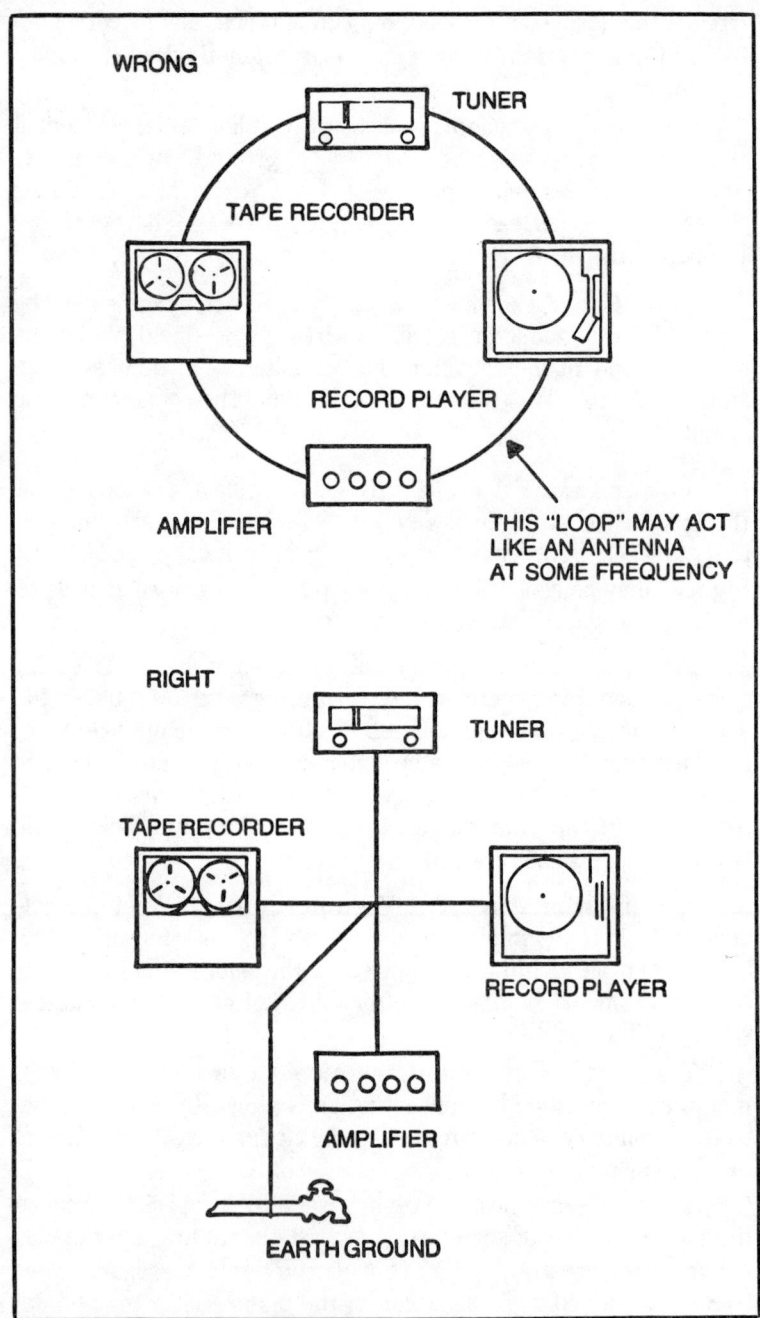

Fig. C-14. The right and the wrong way to ground components.

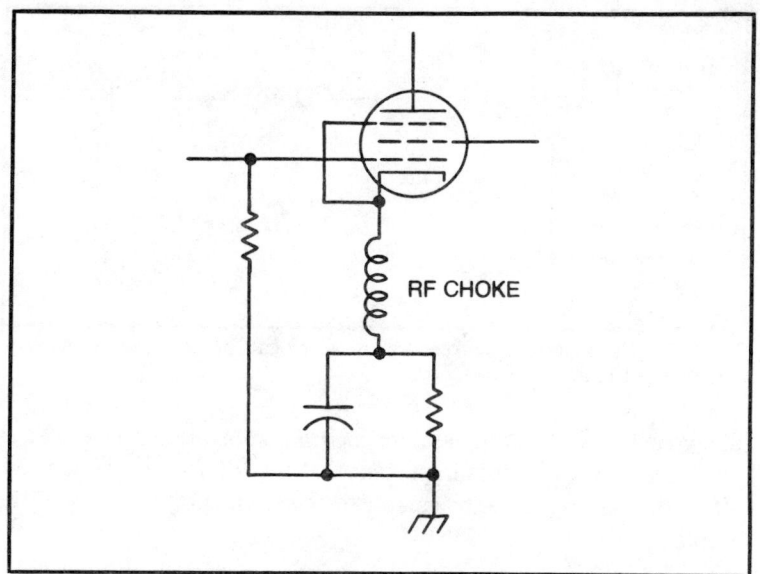

Fig. C-15. Removing interference in tube type equipment by connecting an rf choke.

A grid-stopping or swamping resistor can also be employed. A resistor, ranging in value from 1K to 75K ohms, can be connected in series with the grid as shown in Fig. C-16.

Capacitors, **rf** chokes, and resistors can be used in combination to make filters to eliminate the interference. For circuits such as those shown in Fig. C-17, use a choke of 2 to 6 microhenries and a capacitor of about 10 picofarads. A combination rf filter is shown in Fig. C-18 with the recommended values.

Transistor Equipment. Interference in transistor equipment can usually be eliminated with the use of a shunt capacitor as

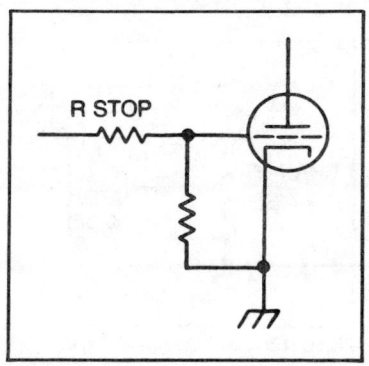

Fig. C-16. Removing interference with a grid-stopping resistor.

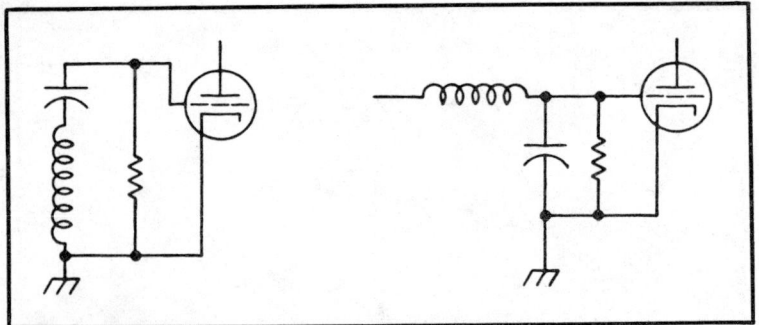

Fig. C-17. Capacitors, rf chokes, and resistors can be combined to make interference filters.

shown in Fig. C-19. A resistor/capacitor combination can be used as shown in Fig. C-20. It is important that the filter network does not affect the biasing of the transistor or the frequency response of the amplifier.

The values of the capacitors used are not critical, but there are some pitfalls to look out for in using capacitors. For example, ceramic caps are best, whereas paper caps do not work at radio frequencies.

Leads should be kept as short as possible. Grounds should be made directly to the emitter and not to the chassis or other grounds, since they may have more **rf** than the signal lead. If the signal increases, then a ground loop has been created, and the inductor method should be tried.

In areas of high **rf** energy, the inductor approach is more effective than the shunt capacitor. An **rf** choke can be used in series with the input and output leads of the amplifier stage since the **rf** can enter a stage through either. This method and the values are shown in Fig. C-21.

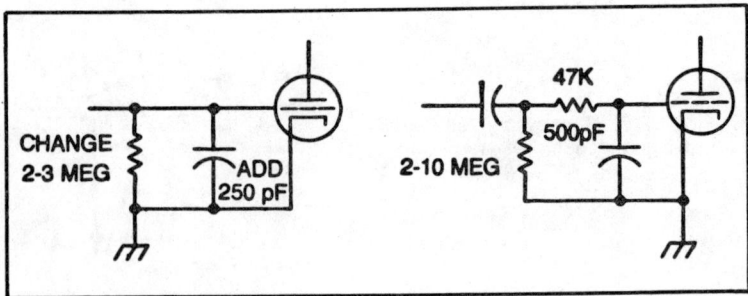

Fig. C-18. Combination of rf filters.

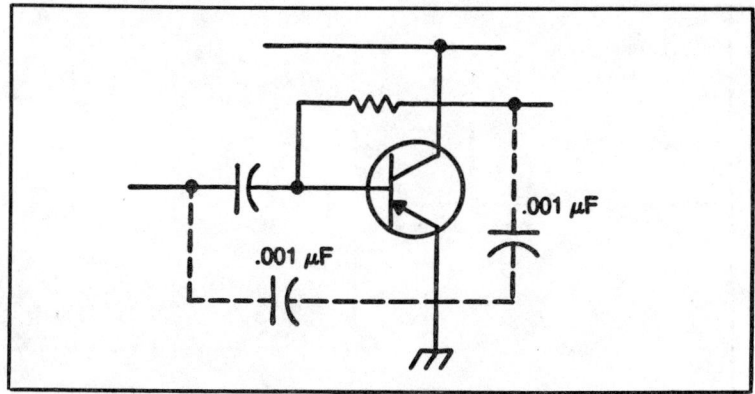

Fig. C-19. Using a shunt capacitor to remove interference in transistor equipment.

Electronic Organs. Organ circuits can be isolated by the use of the swell pedal, band box volume, or tabs (draw bars). By adjusting each one of these different controls, the effect on the interference can be noted. If the volume of the interference changes, the rf is being detected by the amplifier at a point *before* that particular control. If the volume of the interference does not change then the interference is being detected *after* that control. Using this method, the point at which the **rf** is entering the organ can be determined, and the appropriate filter, as described above, can be inserted into the circuit.

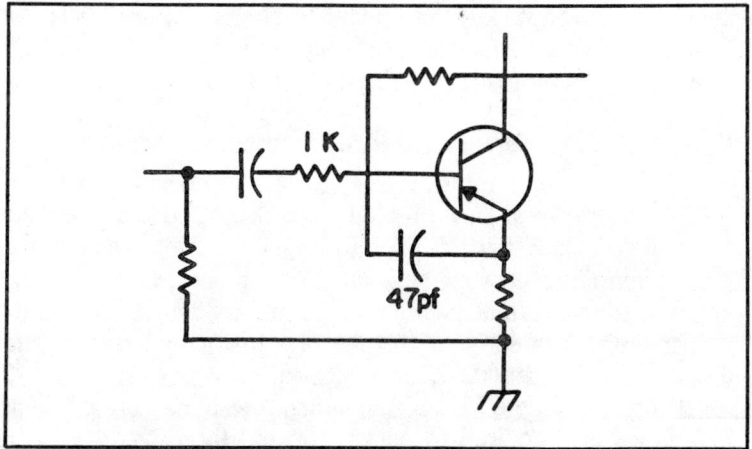

Fig. C-20. A resistor/capacitor combination for removing interference in transistor equipment.

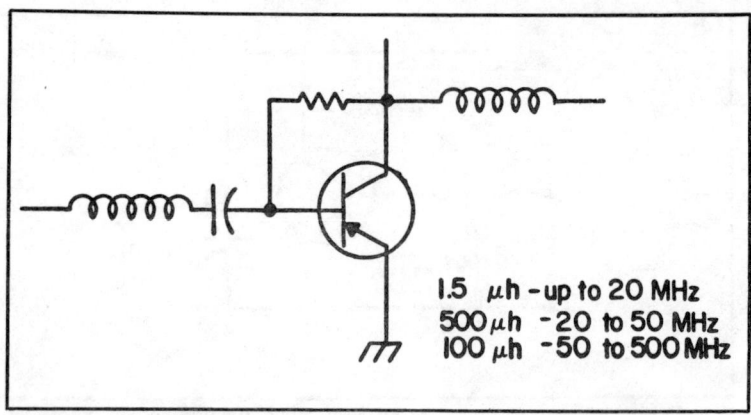

Fig. C-21. An rf choke can be used in series with the input and output leads of an amplifier stage.

Telephones. Telephone rf interference can be eliminated by the use of a 1542A or similar inductor. This inductor must be installed inside the phone and not at the baseboard. To install the inductor inside the phone, the corners of the plastic container will have to be removed. If the phone is too small for the inductor, such as the Princess telephone, then a pair of 2.5-mh chokes (75 mA or higher) must be installed inside the phone, one on each side of the line and as close to the 211A equalizing network as possible.

The information provided here applies primarily to privately owned equipment and should not be applied to equipment owned by the telephone company. Telephone company-owned equipment should be modified only by telephone company personnel. Bell system personnel can obtain additional data in Section 500-150-100 of the *Bell System Practices—Plant Series* manual.

Resolution of Interference For Radio Transmitter Operators

Although some interference problems can be attributed to television receivers, such problems can also be traced to CB and amateur radio transmitter. Therefore, upon receipt of an interference complaint from your neighbor(s), you should take all steps possible to insure that your radio transmitter is not causing the interference. Voluntary installation of a low-pass filter or other steps might eliminate the interference and prevent you from receiving an order from the FCC to implement these measures. You are not, however, required to service or add filtering to the complainant's television, and should not take any such action without the full cooperation of your neighbors.

You are cautioned that the use of an amateur transceiver on the Citizens Band is illegal. Further, the use of external **rf** power amplifiers with CB transceivers is illegal. Both actions may subject you to FCC actions or criminal penalties.

Generally, transmitter equipment that is commercially manufactured and type-accepted by the FCC has precautions built into the set to reduce harmonic radiation. Harmonics are radiations that are multiples of the operating frequency. However, you should follow the steps outlined below to insure that your radio equipment is operating properly.

1. If television interference is occurring, note which channels are affected.

Lower harmonics of CB generally affect TV channels 2, 5, 6, and 9. Therefore, if one or more of these channels are affected, your transmitter is probably radiating harmonics.

If all TV channels are affected, the problem is more likely to be in the TV receiver.

2. If the interference is caused by harmonics, a spectrum analyzer, a calibrated field intensity meter, or frequency selective voltmeter, can be used to accurately measure harmonic and spurious radiations from your transmitter. If any lead-in devices, such as standing-wave ration (SWR) meters are used, measurements should be made with the inline device both installed and removed. This may help identify the interference and lead you to the source. These are complex measurements and should normally be made only by experienced technicians.

3. If it appears that your transmitter is at fault, you should first make sure the chassis of the set is secured to the metal case of the radio by tightening the screws holding the chassis and case together. Then assure that the case of the transmitter is grounded to a good earth ground (metallic cold water pipe or 8-foot ground rod). Solid conductor wire of at least No. 10 gauge or copper ribbon should be used as a ground lead. The lead should be as short as possible.

4. By installing one or more low-pass filters in the transmitter antenna lead, you will reduce the chances of unnecessary harmonic radiation. A low-pass filter allows frequencies up to 30 or 50 MHz, depending on brand, to pass through unattenuated to the antenna while effectively shorting out harmonic radiation. To make this test, connect the equipment as shown in Fig. C-22 and take a power reading. If only a SWR bridge is available, calibrate it in the forward direction to the calibrate line in the meter. Then insert the low-pass

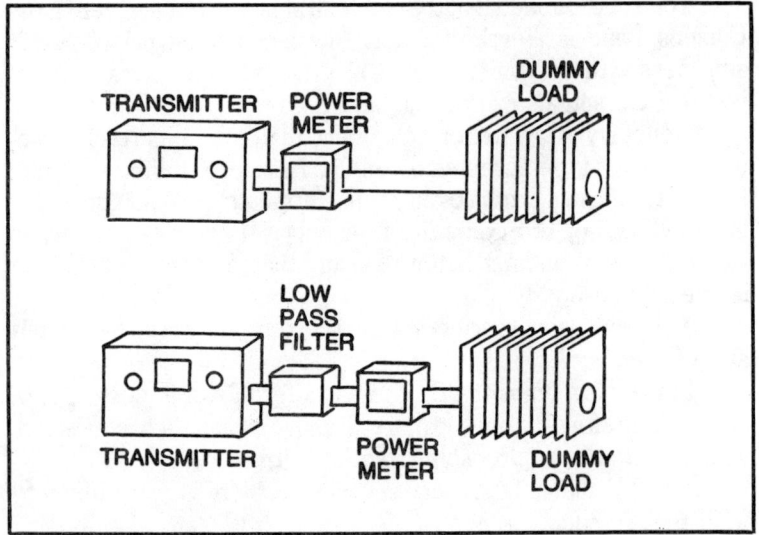

Fig. C-22. Determining if a transmitter is causing interference by using a low-pass filter.

filter and make another power measurement. Do not return the transmitter.

5. If you notice a decrease in output power on a power meter, operating to a properly matched load, with the low-pass filter installed, this is an indication that harmonic content may be present. Even though the meter reading may be lower with the filter installed, it does not mean that the transmitter absolutely has harmonic radiation. Slight detuning of the transmitter by the filter may cause a lower indication.

6. At amateur power levels, corroded metal connections in the area of the transmitting antenna may act like diodes and generate harmonics which may radiate. This type of problem can be found by vibrating suspected offenders such as galvanized downspouts, metal fences, clothes lines, etc., while viewing the affected television set. Sudden changes in the interference pattern which correspond to the vibration should be noted. This test requires on observer at the TV receiver, someone to "shake" suspicious metal objects in the area, and another person to key (but not modulate) the transmitter involved.

7. Finally, some transmitters may actually be radiating harmonic and spurious energy from their cabinet or through the power lines. Try operating the transmitter into a shielded dummy load. If the interference is still present, then cabinet or power line radiation is indicated. A power line filter should be installed. Several types are commercially available. For low power transmitters, the filter in Fig. C-6 may be used.

8. Continued interference with the power line filter installed points toward cabinet radiation. An earth ground should eliminate cabinet radiation.

9. Local television interference (TVI) committees dedicated to resolving CB-TVI problems are now being established. For assistance in locating a TVI Committee in your area, contact: International CB Radio Operator's Association (CBA), P.O. Box 1020, Roanoke, VA 24005.

Resolution Of Interference For Amateur Transmitter Operators

If you have a linear amplifier on your amateur transmitting equipment, use two low-pass filters. One filter should be installed between the actual transmitter (exciter) and the input to the linear amplifier. This prevents harmonics generated in the exciter from reaching the linear amplifier. The second filter should be installed at the output of the linear amplifier to reduce harmonic and spurious contents.

One unique interference problem to TV channel 2 is from an amateur transmitter operating on the 6-meter band. This is due to the close proximity of the frequencies involved. You are not required to service or add filtering to the complainant's television, and should not take any such action without the full cooperation of your neighbor.

Local television interference (TVI) committees are available to assist you in resolving interference problems. Contact the nearest FCC district office (see the Appendix) or the American Radio Relay League, Newington, CT, for assistance in locating a TVI committee in your area.

Transmitter Operator Guidelines For Resolving Audio Interference

Although audio interference (often called audio rectification) is usually resolved by modification of the affected device, you as a radio operator can take certain steps to reduce the possibility of audio rectification by eliminating circulating radio frequency (rf) currents in ground and metal objects in the area. Your radio transmitting equipment should be effectively grounded to a metallic cold water pipe or a ground rod driven into the ground at least 8 feet. The ground lead must be at least No. 10 wire or copper ribbon. The greater the surface area of the ground lead, the more effective it will be. Also, the ground lead should be as short as possible.

You are licensed to use only the amount of power necessary to establish communications. Operating with excessive power is likely to cause audio interference problems.

If you need assistance in performing the above modifications to your equipment, you can contact the dealer or manufacturer representatives. Also, an FCC-licensed service representative may be able to assist you.

Appendix

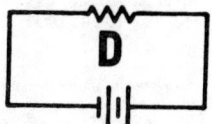

Additional Reading

No.	973	*How To Be A Ham*
No.	989	*Ham Radio Incentive Licensing Guide—2nd Edition*
No.	1073	*Modern Amateur Radio License Study Guide for Novice, Technician & General Class*
No.	1182	*The Complete Handbook of Radio Receivers*
No.	1194	*How To Troubleshoot & Repair Amateur Radio Equipment*
No.	1224	*The Complete Handbook of Radio Transmitters*
No.	1289	*Amateur Radio License Study Guide . . . for all classes*

Appendix

Part 97 FCC Rules and Regulations

Subpart A—General

Sec.
97.1 Basis and purpose.
97.3 Definitions.

Subpart B—Amateur Operator and Station Licenses

OPERATOR LICENSES

97.5 Classes of operator licenses.
97.7 Privileges of operator licenses.
97.9 Eligibility for new operator license.
97.11 Application for operator license.
97.13 Renewal or modification of operator license.

OPERATOR LICENSE EXAMINATIONS

97.19 When examination is required.
97.21 Examination elements.
97.23 Examination requirements.
97.25 Examination credit.
97.27 Mail examinations for applicants unable to travel.
97.28 Manner of conducting examinations.
97.31 Grading of examinations.
97.32 Interim Amateur Permits.
97.33 Eligibility for re-examination.

STATION LICENSES

97.37 General eligibility for station license.
97.39 Eligibility of corporations or organizations to hold station license.
97.40 Station license required.

97.41 Application for station license.
97.42 Mailing address furnished by licensee.
97.43 Location of station.
97.45 Limitations on antenna structures.
97.47 Renewal and/or modification of amateur station license.
97.49 Commission modification of station license.

CALL SIGNS

97.51 Assignment of call signs.

DUPLICATE LICENSES AND LICENSE TERM

97.57 Duplicate license.
97.59 License term.

Subpart C—Technical Standards

97.61 Authorized frequencies and emissions.
97.63 Selection and use of frequencies.
97.65 Emission limitations.
97.67 Maximum authorized power.
97.69 Radio teleprinter transmissions.
97.71 Transmitter power supply.
97.73 Purity of emissions.
97.74 Frequency measurement and regular check.
97.75 Use of external radio frequency (RF) power amolifiers.
97.76 Requirements for type acceptance of external radio frequency (RF) power amplifiers and external radio frequency power amplifier kits.
97.77 Standards for type-acceptance of external radio frequency (RF) power amplifiers and external radio frequency power amplifier kits.

Subpart D—Operating Requirements and Procedures

GENERAL

- 97.78 Practice to be observed by all licensees.
- 97.79 Control operator requirements.
- 97.81 Authorized apparatus.
- 97.82 Availability of operator license.
- 97.83 Availability of station license.
- 97.84 Station identification.
- 97.85 Repeater operation.
- 97.86 Auxiliary operation.
- 97.88 Operation of a remotely controlled station.
- 97.89 Points of communications.
- 97.91 One-way communications.
- 97.93 Modulation of carrier.

STATION OPERATION AWAY FROM AUTHORIZED LOCATION

- 97.95 Operation away from the authorized fixed station location.

SPECIAL PROVISIONS

- 97.99 Stations used only for radio control of remote model crafts and vehicles.
- 97.101 Mobile stations aboard ships or aircraft.

LOGS

- 97.103 Station log requirements.
- 97.105 Retention of logs.

EMERGENCY OPERATIONS

- 97.107 Operation in emergencies.

Subpart E—Prohibited Practices and Administrative Sanctions

PROHIBITED TRANSMISSIONS AND PRACTICES

- 97.112 No remuneration for use of station.
- 97.113 Broadcasting prohibited.
- 97.114 Third party traffic.
- 97.115 Music prohibited.
- 97.116 Amateur radiocommunication for unlawful purposes prohibited.
- 97.117 Codes and ciphers prohibited.
- 97.119 Obscenity, indecency, profanity.
- 97.121 False signals.
- 97.123 Unidentified communications.
- 97.125 Interference.
- 97.126 Retransmitting radio signals.
- 97.127 Damage to apparatus.
- 97.129 Fraudulent licenses.

ADMINISTRATIVE SANCTIONS

- 97.131 Restricted operation.
- 97.133 Second notice of same violation.
- 97.135 Third notice of same violation.
- 97.137 Answers to notices of violations.

Subpart F—Radio Amateur Civil Emergency Service (RACES)

GENERAL

97.161 Basis and purpose.
97.163 Definitions.
97.165 Applicability of rules.

STATION AUTHORIZATIONS

97.169 Station license required.
97.171 Eligibility for RACES station license.
97.173 Application for RACES station license.
97.175 Amateur radio station registration in civil defense organization.

OPERATING REQUIREMENTS

97.177 Operator requirements.
97.179 Operator privileges.
97.181 Availability of RACES station license and operator licenses

TECHNICAL REQUIREMENTS

97.185 Frequencies available.
97.189 Points of communications.
97.191 Permissible communications.
97.193 Limitations on the use of RACES stations.

Subpart G—Operation of Amateur Radio Stations in the United States by Aliens Pursuant to Reciprocal Agreements

97.301 Basis, purpose, and scope.
97.303 Permit required.
97.305 Application for permit.
97.307 Issuance of permit.
97.309 Modification, suspension, or cancellation of permit.
97.311 Operating conditions.
97.313 Station identification.

Subpart H—(Reserved)

APPENDICES

1 Examination points.
2 Extracts from Radio Regulations Annexed to the International Telecommunication Convention (Geneva, 1959).
3 Classification of emissions.
4 Convention between the United States of America and Canada, Relating to the Operation by Citizens of Either Country of Certain Radio Equipment or Stations in the Other Country (Effective May 15, 1952).
5 Determination of Antenna Height above Average Terrain.

AUTHORITY: §§ 97.1 to 97.313 issued under 48 Stat. 1066, 1082, as amended; 47 U.S.C. 154, 303. Interpret or apply 48 Stat. 1064-1068, 1081-1105, as amended; 47 U.S.C. Sub-chap. I, III-VI.

443

SUBPART A—GENERAL

§ 97.1 Basis and purpose.

The rules and regulations in this part are designed to provide an amateur radio service having a fundamental purpose as expressed in the following principles:

(a) Recognition and enhancement of the value of the amateur service to the public as a voluntary noncommercial communication service, particularly with respect to providing emergency communications.

(b) Continuation and extension of the amateur's proven ability to contribute to the advancement of the radio art.

(c) Encouragement and improvement of the amateur radio service through rules which provide for advancing skills in both the communication and technical phases of the art.

(d) Expansion of the existing reservoir within the amateur radio service of trained operators, technicians, and electronics experts.

(e) Continuation and extension of the amateur's unique ability to enhance international good will.

§ 97.3 Definitions.

(a) *Amateur radio service.* A radio communication service of self-training, intercommunication, and technical investigation carried on by amateur radio operators.

(b) *Amateur radio communication.* Noncommercial radio communication by or among amateur radio stations solely with a personal aim and without pecuniary or business interest.

(c) *Amateur radio operator* means a person holding a valid license to operate an amateur radio station issued by the Federal Communications Commission.

(d) *Amateur radio license.* The instrument of authorization issued by the Federal Communications Commission comprised of a station license, and in the case of the primary station, also incorporating an operator license.

Operator license. The instrument of authorization including the class of operator privileges.

Interim Amateur Permit. A temporary operator and station authorization issued to licensees successfully completing Commission supervised examinations for higher class operator licenses.

Station license. The instrument of authorization for a radio station in the Amateur Radio Service.

(e) *Amateur radio station.* A station licensed in the amateur radio service embracing necessary apparatus at a particular location used for amateur radio communication.

(f) *Primary station.* The principal amateur radio station at a specific land location shown on the station license.

(g) *Military recreation station.* An amateur radio station licensed to the person in charge of a station at a land location provided for the recreational use of amateur radio operators, under military auspices of the Armed Forces of the United States.

(h) *Club station.* A separate Amateur radio station licensed to an Amateur radio operator acting as a station trustee for a *bona fide* amateur radio organization or society. A *bona fide* Amateur radio organization or society shall be composed of at least two persons, one of whom must be a licensed Amateur operator, and shall have:

(1) A name,

(2) An instrument of organization (e.g., constitution),

(3) Management, and

(4) A primary purpose which is devoted to Amateur radio activities consistent with § 97.1 and constituting the major portion of the club's activities.

(i) *Space radio station.* An amateur radio station located on an object which is beyond, is intended to go beyond, or has been beyond the major portion of the earth's atmosphere. (Regulations governing this type of station have not yet been adopted and all applications will be considered on an individual basis.)

(j) *Terrestrial location.* Any point within the major portion of the earth's atmosphere, including aeronautical, land, and maritime locations.

(k) *Space location.* (Reserved)

(l) *Amateur radio operation.* Amateur radio communication conducted by amateur radio operators from amateur radio stations, including the following:

Fixed operation. Radio communication conducted from the specific geographical land location shown on the station license.

Portable operation. Radio communication conducted from a specific geographical location other than that shown on the station license.

Mobile operation. Radio communication conducted while in motion or during halts at unspecified locations.

Repeater operation. Radiocommunication, other than auxiliary operation, for retransmitting automatically the radio signals of other amateur radio stations.

Auxiliary operation. Radiocommunication for remotely controlling other amateur radio stations, for automatically relaying the radio signals of other amateur radio stations in a system of stations, or for intercommunicating with other amateur radio stations in a system of amateur radio stations.

(m) *Control* means techniques used for accomplishing the immediate operation of an amateur radio station. Control includes one or more of the following:

(1) *Local control.* Manual control, with the control operator monitoring the operation on duty at the control point located at a station transmitter with the associated operating adjustments directly accessible. (Direct mechanical control, or direct wire control of a transmitter from a control point located on board any aircraft, vessel, or on the same premises on which the transmitter is located, is also considered local control.)

(2) *Remote control.* Manual control, with the control operator monitoring the operation on duty at a control point located elsewhere than at the station transmitter, such that the associated operating adjustments are accessible through a control link.

(3) *Automatic control* means the use of devices and procedures for control so that a control operator does not have to be present at the control point at all times. (Only rules for automatic control of stations in repeater operation have been adopted.)

(n) *Control link.* Apparatus for effecting remote control between a control point and a remotely controlled station.

(o) *Control operator.* An amateur radio operator designated by the licensee of an amateur radio station to also be responsible for the emissions from that station.

(p) *Control point.* The operating position of an amateur radio station where the control operator function is performed.

(q) *Antenna structures.* Antenna structures include the radiating system, its supporting structures, and any appurtenances mounted thereon.

(r) *Antenna height above average terrain.* The height of the center of radiation of an antenna above an averaged value of the elevation above sea level for the surrounding terrain.

(s) *Transmitter.* Apparatus for converting electrical energy received from a source into radio-frequency electromagnetic energy capable of being radiated.

(t) *Effective radiated power.* The product of the radio-frequency power, expressed in watts, delivered to an antenna, and the relative gain of the antenna over that of a half-wave dipole antenna.

(u) *System network diagram.* A diagram showing each station and its relationship to the other stations in a network of stations, and to the control point(s).

(v) *Third-party traffic.* Amateur radio communication by or under the supervision of the control operator at an amateur radio station to another amateur radio station on behalf of anyone other than the control operator.

(w) *Emergency communication.* Any amateur radio communication directly relating to the immediate safety of life of individuals or the immediate protection of property.

(x) *Automatic retransmission.* Retransmission of signals by an amateur radio station whereby the retransmitting station is actuated solely by the presence of a received signal through electrical or electro-mechanical means, i.e., without any direct, positive action by the control operator.

(y) *External radio frequency power amplifier.* Any device which, (1) when used in conjunction with a radio transmitter as a signal source, is capable of amplification of that signal, and (2) is not an integral part of the transmitter as manufactured.

(z) *External radio frequency power amplifier kit.* Any number of electronic parts, usually provided with a schematic diagram or printed circuit board, which, when assembled in accordance with instructions, results in an external radio frequency power amplifier, even if additional parts of any type are required to complete assembly.

SUBPART B—AMATEUR OPERATOR AND STATION LICENSES

OPERATOR LICENSES

§ 97.5 Classes of operator licenses.

Amateur extra class.
Advanced class (previously class A).
General class (previously class B).
Conditional class (previously class C).
Technician class.
Novice class.

§ 97.7 Privileges of operator licenses.

(a) *Amateur Extra Class and Advanced Class.* All authorized amateur privileges including exclusive frequency operating authority in accordance with the following table:

Anyone except a representative of a foreign government is eligible for an amateur operator license.

§ 97.11 Application for operator license.

(a) An application (FCC Form 610) for a new operator license, including an application for change in operating privileges, which will require an examination supervised by Commission personnel at a regular Commission examining office shall be submitted to such office in advance of or at the time of the examination, except that, whenever an examination is to be taken at a designated examination point away from a Commission office, the application, together with the necessary filing fee should be submitted in advance of the examination date to the office which has jurisdiction over the examination point involved.

(b) An application (FCC Form 610) for a new operator license, including an application for change in operating privileges, which requests an examination supervised by a volunteer examiner under the provisions of § 97.27, shall be submitted to the FCC field office nearest the applicant. Applications for the Novice Class license should be sent to the Commission's offices in Gettysburg, Pa. 17325. All applications should be accompanied by any necessary filing fee.

(c) An application (FCC Form 610) for renewal and/or modification of license when no change in operating privileges is involved shall be submitted, together with any necessary filing fee, to the Commission's office at Gettysburg, Pennsylvania, 17325.

Frequencies	Class of license authorized
3500-3525 kHz	
3775-3800 kHz	Amateur Extra Only
7000-7025 kHz	
14,000-14,025 kHz	
21,000-21,025 kHz	
21,250-21,270 kHz	
3800-3890 kHz	
7150-7225 kHz	Amateur Extra and Advanced.
14,200-14,275 kHz	
21,270-21,350 kHz	

(b) *General Class.* All authorized amateur privileges except those exclusive operating privileges which are reserved to the Advanced Class and/or Amateur Extra Class.

(c) *Conditional Class.* Same privileges as General Class. New Conditional Class licensees will not be issued. Present Conditional Class licensees will be issued General Class licenses at time of renewal or modification.

(d) *Technician Class.* All authorized amateur privileges on the frequencies 50.0 MHz and above. Technician Class licenses also convey the full privileges of Novice Class licenses.

(e) *Novice Class.* Radiotelegraphy in the frequency bands 3700-3750 kHz, 7100-7150 kHz (7050-7075 kHz when the terrestrial station location is not within Region 2), 21,100-21,200 kHz, and 28,100-28,200 kHz, using only Type A1 emission.

§ 97.9 Eligibility for new operator license.

§ 97.13 Renewal or modification of operator license.

(a) An Amateur operator license may be renewed upon proper application.

(b) The applicant shall qualify for a new license by examination if the requirements of this section are not fulfilled.

(c) Application for renewal and/or modification of an amateur operator license shall be submitted on FCC Form 610 and shall be accompanied by the applicant's license. Application for renewal of unexpired licenses must be made during the license term and should be filed within 90 days but not later than 30 days prior to the end of the license term. In any case in which the licensee has, in accordance with the provisions of this chapter, made timely and sufficient application for renewal of an unexpired license, no license with reference to any activity of a continuing nature shall expire until such application shall have been finally determined.

(d) If a license is allowed to expire, application for renewal may be made during a period of grace of one year after the expiration date. During this one year period of grace, an expired license is not valid. A license renewed during the grace period will be dated currently and will not be backdated to the date of its expiration. Application for renewal shall be submitted on FCC Form 610 and shall be accompanied by the applicant's expired license.

(e) When the name of a licensee is changed or when the mailing address is changed a formal application for modification of license is not required. However, the licensee shall notify the Commission promptly of these changes. The notice, which may be in letter form, shall contain the name and address of the licensee as they appear in the Commission's records, the new name and/or address, as the case may be, the radio station call sign and class of operator license. The notice shall be sent to Federal Communications Commission, Gettysburg, Pa. 17325 and a copy shall be kept by the licensee until a new license is issued.

OPERATOR LICENSE EXAMINATIONS

§ 97.19 When examination is required.

Examination is required for the issuance of a new amateur operator license, and for a change in class of operating privileges. Credit may be given, however, for certain elements of examination as provided in § 97.25.

§ 97.21 Examination elements.

Examinations for amateur operator privileges will comprise one or more of the following examination elements:

(a) Element 1(A): Beginner's code test at five (5) words per minute;

(b) Element 1(B): General code test at thirteen (13) words per minute;

(c) Element 1(C): Expert's code test at twenty (20) words per minute;

(d) Element 2: Basic law comprising rules and regulations essential to beginners' operation, including sufficient elementary radio theory for the understanding of those rules;

(e) Element 3: General amateur practice and regulations involving radio operation and apparatus and provisions of treaties, statutes, and rules affecting amateur stations and operators;

(f) Element 4(A): Intermediate amateur practice involving intermediate level radio theory and operation as applicable to modern amateur techniques, including, but not limited to, radiotelephony and radiotelegraphy;

(g) Element 4(B): Advanced amateur practice involving advanced radio theory and operation as applicable to modern amateur techniques, including, but not limited to, radiotelephony, radiotelegraphy, and transmissions of energy for measurements and observations applied to propagation, for the radio control of remote objects and for similar experimental purposes.

§ 97.23 Examination requirements.

Applicants for operator licenses will be required to pass the following examination elements:

(a) Amateur Extra Class: Elements 1(C), 2, 3, 4(A) and 4(B);

(b) Advanced Class: Elements 1(B), 2, 3, and 4(A);

(c) General Class: Elements 1(B), 2 and 3;

(d) Technician Class: Elements 1(A), 2, and 3;

(e) Novice Class: Elements 1(A) and 2.

§ 97.25 Examination credit.

(a) An applicant for a higher class of amateur operator license who holds any valid amateur license will be required to pass only those elements of the higher class examination that are not included in the examination for the amateur license held.

(b) Upon presentation of a properly completed Amateur Code Credit Certificate, FCC Form 845, the FCC shall give the applicant for an amateur radio operator license examination credit for the code speed listed on the Amateur Code Credit Certificate. An Amateur Code Credit Certificate is valid for a period of one year from the date of its issuance and will be honored only at the FCC field office that issued the Amateur Code Credit Certificate.

(c) An applicant for an amateur operator license will be given credit for either telegraph code element 1(A) or 1(B) if within 5 years prior to the receipt of his application by the Commission he held a commercial radiotelegraph operator license or permit issued by the Federal Communications Commission. An applicant for an amateur extra class license will be given credit for the telegraph code element 1(C) if he holds a valid first class commercial radiotelegraph operator license or permit issued by the Federal Communications Commission or holds any commercial radiotelegraph operator license or permit issued by the Federal Communications Commission containing an aircraft radiotelegraph endorsement.

(d) An applicant for the amateur extra class operator license will be given credit for examination element 1(C) if he so requests and submits evidence of having held the amateur extra first class license, having continuously held its successor license. An applicant should present his proof in advance of the desired examination time to the Chief, Personal Radio Division, Washington, D.C. 20554 and receive a letter of certification for presentation to the field office where the examination will be taken. No code credit will be given without the letter of certification.

(e) No examination credit, except as herein provided, shall be allowed on the basis of holding or having held any amateur or commercial operator license.

§ 97.27 Mail examinations for applicants unable to travel.

The Commission may permit the examinations for an Amateur Extra, Advanced, General, or Technician Class license to be administered at a location other than a Commission examination point by an examiner chosen by the Commission when it is shown by physician's certification that the applicant is unable to appear at a regular Commission examination point because of a protracted disability preventing travel.

§ 97.28 Manner of conducting examinations.

(a) Except as provided in §97.27, all examinations for Amateur Extra, Advanced, General, and Technician Class operator licenses will be conducted by authorized Commission personnel or representatives at locations and times specified by the Commission. Examination elements given under the provisions of §97.27 will be administered by an examiner selected by the Commission. All applications for consideration of eligibility under §97.27 should be filed on FCC Form 610, and should be sent to the FCC field office nearest the applicant. (A list of these offices appears in §0.121 of the Commission's Rules and can be obtained from the Regional Services Division, Field Operations Bureau, FCC, Washington, D.C. 20554, or any field office.)

(b) The examination for a Novice Class operator license shall be conducted and supervised by a volunteer examiner selected by the applicant, unless otherwise prescribed by the Commission. The volunteer examiner shall be at least 18 years of age, shall be unrelated to the applicant, and shall be the holder of an Amateur Extra, Advanced, or General Class operator license. The written portion of the Novice Class operator examination shall be obtained, administered, and submitted in accordance with the following procedure:

(1) Within 10 days after successfully completing telegraphy examination element 1(A), an applicant shall submit an application (FCC Form 610) to the Commission's office in Gettysburg, Pennsylvania 17325. The application shall include a written request from the volunteer examiner for the examination papers for Element 2. The examiner's written request shall include (i) the names and permanent addresses of the examiner and the applicant, (ii) a description of the examiner's qualifications to administer the examination, (iii) the examiner's statement that the applicant has passed telegraphy element 1(A) under his supervision within the 10 days prior to submission of the request, and (iv) the examiner's written signature. Examination papers will be forwarded only to the volunteer examiner.

(2) The volunteer examiner shall be responsible for the proper conduct and necessary supervision of the examination. Administration of the examination shall be in accordance with the instructions included with the examination papers.

(3) The examination papers, either completed or unopened in the event the examination is not taken, shall be returned by the volunteer examiner to the Commission's office in Gettysburg, Pa., no later than 30 days after the date the papers are mailed by the Commission (the date of mailing is normally stamped by the Commission on the outside of the examination envelope).

(c) The code test required of an applicant for an amateur radio operator license, in accordance with the provisions of §§97.21 and 97.23 shall determine the applicant's ability to transmit by hand key (straight key or, if supplied by the applicant, any other type of hand operated key such as a semiautomatic or electronic key, but not a keyboard keyer) and to receive by ear, in plain language, messages in the international Morse code at not less than the prescribed speed during a five minute test period. Each five characters shall be counted as one word. Each punctuation mark and numeral shall be counted as two characters.

(d) All written portions of the examinations for amateur operator privileges shall be completed by the applicant in legible handwriting or hand printing. Whenever the applicant's signature is required, his normal signature shall be used. Applicants unable to comply with these requirements, because of physical disability, may dictate their answers to the examination questions and the receiving code test. If the examination or any part thereof is dictated, the examiner shall certify the nature of the applicant's disability and the name and address of the person(s) taking and transcribing the applicant's dictation.

§ 97.31 Grading of examinations.

(a) Code tests for sending and receiving are graded separately.

(b) Seventy-four percent (74%) is the passing grade for written examinations. For the purpose of grading, each element required in qualifying for a particular license will be considered as a separate examination. All written examinations will be graded only by Commission personnel.

§ 97.32 Interim Amateur Permits.

(a) Upon successful completion of a Commission supervised Amateur Radio Service operator examination, an applicant already licensed in the Amateur Radio Service may operate his amateur radio station pending issuance of his permanent amateur operator and station licenses under the terms and conditions of an Interim Amateur Permit, evidenced by a properly executed FCC Form 660-B.

(b) An Interim Amateur Permit conveys all operating privileges of the applicant's new operator license classification.

(c) The transmissions of amateur radio stations operated under the authority of Interim Amateur Permits shall be identified in the manner specified in §97.84.

(d) The original Interim Amateur Permit of an amateur radio operator shall be kept in the personal possession of or posted in a conspicuous place in the room occupied by such operator when operating an amateur radio station under the authority of an Interim Amateur Permit.

(e) Interim Amateur Permits are valid for a period of 90 days from the date of issuance or until issuance of the permanent station and operator licenses, whichever comes first, but may be set aside by the Commission within the 90 day term if it appears that the permanent operator and station licenses cannot be granted routinely.

(f) Interim Amateur Permits shall not be renewed.

§ 97.33 Eligibility for re-examination.

An applicant who fails an examination element required for an amateur radio operator license shall not apply to be examined for the same or higher examination element within thirty days of the date the examination element was failed.

STATION LICENSES

§ 97.37 General eligibility for station license.

An amateur radio station license will be issued only to a licensed amateur radio operator, except that a military recreation station license may also be issued to an individual not licensed as an amateur radio operator (other than a representative of a foreign government), who is in charge of a proposed military recreation station not operated by the U.S. Government but which is to be located in approved public quarters.

§ 97.39 Eligibility of corporations or organizations to hold station license.

An amateur station license will not be issued to a school, company, corporation, association, or other organization, except that in the case of a *bona fide* amateur radio organization or society meeting the criteria set forth in Section 97.3, a station license may be issued to a licensed amateur operator, other than the holder of a Novice Class license, as trustee for such society.

§ 97.40 Station license required.

(a) No transmitting station shall be operated in the amateur radio service without being licensed by the Federal Communications Commission.

(b) Every amateur radio operator shall have one, but only one, primary amateur radio station license.

§ 97.41 Application for station license.

(a) Each application for a club or military recreation station license in the Amateur Radio Service shall be made on the FCC Form 610-B. Each application for any other amateur radio license shall be made on the FCC Form 610.

(b) One application and all papers incorporated therein and made a part thereof shall be submitted for each amateur station license. If the application is only for a station license, it shall be filed directly with the Commission's Gettysburg, Pennsylvania office. If the application also contains an application for any class of amateur operator license, it shall be filed in accordance with the provisions of §97.11.

(c) Each applicant in the Safety and Special Radio Services (1) for modification of a station license involving a site change or a substantial increase in tower height or (2) for a license for a new station must, before commencing construction, supply the environmental information, where required, and must follow the procedure prescribed by Subpart 1 of Part 1 of this chapter (§§ 1.1301 through 1.1319) unless Commission action authorizing such construction would be a minor action with the meaning of Subpart 1 of Part I.

§ 97.42 Mailing address furnished by licensee.

Except for applications submitted by Canadian citizens pursuant to agreement between the United States and Canada (TIAS No. 2508 and No. 6931), each application shall set forth and each licensee shall furnish the Commission with an address in the United States to be used by the Commission in serving documents or directing correspondence to that licensee. Unless any licensee advises the Commission to the contrary, the address contained in the licensee's most recent application will be used by the Commission for this purpose.

§ 97.43 Location of station.

Every amateur radio station shall have one land location, the address of which appears on the station license, and at least one control point.

§ 97.45 Limitations on antenna structures.

(a) Except as provided in paragraph (b) of this section, an antenna for a station in the Amateur Radio Service which exceeds the following height limitations may not be erected or used unless notice has been filed with both the FAA on FAA Form 7460-1 and with the Commission on Form 714 or on the license application form, and prior approval by the Commission has been obtained for:

(1) Any construction or alteration of more than 200 feet in height above ground level at its site (§17.7(a) of this chapter).

(2) Any construction or alteration of greater height than an imaginary surface extending outward and upward at one of the following slopes (§17.7(b) of this chapter):

(i) 100 to 1 for a horizontal distance of 20,000 feet from the nearest point of the nearest runway of each airport with at least one runway more than 3,200 feet in length, excluding heliports and seaplane bases without specified boundaries, if that airport is either listed in the Airport Directory of the current Airman's Information Manual or is operated by a Federal military agency.

(ii) 50 to 1 for a horizontal distance of 10,000 feet from the nearest point of the nearest runway of each airport with its longest runway no more than 3,200 feet in length, excluding heliports and seaplane bases without specified boundaries, if that airport is either listed in the Airport Directory or is operated by a Federal military agency.

(iii) 25 to 1 for a horizontal distance of 5,000 feet from the nearest point of the nearest landing and takeoff area of each heliport listed in the Airport Directory or operated by a Federal military agency.

(3) Any construction or alteration on an airport listed in the Airport Directory of the Airman's Information Manual (§17.7(c) of this chapter).

(b) A notification to the Federal Aviation Administration is not required for any of the following construction or alteration:

(1) Any object that would be shielded by existing structures of a permanent and substantial character or by natural terrain or topographic features of equal or greater height, and would be located in the congested area of a city, town, or settlement where it is evident beyond all reasonable doubt that the structure so shielded will not adversely affect safety in air navigation. Applicants claiming such exemption shall submit a statement with their application to the Commission explaining the basis in detail for their finding (§17.14(a) of this chapter).

(2) Any antenna structure of 20 feet or less in height except one that would increase the height of another antenna structure (§17.14(b) of this chapter).

(c) Further details as to whether an aeronautical study and/or obstruction marking and lighting may be required, and specifications for obstruction marking and lighting when required, may be obtained from Part 17 of this chapter, "Construction, Marking, and Lighting of Antenna Structures." Information regarding the inspection and maintenance of antenna structures requiring obstruction marking and lighting is also contained in Part 17 of this chapter.

§ 97.47 Renewal and/or modification of amateur station license.

(a) Application for renewal and/or modification of an individual station license shall be submitted on FCC Form 610, and application for renewal and/or modification of an amateur club or military recreation station shall be submitted on FCC Form 610-B. In every case the application shall be accompanied by the applicant's license or photocopy thereof. Applications for renewal of unexpired licensees must be made during the license term and should be filed not later than 60 days prior to the end of the license term. In any case in which the licensee has in accordance with the provisions of this chapter, made timely and sufficient application for renewal of an unexpired license, no license shall expire until such application shall have been finally determined.

(b) If a license is allowed to expire, application for renewal may be made during a period of grace of 1 year after the expiration date. During this 1-year period of grace, an expired license is not valid. A license renewed during the grace period will be dated currently and will not be backdated to the date of expiration. An application for an individual station license shall be submitted on FCC Form 610. An application for an amateur club or military recreation station license shall be submitted on FCC Form 610-B. In every case the application shall be accompanied by the applicant's expired license or a photocopy thereof.

(c) When the name of a licensee is changed (without changes in the ownership, control, or corporate structure), or when the mailing address is changed (without changing the authorized location of the amateur radio station) a formal application for modification of license is not required. However, the licensee shall notify the Commission promptly of these changes. The notice, which may be in letter form, shall contain the name and address of the licensee as they appear in the Commission's records, the new name and/or address, as the case may be, and the call sign and the class of operator license. The notice shall be sent to Federal Communications Commission, Gettysburg, Pa., 17325, and a copy shall be maintained with the license of each station until a new license is issued.

§ 97.49 Commission modification of station licence.

(a) Whenever the Commission shall determine that public interest, convenience, and necessity would be served, or any treaty ratified by the United States will be more fully complied with, by the modification of any radio station license either for a limited time, or for the duration of the term thereof, it shall issue an order for such licensee to show cause why such license should not be modified.

(b) Such order to show cause shall contain a statement of the grounds and reasons for such proposed modification, and shall specify wherein the said license is required to be modified. It shall require the licensee against whom it is directed to appear at a place and time therein named, in no event to be less than 30 days from the date of receipt of the order, to show cause why the proposed modification should not be made and the order of modification issued.

(c) If the licensee against whom the order to show cause is directed does not appear at the time and place provided in said order, a final order of modification shall issue forthwith.

CALL SIGNS

§ 97.51 Assignment of call signs.

(a) The Commission shall assign the call sign of an amateur radio station on a systematic basis.

(b) The Commission shall not grant any request for a specific call sign.

(c) From time to time the Commission will issue public announcements detailing the policies and procedures governing the systematic assignment of call signs and any changes in those policies and procedures.

DUPLICATE LICENSES AND LICENSE TERM

§ 97.57 Duplicate license.

Any licensee requesting a duplicate license to replace an original which has been lost, mutilated, or destroyed, shall submit a statement setting forth the facts regarding the manner in which the original license was lost, mutilated, or destroyed. If, subsequent to receipt by the licensee of the duplicate license, the original license is found, either the duplicate or the original license shall be returned immediately to the Commission.

§ 97.59 License term.

(b) Amateur station licenses are normally valid for a period of five years from the date of issuance of a new or renewed license. All amateur station licenses, regardless of when issued, will expire on the same date as the licensee's amateur operator license.

(c) A duplicate license or a modified license which is not being renewed shall bear the same expiration date as the license for which it is a modification or duplicate.

SUBPART C—TECHNICAL STANDARDS

§ 97.61 Authorized frequencies and emissions.

(a) The following frequency bands and associated emissions are available to amateur radio stations for amateur radio operation, other than repeater operation and auxiliary operation, subject to the limitations of §97.65 and paragraph (b) of this section:

Frequency band	Emissions	Limitations (See paragraph (b))
kHz		
1800-2000	A1, A3	1,2
3500-4000	A1	
3500-3775	F1	
3775-3890	A5, F5	
3775-4000	A3, F3	4
4383.8	A3J/A3A	13
7000-7300	A1	3,4
7000-7150	F1	3,4
7075-7100	A3, F3	11
7150-7225	A5, F5	3,4
7150-7300	A3, F3	3,4
14000-14350	A1	
14000-14200	F1	
14200-14275	A5, F5	
14200-14350	A3, F3	

MHz

Frequency band	Emissions	Limitations (See paragraph (b))
21.000-21.450	A1	
21.000-21.250	F1	
21.250-21.350	A5, F5	
21.250-21.450	A3, F3	
28.000-29.700	A1	
28.000-28.500	F1	
28.500-29.700	A3, F3, A5, F5	
50.0-54.0	A1	
50.1-54.0	A2, A3, A4, A5, F1, F2, F3, F5	
51.0-54.0	A0	
144-148	A1	

144.1-148.0	A0, A2, A3, A4, A5, F0, F1, F2, F3, F5	
220-225	A0, A1, A2, A3, A4, A5, F0, F1, F2, F3, F4, F5	
420-450	A0, A1, A2, A3, A4, A5, F0, F1, F2, F3, F4, F5	6,7
1215-1300	A0, A1, A2, A3, A4, A5, F0, F1, F2, F3, F4, F5	5
2300-2450	A0, A1, A2, A3, A4, A5, F0, F1, F2, F3, F4, F5, P	5,8
3300-3500	A0, A1, A2, A3, A4, A5, F0, F1, F2, F3, F4, F5, P	5,12
5650-5925	A0, A1, A2, A3, A4, A5, F0, F1, F2, F3, F4, F5, P	5,9

GHz

10.000-10.500	A0, A1, A2, A3, A4, A5, F0, F1, F2, F3, F4, F5, P	5
24.000-24.250	A0, A1, A2, A3, A4, A5, F0, F1, F2, F3, F4, F5, P	5,10
48.000-50.000	A0, A1, A2, A3, A4, A5, F0, F1, F2, F3, F4, F5, P	
71.000-76.000	A0, A1, A2, A3, A4, A5, F0, F1, F2, F3, F4, F5, P	
165.000-170.000	A0, A1, A2, A3, A4, A5, F0, F1, F2, F3, F4, F5, P	
240.000-250.000	A0, A1, A2, A3, A4, A5, F0, F1, F2, F3, F4, F5, P	
Above 300.000	A0, A1, A2, A3, A4, A5, F0, F1, F2, F3, F4, F5, P	

(b) Limitations:

(1) The use of frequencies in this band is on a shared basis with the LORAN-A radionavigation system and is subject to cancellation or revision, in whole or in part, by order of the Commission, without hearing, whenever the Commission shall determine such action is necessary in view of the priority of the LORAN-A radionavigation system. The use of these frequencies by amateur stations shall not cause harmful interference to LORAN-A system. If an amateur station causes such interference, operation on the frequencies involved must cease if so directed by the Commission.

(2) Operation shall be limited to:

Maximum DC plate input power in watts

Area	1800-1825 kHz Day/Night	1825-1850 kHz Day/Night	1850-1875 kHz Day/Night	1875-1900 kHz Day/Night	1900-1925 kHz Day/Night	1925-1950 kHz Day/Night	1950-1975 kHz Day/Night	1975-2000 kHz Day/Night
Alabama	500/100	100/25	0	0	0	0	0	500/100
Alaska	1000/200	500/100	500/100	100/25	0	0	100/25	0
Arizona	1000/200	500/100	500/100	0	0	0	0	0
Arkansas	1000/200	500/100	100/25	0	0	100/25	100/25	500/100
California	1000/200	500/100	500/100	100/25	0	0	0	0
Colorado	1000/200	500/100	200/50	0	0	0	0	200/50
Connecticut	500/100	100/25	0	0	0	0	0	0
Delaware	500/100	100/25	0	0	0	0	0	100/25
District of Columbia	500/100	100/25	0	0	0	0	100/25	100/25
Florida	500/100	100/25	0	0	200/50	100/25	100/25	500/100
Georgia	500/100	100/25	0	0	100/25	100/25	0	500/100
Hawaii	0	0	0	0	0	0	0	200/50
Idaho	1000/200	500/100	500/100	100/25	0	0	100/25	500/100
Illinois	1000/200	500/100	100/25	0	0	100/25	100/25	200/50
Indiana	1000/200	500/100	100/25	0	0	0	0	200/50
Iowa	1000/200	500/100	200/50	0	100/25	100/25	100/25	500/100
Kansas	1000/200	500/100	100/25	100/25	0	100/25	100/25	500/100
Kentucky	1000/200	500/100	100/25	0	0	0	0	500/100
Louisiana	500/100	100/25	0	0	0	0	100/25	200/50
Maine	500/100	100/25	0	0	0	0	0	500/100
Maryland	500/100	100/25	100/25	0	0	0	0	0
Massachusetts	1000/200	500/100	500/100	100/25	100/25	100/25	100/25	100/25
Michigan	1000/200	500/100	0	0	0	0	0	0
Minnesota	500/100	100/25	0	0	0	0	100/25	100/25
Mississippi	1000/200	500/100	100/25	100/25	0	100/25	100/25	500/100
Missouri	1000/200	500/100	100/25	0	0	100/25	100/25	500/100
Montana	1000/200	500/100	500/100	100/25	0	100/25	100/25	500/100
Nebraska	1000/200	500/100	200/50	0	0	100/25	100/25	200/50
Nevada	1000/200	500/100	500/100	100/25	0	100/25	100/25	500/100
New Hampshire	500/100	100/25	0	0	0	0	0	0
New Jersey	500/100	100/25	0	0	0	0	0	100/25
New Mexico	1000/200	500/100	100/25	0	0	0	0	0
New York	500/100	100/25	0	0	0	100/25	500/100	500/100
North Carolina	500/100	100/25	0	0	0	0	0	500/100
North Dakota	1000/200	500/100	500/100	100/25	0	100/25	100/25	1000/200
Ohio	1000/200	500/100	100/25	0	0	0	0	100/25
Oklahoma	1000/200	500/100	100/25	0	0	0	100/25	500/100
Oregon	1000/200	500/100	500/100	100/25	0	0	0	500/100

Area	1800-1825 kHz Day/Night	1825-1850 kHz Day/Night	1850-1875 kHz Day/Night	1875-1900 kHz Day/Night	1900-1925 kHz Day/Night	1925-1950 kHz Day/Night	1950-1975 kHz Day/Night	1975-2000 kHz Day/Night
				Maximum DC plate input power in watts				

Area	1800-1825	1825-1850	1850-1875	1875-1900	1900-1925	1925-1950	1950-1975	1975-2000
Pennsylvania	500/100	100/25	0	0	0	0	0	0
Rhode Island	500/100	100/25	0	0	0	0	0	0
South Carolina	500/100	100/25	0	0	0	0	0	200/50
South Dakota	1000/200	500/100	500/100	100/25	100/25	100/25	100/25	500/100
Tennessee	1000/200	100/25	100/25	0	0	0	0	200/50
Texas	500/100	500/100	0	100/25	100/25	0	0	200/50
Utah	1000/200	500/100	500/100	0	0	0	0	100/25
Vermont	500/100	100/25	0	0	0	0	0	0
Virginia	500/100	500/100	0	100/25	0	0	0	100/25
Washington	1000/200	500/100	500/100	0	0	0	0	100/25
West Virginia	1000/200	500/100	100/25	0	0	0	0	200/50
Wisconsin	1000/200	500/100	200/50	0	0	0	0	200/50
Wyoming	1000/200	500/100	500/100	100/25	100/25	0	0	200/50
Puerto Rico	500/100	100/25	0	0	0	0	0	200/50
Virgin Islands	500/100	100/25	0	0	0	0	100/25	500/100
Swan Island	500/100	100/25	0	0	0	0	100/25	500/100
Serrana Bank	500/100	100/25	0	0	0	0	100/25	500/100
Roncador Key	500/100	100/25	0	0	0	0	0	200/50
Navassa Island	0	0	0	0	0	0	0	0
Baker, Canton, Enderbury, Howland	100/25	0	0	100/25	100/25	0	0	100/25
Guam, Johnston, Midway	0	0	0	0	100/25	0	0	100/25
American Samoa	200/50	0	0	200/50	200/50	0	0	200/50
Wake	100/25	0	0	100/25	0	0	0	0
Palmyra, Jarvis	0	0	0	0	200/50	0	0	200/50

(3) Where, in adjacent regions or subregions, a band of frequencies is allocated to different services of the same category, the basic principle is the equality of right to operate. Accordingly, the stations of each service in one region or subregion must operate so as not to cause harmful interference to services in the other regions or subregions (No. 117, the Radio Regulations, Geneva, 1959).

(4) 3900-4000 kHz and 7100-7300 kHz are not available in the following U.S. possessions: Baker, Canton, Enderbury, Guam, Howland, Jarvis, Palmyra, American Samoa, and Wake Islands.

(5) Amateur stations shall not cause interference to the Government radiolocation service.

(6) (Reserved)

(7) In the following areas the d.c. plate input power to the final transmitter stage shall not exceed 50 watts, except when authorized by the appropriate Commission Engineer in Charge and the appropriate Military Area Frequency Coordinator

(i) Those portions of Texas and New Mexico bounded by latitude 33°24′ N, 31°53′ N, and longitude 105°40′ W. and 106°40′ W

(ii) The State of Florida, including the Key West area and the areas enclosed within circles of 200-mile radius centered at 28°21′ N., 80°43′W. and 30°30′ N., 86°30′ W.

(iii) The State of Arizona.

(iv) Those portions of California and Nevada south of latitude 37°10′ N. and the area within a 200-mile radius of 34°09′ N., 119°11′ W.

(8) No protection in the band 2400-2500 MHz is afforded from interference due to the operation of industrial, scientific, and medical devices on 2450 MHz.

(9) No protection in the band 5725-5875 MHz is afforded from interference due to the operation of industrial, scientific and medical devices on 5800 MHz.

(10) No protection in the band 24.00-24.25 GHz is afforded from interference due to the operation of industrial, scientific and medical devices on 24.125 GHz.

(11) The use of A3 and F3 in this band is limited to amateur radio stations located outside Region 2.

(12) Amateur stations shall not cause interference to the Fixed-Satellite Service operating in the band 3400-3500 MHz.

(13) The frequency 4383.8 kHz, maximum power 150 watts, may be used by any station authorized under this part to communicate with any other station authorized in the State of Alaska for emergency communications. No airborne operations will be permitted on this frequency. Additionally, all stations operating on this frequency must be located in or within 50 nautical miles of the State of Alaska.

(c) All amateur frequency bands above 29.5 MHz are available for repeater operation, except 50.0-52.0 MHz, 144.0-144.5 MHz, 145.5-146.0 MHz, 220.0-220.5 MHz, 431.0-433.0 MHz, and 435.0-438.0 MHz. Both the input (receiving) and output (transmitting) frequencies of a station in repeater operation shall be frequencies available for repeater operation.

(d) All amateur frequency bands above 220.5 MHz, except 431-433 MHz, and 435-438 MHz, are available for auxiliary operation.

§ 97.63 Selection and use of frequencies.

(a) An amateur station may transmit on any frequency within any authorized amateur frequency band.

(b) Sideband frequencies resulting from keying or modulating a carrier wave shall be confined within the authorized amateur band.

(c) The frequencies available for use by a control operator of an amateur station are dependent on the operator license classification of the control operator and are listed in §97.7.

§ 97.65 Emission limitations.

(a) Type A0 emission, where not specifically designated in the bands listed in §97.61, may be used for short periods of time when required for authorized remote control purposes or for experimental purposes. However, these limitations do not apply where type A0 emission is specifically designated.

(b) Whenever code practice, in accordance with §97.91(d), is conducted in bands authorized for A3 emission, tone modulation of the radiotelephone transmitter may be utilized when interspersed with appropriate voice instructions.

(c) On frequencies below 29.0 MHz and between 50.1 and 52.5 MHz, the bandwidth of an F3 emission (frequency or phase modulation) shall not exceed that of an A3 emission having the same audio characteristics; and the purity and stability of emissions shall comply with the requirements of §97.73.

(d) On frequencies below 50 MHz, the bandwidth of A5 and F5 emissions shall not exceed that of an A3 single sideband emission.

(e) On frequencies between 50 MHz and 225 MHz, single sideband or double sideband A5 emission may be used and the bandwidth shall not exceed that of an A3 single sideband or double sideband signal respectively. The bandwidth of F5 emission shall not exceed that of an A3 single sideband emission.

(f) Below 225 MHz, A3 and A5 emissions may be used simultaneously on the same carrier frequency provided the total bandwidth does not exceed that of an A3 double sideband emission.

§ 97.67 Maximum authorized power.

(a) Except for power restrictions as set forth in §97.61 and paragraph (d) below each amateur transmitter may be operated with a power input not exceeding one kilowatt to the plate circuit of the final amplifier stage of an amplifier oscillator transmitter or to the plate circuit of an oscillator transmitter. An amateur transmitter operating with a power input exceeding 900 watts to the plate circuit shall provide means for accurately measuring the plate power input to the vacuum tube or tubes supplying power to the antenna.

(b) Notwithstanding the provisions of paragraph (a) of this section, amateur stations shall use the minimum amount of transmitter power necessary to carry out the desired communications.

(c) Within the limitations of paragraphs (a) and (b) of this section, the effective radiated power of an amateur radio station in repeater operation shall not exceed the power specified for the antenna height above average terrain in the following table:

Antenna height above average terrain	Maximum effective radiated power for frequency bands above:			
	52 MHz	144.5 MHz	420 MHz	1215 MHz
			Paragraphs (a) and (b)	Paragraphs (a) and (b)
Below 50 feet	100 watts	800 watts	..do...	..do...
50-99 feet	100 watts	400 watts	..do...	..do...
100-499 feet	50 watts	400 watts	800 watts	..do...
500-999 feet	25 watts	200 watts	800 watts	..do...
Above 1000 feet	25 watts	100 watts	400 watts	..do...

(d) In the frequency bands 3700-3750 kHz, 7100-7150 kHz (7050-7075 kHz when the terrestrial location of the station is not within Region 2), 21,100-21,200 kHz and 28,100-28,200 kHz, the power input to the transmitter final amplifying stage supplying radio frequency energy to the antenna shall not exceed 250 watts, exclusive of power for heating the cathode of a vacuum tube(s).

§ 97.69 Radio teleprinter transmissions.

The following special conditions shall be observed during the transmission of radio teleprinter signals on authorized frequencies by amateur stations:

(a) A single channel five-unit (start-stop) teleprinter code shall be used which shall correspond to the International Telegraphic Alphabet No. 2 with respect to all letters and numerals (including the slant sign or fraction bar) but special signals may be employed for the remote control of receiving printers, or for other purposes, in "figures" positions not utilized for numerals. In general, this code shall conform as nearly as possible to the teleprinter code or codes in common commercial usage in the United States.

(b) The normal transmitting speed of the radio teleprinter signal keying equipment shall be adjusted as closely as possible to one of the standard teleprinter speeds, namely, 60 (45 bauds), 67 (50 bauds), 75 (56.25 bauds) or 100 (75 bauds) words per minute, and in any event, within the range of ± 5 words per minute of the selected standard speed.

(c) When frequency shift keying (type F1 emission) is utilized, the deviation in frequency from the mark signal to space signal, or from the space signal to the mark signal, shall be less than 900 Hertz.

(d) When audio frequency shift keying (type A2 or type F2 emission) is utilized, the highest fundamental modulating audio frequency shall not exceed 3000 hertz, and the difference between the modulating audio frequency for the mark signal and that for the space signal shall be less than 900 hertz.

§ 97.71 Transmitter power supply.

The licensee of an amateur station using frequencies below 144 megahertz shall use adequately filtered direct-current plate power supply for the transmitting equipment to minimize modulation from this source.

§ 97.73 Purity of emissions.

(a) Except for a transmitter or transceiver built before April 15, 1977 or first marketed before January 1, 1978, the mean power of any spurious emission or radiation from an amateur transmitter, transceiver, or external radio frequency power amplifier being operated with a carrier frequency below 30 MHz shall be at least 40 decibels below the mean power of the fundamental without exceeding the power of 50 milliwatts. For equipment of mean power less than five watts, the attenuation shall be at least 30 decibels.

(b) Except for a transmitter or transceiver built before April 15, 1977 or first marketed before January 1, 1978, the mean power of any spurious emission or radiation from an amateur transmitter, transceiver, or external radio frequency power amplifier being operated with a carrier frequency above 30 MHz but below 235 MHz shall be at least 60 decibels below the mean power of the fundamental. For a transmitter having a mean power of 25 watts or less, the mean power of any spurious radiation supplied to the antenna transmission line shall be at least 40 decibels below the mean power of the fundamental without exceeding the power of 25 microwatts, but need not be reduced below the power of 10 microwatts.

(c) Paragraphs (a) and (b) of this section notwithstanding, all spurious emissions or radiation from an amateur transmitter, transceiver, or external radio frequency power amplifier shall be reduced or eliminated in accordance with good engineering practice.

(d) If any spurious radiation, including chassis or power line radiation, causes harmful interference to the reception of another radio station, the licensee may be required to take steps to eliminate the interference in accordance with good engineering practice.

NOTE: For the purposes of this section, a spurious emission or radiation means any emission or radiation from a transmitter, transceiver, or external radio frequency power amplifier which is outside of the authorized Amateur Radio Service frequency band being used.

§ 97.74 Frequency measurement and regular check.

The licensee of an amateur station shall provide for measurement of the emitted carrier frequency or frequencies and shall establish procedures for making such measurement regularly. The measurement of the emitted carrier frequency or frequencies shall be made by means independent of the means used to control the radio frequency or frequencies generated by the transmitting apparatus and shall be of sufficient accuracy to assure operation within the amateur frequency band used.

§ 97.75 Use of external radio frequency (RF) power amplifiers.

(a) Until April 28, 1981, any external radio frequency (RF) power amplifier used or attached at any amateur radio station shall be type accepted in accordance with Subpart J of Part 2 of the FCC's Rules for operation in the Amateur Radio Service, unless one or more of the following conditions are met:

(1) The amplifier is not capable of operation on any frequency or frequencies below 144 MHz (the amplifier shall be considered incapable of operation below 144 MHz if the mean output power decreases, as frequency decreases from 144 MHz, to a point where 0 decibels or less gain is exhibited at 120 MHz and below and the amplifier is not capable of being easily modified to provide amplification below 120 MHz);

(2) The amplifier was originally purchased before April 28, 1978;

(3) The amplifier was—

(i) Constructed by the licensee, not from an external RF power amplifier kit, for use at his amateur radio station;

(ii) Purchased by the licensee as an external RF power amplifier kit before April 28, 1978 for use at his amateur radio station; or

(iii) Modified by the licensee for use at his amateur radio station in accordance with §2.1001 of the FCC's Rules;

(4) The amplifier was purchased by the licensee from another amateur radio operator who—

(i) Constructed the amplifier, but not from an external RF power amplifier kit;

(ii) Purchased the amplifier as an external RF power amplifier kit before April 28, 1978 for use at his amateur radio station; or

(iii) Modified the amplifier for use at his amateur radio station in accordance with §2.1001 of the FCC's Rules;

(5) The external RF power amplifier was purchased from a dealer who obtained it from an amateur radio operator who—

(i) Constructed the amplifier, but not from an external RF power amplifier kit;

(ii) Purchased the amplifier as an external RF power amplifier kit before April 28, 1978 for use at his amateur radio station; or

(iii) Modified the amplifier for use at his amateur radio station in accordance with §2.1001 of the FCC's Rules; or

(6) The amplifier was originally purchased after April 27, 1978 and has been issued a marketing waiver by the FCC.

(b) A list of type accepted equipment may be inspected at FCC headquarters in Washington, D.C. or at any FCC field office. Any external RF power amplifier appearing on this list as type accepted for use in the Amateur Radio Service may be used in the Amateur Radio Service.

NOTE: No more than one unit of one model of an external RF power amplifier shall be constructed or modified during any calendar year by an amateur radio operator for use in the Amateur Radio Service without a grant of type acceptance.

§ 97.76 Requirements for type acceptance of external radio frequency (RF) power amplifiers and external radio frequency power amplifier kits.

(a) Until April 28, 1981, any external radio frequency (RF) power amplifier or external RF power amplifier kit marketed (as defined in §2.815), manufactured, imported or modified for use in the Amateur Radio Service shall be type accepted for use in the Amateur Radio Service in accordance with Subpart J or Part 2 of the FCC's Rules. This requirement does not apply if one or more of the following conditions are met:

(1) The amplifier is not capable of operation on any frequency or frequencies below 144 MHz (the amplifier shall be considered incapable of operation below 144 MHz if the mean output power decreases, as frequency decreases from 144 MHz, to a point where 0 decibels or less gain is exhibited at 120 MHz and below and the amplifier is not capable of being easily modified to provide amplification below 120 MHz).

(2) The amplifier was originally purchased before April 28, 1978 by an amateur radio operator for use at his amateur radio station;

(3) The amplifier was constructed or modified by an amateur radio operator for use at his amateur radio station in accordance with §2.1001 of the FCC's Rules;

(4) The amplifier was constructed or modified by an amateur radio operator in accordance with §2.1001 of the FCC's Rules and sold to another amateur radio operator or to a dealer;

(5) The amplifier was constructed or modified by an amateur radio operator in accordance with §2.1001 of the FCC's Rules and sold by a dealer to an amateur radio operator for use at his amateur radio station; or

(6) The amplifier was manufactured before April 28, 1978 and has been issued a marketing waiver by the FCC.

(b) No more than one unit of one model of an external RF power amplifier shall be constructed or modified during any calendar year by an amateur radio operator for use in the Amateur Radio Service without a grant of type acceptance.

(c) A list of type accepted equipment may be inspected at FCC headquarters in Washington, D.C. or at any FCC field office. Any external RF power amplifier appearing on this list as type accepted for use in the Amateur Radio Service may be marketed for use in the Amateur Radio Service.

§ 97.77 Standards for type acceptance of external radio frequency (RF) power amplifiers and external radio frequency power amplifier kits.

(a) An external radio frequency (RF) power amplifier or external RF power amplifier kit will receive a grant of type acceptance under this Part only if a grant of type acceptance would serve the public interest, convenience or necessity.

(b) To receive a grant of type acceptance under this Part, an external RF power amplifier shall meet the emission limitations of §97.73 when the amplifier is—

(1) Operated at its full output power;

(2) Placed in the "standby" or "off" positions. but still connected to the transmitter; and

(3) Driven with at least 50 watts mean radio frequency input power (unless a higher drive level is specified).

(c) To receive a grant of type acceptance under this part, an external RF power amplifier shall not be capable of operation on any frequency or frequencies between 24.00 MHz and 35.00 MHz. The amplifier will be deemed incapable of operation between 24.00 MHz and 35.00 MHz if—

(1) The amplifier has no more than 6 decibels of gain between 24.00 MHz and 26.00 MHz and between 28.00 MHz and 35.00 MHz. (This gain is determined by the ratio of the input RF driving signal (mean power measurement) to the mean RF output power of the amplifier.); and

(2) The amplifier exhibits no amplification (0 decibels of gain) between 26.00 MHz and 28.00 MHz.

(d) Type acceptance of external radio frequency power amplifiers or amplifier kits may be denied when denial serves the public interest, convenience or necessity by preventing the use of these amplifiers in services other than the Amateur Radio Service. Other uses of these amplifiers, such as in the Citizens Band Radio Service, is prohibited (Section 95.509). Examples of features which may result in dismissal or denial of an application for type acceptance of an external RF power amplifier include, but are not limited to, the following:

(1) Any accessible wiring which, when altered, would permit operation of the amplifier in a manner contrary to the FCC's Rules;

(2) Circuit boards or similar circuitry to facilitate the addition of components to change the amplifier's operating characteristics in a manner contrary to the FCC's Rules;

(3) Instructions for operation or modification of the amplifier in a manner contrary to the FCC's Rules;

(4) Any internal or external controls or adjustments to facilitate operation of the amplifier in a manner contrary to the FCC's Rules.

(5) Any internal radio frequency sensing circuitry or any external switch, the purpose of which is to place the amplifier in the transmit mode;

(6) The incorporation of more gain in the amplifier than is necessary to operate in the Amateur Radio Service. For purposes of this paragraph, an amplifier must meet the following requirements:

(i) No amplifier shall be capable of achieving designed output (or designed d.c. input) power when driven with less than 50 watts mean radio frequency input power;

(ii) No amplifier shall be capable of amplifying the input RF driving signal by more than 13 decibels. (This gain limitation is determined by the ratio of the input RF driving signal (mean power) to the mean RF output power of the amplifier). If the amplifier has a designed d.c. input power of less than 1000 watts, the gain allowance is reduced accordingly. (For example, an amplifier with a designed d.c. input power of 500 watts shall not be capable of amplifying the input RF driving signal (mean power measurement) by more than 10 decibels, compared to the mean RF output power of the amplifier.);

(iii) The amplifier shall not exhibit more gain than permitted by paragraph (d)(6)(ii) of this section when driven by a radio frequency input signal of less than 50 watts mean power; and

(iv) The amplifier shall be capable of sustained operation at its designed power level.

(7) Any attenuation in the input of the amplifier which, when removed or modified, would permit the amplifier to function at its designed output power when driven by a radio frequency input signal of less than 50 watts mean power.

SUBPART D—OPERATING REQUIREMENTS AND PROCEDURES

GENERAL

§ 97.78 Practice to be observed by all licensees.

In all respects not specifically covered by these regulations each amateur station shall be operated in accordance with good engineering and good amateur practice.

§ 97.79 Control operator requirements.

(a) The licensee of an amateur station shall be responsible for its proper operation.

(b) Every amateur radio station, when in operation, shall have a control operator at an authorized control point. The control operator shall be on duty, except where the station is operated under automatic control. The control operator may be the station licensee, if a licensed amateur radio operator, or may be another amateur radio operator with the required class, of license and designated by the station licensee. The control operator shall also be responsible, together with the station licensee, for the proper operation of the station.

(c) An amateur station may only be operated in the manner and to the extent permitted by the operator privileges authorized for the class of license held by the control operator, but may exceed those of the station licensee provided proper station identification procedures are performed.

(d) The licensee of an amateur radio station may permit any third party to participate in amateur radio communication from his station, provided that a control operator is present and continuously monitors and supervises the radio communication to insure compliance with the rules.

§ 97.81 Authorized apparatus.

An amateur station license authorizes the use under control of the licensee of all transmitting apparatus at the fixed location specified in the station license which is operated on any frequency, or frequencies allocated to the amateur service, and in addition authorizes the use, under control of the licensee, of portable and mobile transmitting apparatus operated at other locations.

§ 97.82 Availability of operator license.

The original operator license of each operator shall be kept in the personal possession of the operator while operating an amateur station. When operating an amateur station at a fixed location, however, the license may be posted in a conspicuous place in the room occupied by the operator. The license shall be available for inspection by any authorized Government official whenever the operator is operating an amateur station and at other times upon request made by an authorized representative of the Commission, except when such license has been filed with application for modification or renewal thereof, or has been mutilated, lost or destroyed, and request has been made for a duplicate license in accordance with §97.57. No recognition shall be accorded to any photocopy of an operator license; however, nothing in this section shall be construed to prohibit the photocopying for other purposes of any amateur radio operator license.

§ 97.83 Availability of station license.

The original license of each amateur station or a photocopy thereof shall be posted in a conspicuous place in the room occupied by the licensed operator r⋯'le the station is being operated at a fixed location or shall be kept in his personal possession. When the station is operated at other than a fixed location, the original station license or a photocopy thereof shall be kept in the personal possession of the station licensee (or a licensed representative) who shall be present at the station while it is being operated as a portable or mobile station. The original station license shall be available for inspection by any authorized Government official at all times while the station is being operated and at other times upon request made by an authorized representative of the Commission, except when such license has been filed with application for modification or renewal thereof, or has been mutilated, lost, or destroyed, and request has been made for a duplicate license in accordance with §97.57.

§ 97.84 Station identification.

(a) An amateur station shall be identified by the transmission of its call sign at the beginning and end of each single transmission or exchange of transmissions and at intervals not to exceed 10 minutes during any single transmission or exchange of transmissions of more than 10 minutes duration. Additionally, at the end of an exchange of telegraphy (other than teleprinter) or telephony transmissions between amateur stations, the call sign (or the generally accepted network identifier) shall be given for the station, or for at least one of the group of stations, with which communication was established.

(b) Under conditions when the control operator is other than the station licensee, the station identification shall be the assigned call sign for that station. However, when a station is operated within the privileges of the operator's class of license but which exceeds those of the station licensee, station identification shall be made by following the station call sign with the operator's primary station call sign (i.e. WN4XYZ/W4XX).

(c) An amateur radio station in repeater operation or a station in auxiliary operation used to relay automatically the signals of other stations in a system of stations shall be identified by radiotelephony or radiotelegraphy at a level of modulation sufficient to be intelligible through the repeated transmission at intervals not to exceed ten minutes.

(d) When an amateur radio station is in repeater or auxiliary operation, the following additional identifying information shall be transmitted:

(1) When identifying by radiotelephony, a station in repeater operation shall transmit the word "repeater" at the

end of the station call sign. When identifying by radiotelegraphy, a station in repeater operation shall transmit the fraction bar $\overline{DN}$ followed by the letters "RPT" or "R" at the end of the station call sign. (The requirements of this subparagraph do not apply to stations having call signs prefixed by the letters "WR".)

(2) When identifying by radiotelephony, a station in auxillary operation shall transmit the word "auxiliary" at the end of the station call sign. When identifying by radiotelegraphy, a station in auxiliary operation shall transmit the fraction bar $\overline{DN}$ followed by the letters "AUX" or "A" at the end of the station call sign.

(e) A station in auxiliary operation may be identified by the call sign of its associated station.

(f) When operating under the authority of an Interim Amateur Permit with privileges authorized by the Permit, but which exceed the privileges of the licensee's permanent operator license, the station must be identified in the following manner:

(1) On radiotelephony, by the transmission of the station call sign, followed by the word "interim", followed by the special identifier shown on the Interim Permit;

(2) On radiotelegraphy, by the transmission of the station call sign, followed by the fraction bar $\overline{DN}$, followed by the special identifier shown on the interim permit.

(g) The identification required by this section shall be given on each frequency being utilized for transmission and shall be transmitted either by telegraphy using the international Morse code, or by telephony, using the English language. If the identification required by this section is made by an automatic device used only for identification by telegraphy, the code speed shall not exceed 20 words per minute. The Commission encourages the use of a nationally or internationally recognized standard phonetic alphabet as an aid for correct telephone identification.

§ 97.85 Repeater operation.

(a) Emissions from a station in repeater operation shall be discontinued within five seconds after cessation of radiocommunications by the user station. Provisions to limit automatically the access to a station in repeater operation may be incorporated but are not mandatory.

(b) Except for operation under automatic control, as provided in paragraph (e) of this section, the transmitting and receiving frequencies used by a station in repeater operation shall be continuously monitored by a control operator immediately before and during periods of operation.

(c) A station in repeater operation shall not concurrently retransmit amateur radio signals on more than one frequency in the same amateur frequency band, from the same location.

(d) A station in repeater operation shall be operated in a manner ensuring that it is not used for one-way communications, except as provided in §97.91.

(e) A station in repeater operation, either locally controlled or remotely controlled, may also be operated by automatic control when devices have been installed and procedures have been implemented to ensure compliance with the rules when a duty control operator is not present at a control point of the station. Upon notification by the Commission of improper operation of a station under automatic control, operation under automatic control shall be immediately discontinued until all deficiencies have been corrected.

§ 97.86 Auxiliary operation.

(a) A station in auxiliary operation, either locally controlled or remotely controlled, may also be operated by automatic control when it is operated as part of a system of stations in repeater operation operated under automatic control.

(b) If a station in auxiliary operation is relaying signals of another amateur radio station(s) to a station in repeater operation, the station in auxiliary operation may use an input (receiving) frequency in frequency bands reserved for auxiliary operation, repeater operation, or both.

(c) A station in auxiliary operation shall be used only to communicate with stations shown in the system network diagram.

§ 97.88 Operation of a station by remote control.

An amateur radio station may be operated by remote control only if there is compliance with the following:

(a) A photocopy of the remotely controlled station license shall be—

(1) posted in a conspicuous place at the remotely controlled transmitter location, and

(2) placed in the log of each authorized control operator.

(b) The name, address, and telephone number of the remotely controlled station licensee and at least one control operator shall be posted in a conspicuous place at the remotely controlled transmitter location.

(c) Except for operation under automatic control, a control operator shall be on duty when the station is being remotely controlled. Immediately before and during the periods the remotely controlled station is in operation, the frequencies used for emission by the remotely controlled station shall be monitored by the control operator. The control operator shall terminate all transmissions upon any deviation from the rules.

(d) Provisions must be incorporated to limit transmission to a period of no more than 3 minutes in the event of malfunction in the control link.

(e) A station in repeater operation shall be operated by radio remote control only when the control link uses frequencies other than the input (receiving) frequencies of the station in repeater operation.

§ 97.89 Points of Communications.

(a) Amateur stations may communicate with:

(1) Other amateur stations, excepting those prohibited by Appendix 2.

(2) Stations in other services licensed by the Commission and with U.S Government stations for civil defense purposes in accordance with Subpart F of this part, in emergencies and, on a temporary basis, for test purposes.

(3) Any station which is authorized by the Commission to communicate with amateur stations.

(b) Amateur stations may be used for transmitting signals, or communications, or energy, to receiving apparatus for the measurement of emissions, temporary observation of transmission phenomena, radio control of remote objects, and similar experimental purposes and for the purposes set forth in §97.91.

§ 97.91 One-way communications.

In addition to the experimental one-way transmission permitted by §97.89, the following kinds of one-way communications, addressed to amateur stations, are authorized and will not be construed as broadcasting: (a) Emergency communications, including bona fide emergency drill practice transmissions; (b) Information bulletins consisting solely of subject matter having direct interest to the amateur radio service as such; (c) Round-table discussions or net-type operations where more than two amateur stations are in communication, each station taking a turn at transmitting to other station(s) of the group; and (d) Code practice transmissions intended for persons learning or improving proficiency in the international Morse code.

§ 97.93 Modulation of carrier.

Except for brief tests or adjustments, an amateur radiotelephone station shall not emit a carrier wave on frequencies below 51 megahertz unless modulated for the purpose of communication. Single audiofrequency tones may be transmitted for test purposes of short duration for the development and perfection of amateur radio telephone equipment.

STATION OPERATION AWAY FROM AUTHORIZED LOCATION

§ 97.95 Operation away from the authorized fixed station location.

(a) Operation within the United States, its territories or possessions is permitted as follows:

(1) When there is no change in the authorized fixed operation station location, an amateur radio station, other than a military recreation station, may be operated portable or mobile under its station license anywhere in the United States, its territories or possessions, subject to §97.61.

(2) When the authorized fixed station location is changed, the licensee shall submit an application for modification of the station license in accordance with §97.47.

(b) When outside the continental limits of the United States, its territories, or possessions, an amateur radio station may be operated as portable or mobile only under the following conditions:

(1) Operation may not be conducted within the jurisdiction of a foreign government except pursuant to, and in accordance with express authority granted to the licensee by such foreign government. When a foreign government permits Commission licensees to operate within its territory, the amateur frequency bands which may be used shall be as prescribed or limited by that government. (See Appendix 4 of this Part for the text of treaties or agreements between the United States and foreign governments relative to reciprocal amateur radio operation.)

(2) When outside the jurisdiction of a foreign government, amateur operation may be conducted within ITU Region 2 subject to the limitations of, and on those frequency bands listed in, §97.61.

(3) When outside the jurisdiction of a foreign government, amateur operation may be conducted within ITU Regions 1 and 3 on the following frequencies, subject to the limitations and provisions of Section IV of Article 5 of the Radio Regulations of the ITU:

(i)

REGION 1	REGION 3
3.5-3.8 MHz	1.8-2.0 MHz
7.0-7.1 MHz	3.5-3.9 MHz
14.0-14.35 MHz	7.0-7.1 MHz
21.0-21.45 MHz	14.0-14.35 MHz
28.0-29.7 MHz	21.0-21.45 MHz
144-146 MHz	28.0-29.7 MHz
430-440 MHz	50.0-54.0 MHz
1215-1300 MHz	144-148 MHz
2300-2450 MHz	420-450 MHz
	1215-1300 MHz
	2300-2450 MHz

(ii) Operation on amateur frequency bands above 2450 MHz may be conducted subject to the limitations and provisions of Section IV of Article 5 of the Radio Regulations of the ITU.

(4) Except as otherwise provided, amateur operation conducted outside the jurisdiction of a foreign government shall comply with all requirements of Part 97 of this Chapter.

SPECIAL PROVISIONS

§ 97.99 Stations used only for radio control of remote model crafts and vehicles.

An amateur transmitter when used for the purpose of transmitting radio signals intended only for the control of a remote model craft or vehicle and having mean output power not exceeding one watt may be operated under the special provisions of this section provided an executed Transmitter Identification Card (FCC Form 452-C) or a plate made of a durable substance indicating the station call sign and licensee's name and address is affixed to the transmitter.

(a) Station identification is not required for transmissions directed only to a remote model craft or vehicle.

(b) Transmissions containing only control signals directed only to a remote model craft or vehicle are not considered to be codes or ciphers in the context of the meaning of §97.117.

(c) Station logs need not indicate the times of commencing and terminating each transmission or series of transmissions.

§ 97.101 Mobile stations aboard ships or aircraft.

In addition to complying with all other applicable rules, an amateur mobile station operated on board a ship or aircraft must comply with all of the following special conditions: (a) The installation and operation of the amateur mobile station shall be approved by the master of the ship or captain of the aircraft; (b) The amateur mobile station shall be separate from and independent of all other radio equipment, if any, installed on board the same ship or aircraft; (c) The electrical installation of the amateur mobile station shall be in accord with the rules applicable to ships or aircraft as promulgated by the appropriate government agency; (d) The operation of the amateur mobile station shall not interfere with the efficient operation of any other radio equipment installed on board the same ship or aircraft; and (e) The amateur mobile station and its associated

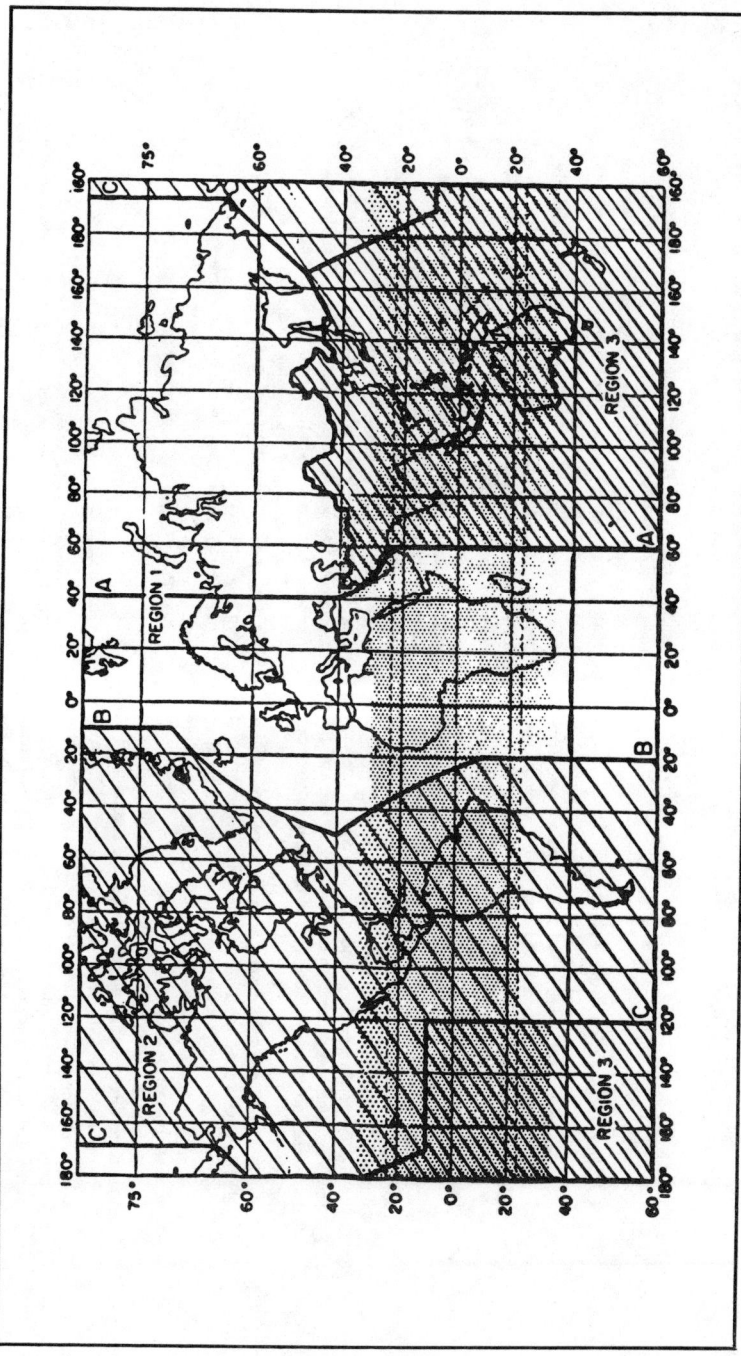

equipment, either in itself or in its method of operation, shall not constitute a hazard to the safety of life or property.

LOGS

§ 97.103 Station log requirements.

An accurate legible account of station operation shall be entered into a log for each amateur radio station. The following items shall be entered as a minimum:

(a) The call sign of the station, the signature of the station licensee, or a photocopy of the station license.

(b) The locations and dates upon which fixed operation of the station was initiated and terminated. If applicable, the location and dates upon which portable operation was initiated and terminated at each location.

(1) The date and time periods the duty control operator for the station was other than the station licensee, and the signature and primary station call sign of that duty control operator.

(2) A notation of third party traffic sent or received, including names of all third parties, and a brief description of the traffic content. This entry may be in a form other than written, but one which can be readily transcribed by the licensee into written form.

(3) Upon direction of the Commission, additional information as directed shall be recorded in the station log.

(c) In addition to the other information required by this section, the log of a remotely controlled station shall have entered the names, addresses, and call signs of all authorized control operators and a functional block diagram of, and a technical explanation sufficient to describe the operation of the control link. Additionally, the following information shall be entered:

(1) A description of the measures taken for protection against access to the remotely controlled station by unauthorized persons;

(2) A description of the measures taken for protection against unauthorized station operation, either through activation of the control link, or otherwise;

(3) A description of the provisions for shutting down the station in the case of control link malfunction; and

(4) A description of the means used for monitoring the transmitting frequencies.

(d) When a station has one or more associated stations, that is, stations in repeater or auxiliary operation, a system network diagram shall be entered in the station log.

(e) In addition to the other information required by this section, the log of a station in repeater operation transmitting with an effective radiated power greater than the minimum effective radiated power listed in §97.67(c) for the frequency band in use shall contain the following:

(1) The location of the station transmitting antenna, marked upon a topographic map having a scale of 1:250,000 and contour intervals[1];

(2) The antenna transmitting height above average terrain[2];

(3) The effective radiated power in the horizontal plane

---

[1]Indexes and ordering information for suitable maps are available from the U.S. Geological Survey, Washington, D.C. 20242, or from the Federal Center, Denver, Colorado 80255.

[2]See Appendix 5.

for the main lobe of the antenna pattern, calculated for maximum transmitter output power;

(4) The transmitter output power;

(5) The loss in the transmission line between the transmitter and the antenna, expressed in decibels;

(6) The relative gain in the horizontal plane of the transmitting antenna; and

(7) The horizontal and vertical radiation patterns of the transmitting antenna, with reference to true north (for horizontal pattern only), upon polar coordinate graph paper, and the method used in determining these patterns.

(f) In addition to the other information required by this section, the log of a station in auxiliary operation shall have the following information entered:

(1) A system network diagram for each system with which the station is associated;

(2) The station transmitting band(s);

(3) The transmitter input power; and

(4) If operated by remote control, the information required by paragraph (c) of this section.

(g) Notwithstanding the provisions of §97.105, the log entries required by paragraphs (c), (d), (e), and (f) of this section shall be retained in the station log as long as the information contained in those entries is accurate.

§ 97.105 Retention of logs.

The station log shall be preserved for a period of at least 1 year following the last date of entry and retained in the possession of the licensee. Copies of the log, including the sections required to be transcribed by §97.103, shall be available to the Commission for inspection.

EMERGENCY OPERATIONS

§ 97.107 Operation in emergencies.

In the event of an emergency disrupting normally available communication facilities in any widespread area or areas, the Commission, in its discretion, may declare that a general state of communications emergency exists, designate the area or areas concerned, and specify the amateur frequency bands, or segments of such bands, for use only by amateurs participating in emergency communication within or with such affected area or areas. Amateurs desiring to request the declaration of such a state of emergency should communicate with the Commission's Engineer in Charge of the area concerned. Whenever such declaration has been made, operation of and with amateur stations in the area concerned shall be only in accordance with the requirements set forth in this section. but such requirements shall in nowise affect other normal amateur communication in the affected area when conducted on frequencies not designated for emergency operation.

(a) All transmissions within all designated amateur communications bands[1] other than communications relating directly to relief work, emergency service, or the establishment and maintenance of efficient amateur radio networks for the handling of such communications shall be suspended.

---

[1]The frequency 4383.8 kHz may be used by any station authorized under this part to communicate with any other station in the State of Alaska for emergency communications. No airborne operations will be permitted on this frequency. Additionally, all stations operating on this frequency must be located in or within 50 nautical miles of the State of Alaska.

Incidental calling, answering, testing or working (including casual conversations, remarks or messages) not pertinent to constructive handling of the emergency situation shall be prohibited within these bands.

(b) The Commission may designate certain amateur stations to assist in the promulgation of information relating to the declaration of a general state of communications emergency, to monitor the designated amateur emergency communications bands, and to warn non-complying stations observed to be operating in those bands. Such station, when so designated, may transmit for that purpose on any frequency or frequencies authorized to be used by that station, provided such transmissions do not interfere with essential emergency communications in progress; however, such transmissions shall preferably be made on authorized frequencies immediately adjacent to those segments of the amateur bands being cleared for the emergency. Individual transmissions for the purpose of advising other stations of the existence of the communications emergency shall refer to this section by number (§97.107) and shall specify, briefly and concisely, the date of the Commission's declaration, the area and nature of the emergency, and the amateur frequency bands or segments of such bands which constitute the amateur emergency communications bands at the time. The designated stations shall not enter into discussions with other stations beyond furnishing essential facts relative to the emergency, or acting as advisors to stations desiring to assist in the emergency, and the operators of such designated stations shall report fully to the Commission the identity of any stations failing to comply, after notice, with any of the pertinent provisions of this section.

(c) The special conditions imposed under the provisions of this section shall cease to apply only after the Commission or its authorized representative, shall have declared such general state of communications emergency to be terminated: however, nothing in this paragraph shall be deemed to prevent the Commission from modifying the terms of its declaration from time to time as may be necessary during the period of a communications emergency, or from removing those conditions with respect to any amateur frequency band or segment of such band which no longer appears essential to the conduct of the emergency communications.

SUBPART E—PROHIBITED PRACTICES AND ADMINISTRATIVE SANCTIONS

PROHIBITED TRANSMISSIONS AND PRACTICES

§ 97.112 No remuneration for use of station.

(a) An amateur station shall not be used to transmit or receive messages for hire, nor for communication for material compensation, direct or indirect, paid or promised.

(b) Control operators of a club station may be compensated when the club station is operated primarily for the purpose of conducting amateur radiocommunication to provide telegraphy practice transmissions intended for persons learning or improving proficiency in the international Morse code, or to disseminate information bulletins consisting solely of subject matter having direct interest to the Amateur Radio Service provided:

(1) The station conducts telegraphy practice and bulletin transmission for at least 40 hours per week.

(2) The station schedules operations on all allocated medium and high frequency amateur bands using reasonable measures to maximize coverage.

(3) The schedule of normal operating times and frequencies is published at least 30 days in advance of the actual transmissions.

Control operators may accept compensation only for such periods of time during which the station is transmitting telegraphy practice or bulletins. A control operator shall not accept any direct or indirect compensation for periods during which the station is transmitting material other than telegraphy practice or bulletins.

§ 97.113 Broadcasting prohibited.

Subject to the provisions of §97.91, an amateur station shall not be used to engage in any form of broadcasting, that is, the dissemination of radio communications intended to be received by the public directly or by the intermediary of relay stations, nor for the retransmission by automatic means of programs or signals emanating from any class of station other than amateur. The foregoing provisions shall not be construed to prohibit amateur operators from giving their consent to the rebroadcast by broadcast stations of the transmissions of their amateur stations, provided, that the transmissions of the amateur stations shall not contain any direct or indirect reference to the rebroadcast.

§ 97.114 Third party traffic.

The transmission or delivery of the following amateur radiocommunication is prohibited:

(a) International third party traffic except with countries which have assented thereto.

(b) Third party traffic involving material compensation, either tangible or intangible, direct or indirect, to a third party, a station licensee, a control operator, or any other person.

(c) Except for an emergency communication as defined in this part, third party traffic consisting of business communications on behalf of any party. For the purpose of this section business communication shall mean any transmission or communication the purpose of which is to facilitate the regular business or commercial affairs of any party.

§ 97.115 Music prohibited.

The transmission of music by an amateur station is forbidden.

§ 97.116 Amateur radiocommunication for unlawful purposes prohibited.

The transmission of radiocommunication or messages by an amateur radio station for any purpose, or in connection with any activity, which is contrary to Federal, State or local law is prohibited.

§ 97.117 Codes and ciphers prohibited.

The transmission by radio of messages in codes or ciphers in domestic and international communications to or between amateur stations is prohibited. All communications regardless of type of emission employed shall be in plain language except that generally recognized abbreviations established by regulation or custom and usage are permissible i

475

as are any other abbreviations or signals where the intent is not to obscure the meaning but only to facilitate communications.

§ 97.119 Obscenity, indecency, profanity.

No licensed radio operator or other person shall transmit communications containing obscene, indecent, or profane words, language, or meaning.

§ 97.121 False signals.

No licensed radio operator shall transmit false or deceptive signals or communications by radio, or any call letter or signal which has not been assigned by proper authority to the radio station he is operating.

§ 97.123 Unidentified communications.

No licensed radio operator shall transmit unidentified radio communications or signals.

§ 97.125 Interference.

No licensed radio operator shall willfully or maliciously interfere with or cause interference to any radio communication or signal.

§ 97.126 Retransmitting radio signals.

(a) An amateur radio station, except a station in repeater operation or auxiliary operation, shall not automatically retransmit the radio signals of other amateur radio stations.

(b) A remotely controlled station, other than a remotely controlled station in repeater operation or auxiliary operation, shall automatically retransmit only the radio signals of stations in auxiliary operation shown on the remotely controlled station's system network diagram.

§ 97.127 Damage to apparatus.

No licensed radio operator shall willfully damage, or cause or permit to be damaged, any radio apparatus or installation in any licensed radio station.

§ 97.129 Fraudulent licensees.

No licensed radio operator or other person shall obtain or attempt to obtain, or assist another to obtain or attempt to obtain, an operator license by fraudulent means.

ADMINISTRATIVE SANCTIONS

§ 97.131 Restricted operation.

(a) If the operation of an amateur station causes general interference to the reception of transmissions from stations operating in the domestic broadcast service when receivers of good engineering design including adequate selectivity characteristics are used to receive such transmission and this fact is made known to the amateur station licensee, the amateur station shall not be operated during the hours from 8 p.m. to 10:30 p.m., local time, and on Sunday for the additional period from 10:30 a.m. until 1 p.m., local time, upon the frequency or frequencies used when the interference is created.

(b) In general, such steps as may be necessary to minimize interference to stations operating in other services may be required after investigation by the Commission.

§ 97.133 Second notice of same violation.

In every case where an amateur station licensee is cited within a period of 12 consecutive months for the second violation of the provisions of §§97.61, 97.63, 97.65, 97.71, or 97.73, the station licensee, if directed to do so by the Commission, shall not operate the station and shall not permit it to be operated from 6 p.m. to 10:30 p.m., local time, until written notice has been received authorizing the resumption of full-time operation. This notice will not be issued until the licensee has reported on the results of tests which he has conducted with at least two other amateur stations at hours other than 6 p.m. to 10:30 p.m., local time. Such tests are to be made for the specific purpose of aiding the licensee in determining whether the emissions of the station are in accordance with the Commission's rules. The licensee shall report to the Commission the observations made by the cooperating amateur licensee in relation to reported violations. This report shall include a statement as to the corrective measures taken to insure compliance with the rules.

§ 97.135 Third notice of same violation.

In every case where an amateur station licensee is cited within a period of 12 consecutive months for the third violation of §97.61, 97.63, 97.65, 97.71, or 97.73, the station licensee, if directed by the Commission, shall not operate the station and shall not permit it to be operated from 8 a.m. to 12 midnight, local time, except for the purpose of transmitting a prearranged test to be observed by a monitoring station of the Commission to be designated in each particular case. The station shall not be permitted to resume operation during these hours until the licensee is authorized by the Commission, following the test, to resume full-time operation. The results of the test and the licensee's record shall be considered in determining the advisability of suspending the operator license or revoking the station license, or both.

§ 97.137 Answers to notices of violations.

Any licensee receiving official notice of a violation of the terms of the Communications Act of 1934, as amended, any legislative act, Executive order, treaty to which the United States is a party, or the rules and regulations of the Federal Communications Commission, shall, within 10 days from such receipt, send a written answer direct to the office of the Commission originating the official notice: *Provided, however,* That if an answer cannot be sent or an acknowledgement made within such 10-day period by reason of illness or other unavoidable circumstances, acknowledgement and answer shall be made at the earliest practicable date with a satisfactory explanation of the delay. The answer to each notice shall be complete in itself and shall not be abbreviated by reference to other communications or answers to other notices. If the notice relates to some violation that may be due to the physical or electrical characteristics of transmitting apparatus, the answer shall state fully what steps, if any, are taken to prevent future violations, and if any new apparatus is to be installed, the date such apparatus was ordered, the name of the manufacturer, and promised date of delivery. If the notice of violation relates to some lack of attention to or improper operation of the transmitter, the name of the operator in charge shall be given.

SUBPART F—RADIO AMATEUR CIVIL EMERGENCY SERVICE (RACES)

GENERAL

§ 97.161 Basis and purpose.

The Radio Amateur Civil Emergency Service provides for amateur radio operation for civil defense communications purposes only, during periods of local, regional or national civil emergencies, including any emergency which may necessitate invoking of the President's War Emergency Powers under the provisions of section 606 of the Communications Act of 1934, as amended.

§ 97.163 Definitions.

For the purposes of this Subpart, the following definitions are applicable:

(a) *Radio Amateur Civil Emergency Service.* A radiocommunication service conducted by volunteer licensed amateur radio operators, for providing emergency radiocommunications to local, regional, or state civil defense organizations.

(b) *RACES station.* An amateur radio station licensed to a civil defense organization, at a specific land location, for the purpose of providing the facilities for amateur radio operators to conduct amateur radiocommunications in the Radio Amateur Civil Emergency Service.

§ 97.165 Applicability of rules.

In all cases not specifically covered by the provisions contained in this Subpart, amateur radio stations and RACES stations shall be governed by the provisions of the rules governing amateur radio stations and operators (Subpart A through E of this part).

STATION AUTHORIZATIONS

§ 97.169 Station license required.

No transmitting station shall be operated in the Radio Amateur Civil Emergency Service unless:

(a) The station is licensed as a RACES station by the Federal Communications Commission, or

(b) The station is an amateur radio station licensed by the Federal Communications Commission, and is certified by the responsible civil defense organization as registered with that organization.

§ 97.171 Eligibility for RACES station license.

A RACES station will only be licensed to a local, regional, or state civil defense organization.

§ 97.173 Application for RACES station licensee.

(a) Each application for a RACES station license shall be made on the FCC Form 610-B.

(b) The application shall be signed by the civil defense official responsible for the coordination of all civil defense activities in the area concerned.

(c) The application shall be countersigned by the responsible official for the governmental entity served by the civil defense organization.

(d) If the application is for a RACES station to be in any special manner covered by §97.41, those showings specified for non-RACES stations shall also be submitted.

§ 97.175 Amateur radio station registration in civil defense organization.

No amateur radio station shall be operated in the Radio Amateur Civil Emergency Service unless it is registered in a civil defense organization by that organization.

OPERATING REQUIREMENTS

§ 97.177 Operator requirements.

No person shall be the control operator of a RACES station, or shall be the control operator of an amateur radio station conducting communications in the Radio Amateur Civil Emergency Service unless that person holds a valid amateur radio operator license and is certified as enrolled in a civil defense organization by that organization.

§ 97.179 Operator privileges.

Operator privileges in the Radio Amateur Civil Emergency Service are dependent upon, and identical to, those for the class of operator license held in the Amateur Radio Service.

§ 97.181 Availability of RACES station license and operator licenses.

(a) The original license of each RACES station, or a photocopy thereof, shall be attached to each transmitter of such station, and at each control point of such station. Whenever a photocopy of the RACES station license is utilized in compliance with this requirement, the original station license shall be available for inspection by any authorized Government official at all times while the station is being operated and at other times upon request made by an authorized representative of the Commission, except when such license has been filed with application for modification or renewal thereof, or has been mutilated, lost, or destroyed, and request has been made for a duplicate license in accordance with §97.57.

(b) In addition to the operator license availability requirements of §97.82, a photocopy of the control operator's amateur radio operator license shall be posted at a conspicuous place at the control point for the RACES station.

TECHNICAL REQUIREMENTS

§ 97.185 Frequencies available.

(a) All of the authorized frequencies and emissions allocated to the Amateur Radio Service are also available to the Radio Amateur Civil Emergency Service on a shared basis.

(b) In the event of an emergency which necessitates the invoking of §606 of the Communications Act of 1934 as amended, unless otherwise modified or directed, RACES stations and amateur radio stations participating in RACES will be limited in operation to the following:

FREQUENCY OR FREQUENCY BANDS	Limitations
kHz:	
1800-1825	
1975-2000	
3500-3510	

3510-3516	4
3516-3550	2,4
3984-4000	3
3997	4
7097-7103	2,4
7103-7125	2,4
7245-7255	4
14047-14053	2,4
14220-14230	4
21047-21053	4
MHz:	
28.55-28.75	
29.45-29.65	
50.35-50.75	
53.30	3
53.35-53.75	
145.17-145.71	
146.79-147.33	
220-225	5

(c) Limitations: (1) Use of frequencies in the band 1800-2000 kHz is subject to the priority of the LORAN system of radionavigation in this band and to the geographical, frequency, emission, and power limitations contained in §97.61 governing amateur radio stations and operators (Subparts A through E of this part).

(2) The availability of the frequency bands 3515-3550 kHz, 7103-7125 kHz, 7245-7247 kHz, 7253-7255 kHz, 14220-14222 kHz, and 14228-14230 kHz for use during periods of actual civil defense emergency is limited to the initial 30 days of such emergency, unless otherwise ordered by the Commission.

(3) For use in emergency areas when required to make initial contact with a military unit; also, for communications with military stations on matters requiring coordinations.

(4) For use by all authorized stations only in the continental United States, except that the bands 7245-7255 kHz and 14220-14230 kHz are also available in Alaska, Hawaii, Puerto Rico, and the Virgin Islands.

(5) Those stations operating in the band 220-225 MHz shall not cause harmful interference to the government radiolocation service.

§ 97.189 Point of communications.

(a) RACES stations may only be used to communicate with:

(1) Other RACES stations;

(2) Amateur radio stations certified as being registered with a civil defense organization, by that organization;

(3) Stations in the Disaster Communications Service;

(4) Stations of the United States Government authorized by the responsible agency to exchange communications with RACES stations;

(5) Any other station in any other service regulated by the Federal Communications Commission, whenever such station is authorized by the Commission, to exchange communications with stations in the Radio Amateur Civil Emergency Service.

(b) Amateur radio stations registered with a civil defense organization may only be used to communicate with:

(1) RACES stations licensed to the civil defense organization with which the amateur radio station is registered;

(2) Any of the following stations upon authorization of the responsible civil defense official for the organization in which the amateur radio station is registered:

(i) Any RACES station licensed to other civil defense organizations;

(ii) Amateur radio stations registered with the same or another civil defense organization;

(iii) Stations in the Disaster Communications Service;

(iv) Stations of the United States Government authorized by the responsible agency to exchange communications with RACES stations;

(v) Any other station in any other service regulated by the Federal Communications Commission, whenever such station is authorized by the Commission to exchange communications with stations in the Radio Amateur Civil Emergency Service.

§ 97.191 Permissible communications.

All communications in the Radio Amateur Civil Emergency Service must be specifically authorized by the civil defense organization for the area served. Stations in this service may transmit only civil defense communications of the following types:

(a) Communications concerning impending or actual conditions jeopardizing the public safety, or affecting the national defense or security during periods of local, regional, or national civil emergencies:

(1) Communications directly concerning the immediate safety of life or individuals, the immediate protection of property, maintenance of law and order, alleviation of human suffering and need, and the combating of armed attack or sabotage;

(2) Communications directly concerning the accumulation and dissemination of public information or instructions to the civilian population essential to the activities of the civil defense organization or other authorized governmental or relief agencies.

(b) Communications for training drills and tests necessary to ensure the establishment and maintenance of orderly and efficient operation of the Radio Amateur Civil Emergency Service as ordered by the responsible civil defense organization served. Such tests and drills may not exceed a total time of one hour per week.

(c) Brief one way transmissions for the testing and adjustment of equipment.

§ 97.193 Limitations on the use of RACES stations.

(a) No station in the Radio Amateur Civil Emergency Service shall be used to transmit or to receive messages for hire, nor for communications for material compensation, direct or indirect, paid or promised.

(b) All messages which are transmitted in connection with drills or tests shall be clearly identified as such by use of the words "drill" or "test", as appropriate, in the body of the messages.

SUBPART G—OPERATION OF AMATEUR RADIO STATIONS IN THE UNITED STATES BY ALIENS PURSUANT TO RECIPROCAL AGREEMENTS

§ 97.301 Basis, purpose, and scope.

(a) The rules in this subpart are based on, and are applicable solely to, alien amateur operations pursuant to section 303(1)(3) and 310(a) of the Communications Act of 1934, as amended. (See Pub. L. 93-505, 88 Stat. 1576.)

(b) The purpose of this subpart is to implement Public Law 88-313 by prescribing the rules under which an alien, who holds an amateur operator and station license issued by his government (hereafter referred to as an alien amateur), may operate an amateur radio station in the United States, in its possessions, and in the Commonwealth of Puerto Rico (hereafter referred to only as the United States).

§ 97.303 Permit required.

(a) Before he may operate an amateur radio station in the United States, under the provisions of sections 303(1)(2) and 310(a) of the Communications Act of 1934, as amended, an alien amateur licensee must obtain a permit for such operation from the Federal Communications Commission. A permit for such operation shall be issued only to an alien holding a valid amateur operator and station authorization from his government, and only when there is in effect a bilateral agreement between the United States and that government for such operation on a reciprocal basis by United States amateur radio operators.

§ 97.305 Application for permit.

(a) Application for a permit shall be made on FCC Form 610-A. Form 610-A may be obtained from the Commission's Washington, D.C., office, from any of the Commission's field offices and, in some instances, from United States missions abroad.

(b) The application form shall be completed in full in English and signed by the applicant. A photocopy of the applicant's amateur operator and station license issued by his government shall be filed with the application. The Commission may require the applicant to furnish additional information. The application must be filed by mail or in person with the Federal Communications Commission, Gettysburg, Pennsylvania 17325, U.S.A. To allow sufficient time for processing, the application should be filed at least 60 days before the date on which the applicant desires to commence operation.

§ 97.307 Issuance of permit.

(a) The Commission may issue a permit to an alien amateur under such terms and conditions as it deems appropriate. If a change in the terms of a permit is desired, an application for modification of the permit is required. If operation beyond the expiration date of a permit is desired, an application for renewal of the permit is required. In any case in which the permittee has, in accordance with the provisions of this subpart, made a timely and sufficient application for renewal of an unexpired permit, such permit shall not expire until the application has been finally determined. Applications for modification or for renewal of a permit shall be filed on FCC Form 610-A.

(b) The Commission, in its discretion may deny any application for a permit under this subpart. If an application is denied, the applicant will be notified by letter. The applicant may, within 90 days of the mailing of such letter, request the Commission to reconsider its action.

(c) Normally, a permit will be issued to expire 1 year after issuance but in no event after the expiration of the license issued to the alien amateur by his government.

§ 97.309 Modification, suspension, or cancellation of permit.

At any time the Commission may, in its discretion, modify, suspend, or cancel any permit issued under this subpart. In this event, the permittee will be notified of the Commission's action by letter mailed to his mailing address in the United States and the permittee shall comply immediately. A permittee may, within 90 days of the mailing of such letter, request the Commission to reconsider its action. The filing of a request for reconsideration shall not stay the effectiveness of that action, but the Commission may stay its action on its own motion.

§ 97.311 Operating conditions.

(a) The alien amateur may not under any circumstances begin operation until he has received a permit issued by the Commission.

(b) Operation of an amateur station by an alien amateur under a permit issued by the Commission must comply with all of the following:

(1) The terms of the bilateral agreement between the alien amateur's government and the government of the United States;

(2) The provisions of this subpart and of Subparts A through E of this part;

(3) The operating terms and conditions of the license issued to the alien amateur by his government; and

(4) Any further conditions specified on the permit issued by the Commission.

§ 97.313 Station identification.

(a) The alien amateur shall identify his station as follows:

(1) Radio telegraph operation: The amateur shall transmit the call sign issued to him by the licensing country followed by a slant (/) sign and the United States amateur call sign prefix letter(s) and number appropriate to the location of his station.

(2) Radiotelephone operation: The amateur shall transmit the call sign issued to him by the licensing country followed by the words "fixed", "portable" or "mobile", as appropriate, and the United States amateur call sign prefix letter(s) and number appropriate to the location of his station. The identification shall be made in the English language.

(b) At least once during each contact with another amateur station, the alien amateur shall indicate, in English, the geographical location of his station as nearly as possible by city and state, commonwealth, or possession.

SUBPART H—(RESERVED)

APPENDICES
APPENDIX 1

EXAMINATION POINTS

Examinations for amateur radio operator licenses are conducted at the Commission's office in Washington, D.C., and at each field office of the Commission on the days designated by the Engineer in Charge of each office. Specific dates should be obtained from the Engineer in Charge of the nearest field office of the Commission.

Examinations are also given at prescribed intervals in the cities listed in the Commission's current Examination Schedule, copies of which are available from the Federal Communications Commission Regional Services Division, Washington, D.C. 20554, or from any one of the Commission's field offices listed in §0.121.

APPENDIX 2

Extracts From Radio Regulations Annexed to the International Telecommunication Convention (Geneva, 1959)

ARTICLE 41—AMATEUR STATIONS

SECTION 1. Radiocommunications between amateur stations of different countries[1] shall be forbidden if the administration of one of the countries concerned has notified that it objects to such radiocommunications.

SEC. 2.(1) When transmissions between amateur stations of different countries are permitted, they shall be made in plain language and shall be limited to messages of a technical nature relating to tests and to remarks of a personal character for which, by reason of their unimportance, recourse to the public telecommunications service is not justified. It is absolutely forbidden for amateur stations to be used for transmitting international communications on behalf of third parties.

(2) The preceding provisions may be modified by special arrangements between the administrations of the countries concerned.

SEC. 3(1) Any person operating the apparatus of an amateur station shall have proved that he is able to send correctly by hand and to receive correctly by ear, texts in Morse code signals. Administrations concerned may, however, waive this requirement in the case of stations making use exclusively of frequencies above 144 MHz.

(2) Administrations shall take such measures as they judge necessary to verify the technical qualifications of any person operating the apparatus of an amateur station.

SEC. 4. The maximum power of amateur stations shall be fixed by the administrations concerned, having regard to the technical qualifications of the operators and to the conditions under which these stations are to work.

SEC. 5. (1) All the general rules of the Convention and of these Regulations shall apply to amateur stations. In particular, the emitted frequency shall be as stable and as free from spurious emissions as the state of technical development for such stations permits.

(2) During the course of their transmissions, amateur stations shall transmit their call sign at short intervals.

RESOLUTION NO. 10

Relating to the use of the bands 7000 to 7100 kHz and 7100 to 7300 kHz by the Amateur Service and the Broadcasting Service.

The Administrative Radio Conference Geneva, 1959.

Considering—

(a) That the sharing of frequency bands by amateur, fixed, and broadcasting services is undesirable and should be avoided;

(b) That it is desirable to have worldwide exclusive allocations for these services in Band 7;

(c) That the band 7000 to 7100 kHz is allocated on a worldwide basis exclusively to the amateur service;

(d) That the band 7100 to 7300 kHz is allocated in Regions 1 and 3 to the broadcasting service and in Region 2 to the amateur service,

resolves,

that the broadcasting service should be prohibited from the band 7000 to 7100 kHz and that broadcasting stations operating on frequencies in this band should cease such operation;

and noting,

the provisions of No. 117 of the Radio Regulations;

further resolves,

that interregional amateur contacts should be only in the band 7000 to 7100 kHz and that the administrations should make every effort to ensure that the broadcasting service in the band 7100 to 7300 kHz, in Regions 1 and 3, does not cause interference to the amateur service in Region 2; such being consistent with the provisions of No. 117 of the Radio Regulations.

APPENDIX 3

CLASSIFICATION OF EMISSIONS

For convenient reference the tabulation below is extracted from the classification of typical emissions in Part 2 of the Commission's Rules and Regulations and in the Radio Regulations, Geneva, 1959, and it includes only those general classifications which appear most applicable to the Amateur Radio Service.

Type of modulation	Type of transmission	Symbol
Amplitude	With no modulation	A0
	Telegraph without the use of modulating audio frequency (by on-off keying)	A1
	Telegraphy by the on-off keying of an amplitude modulating audio frequency or audio frequencies or by the on-off keying of the modulated emission (special case: an unkeyed emission amplitude modulated).	A2
	Telephony	A3[1]
	Facsimile	A4
	Television	A5
Frequency (or phase)	Telegraphy by frequency shift keying without the use of a modulating audio frequency.	F1
	Telegraphy by the on-off keying of a frequency modulating audio frequency or by the on-off	F2

laying of frequency modulated emission (special case; an unkeyed emission frequency modulated)	
Telephony	F3
Facsimile	F4
Television	F5
Pulse	P

¹(In Part 97) Unless specified otherwise, A3 includes single and double sideband with full, reduced, or suppressed carrier.

APPENDIX 4

Convention Between the United States of America and Canada, Relating to the Operation by Citizens of Either Country of Certain Radio Equipment or Stations in the Other Country (Effective May 15, 1952)

ARTICLE III

It is agreed that persons holding appropriate amateur licenses issued by either country may operate their amateur stations in the territory of the other country under the following conditions:

(a) Each visiting amateur may be required to register and receive a permit before operating any amateur station licensed by his government.

(b) The visiting amateur will identify his station by:

(1) *Radiotelegraphy operation.* The amateur call sign issued to him by the licensing country followed by a slant (/) sign and the amateur call sign prefix and call area number of the country he is visiting.

(2) *Radiotelephone operation.* The amateur call sign in English issued to him by the licensing country followed by the words, "fixed" "portable" or "mobile," as appropriate, and the amateur call sign prefix and call area number of the country he is visiting.

(c) Each amateur station shall indicate at least once during each contact with another station its geographical location as nearly as possible by city and, state or city and province.

(d) In other respects the amateur station shall be operated in accordance with the laws and regulations of the country in which the station is temporarily located.

APPENDIX 5

DETERMINATION OF ANTENNA HEIGHT ABOVE AVERAGE TERRAIN

The effective height of the transmitting antenna shall be the height of the antenna's center of radiation above "average terrain." For this purpose "effective height" shall be established as follows:

(a) On a U.S. Geological Survey Map having a scale of 1:250,000, lay out eight evenly spaced radials, extending from the transmitter site to a distance of 10 miles and beginning at (0°, 45°, 90°, 135°, 180°, 225°, 270°, 315°T.) If preferred, maps of greater scale may be used.

(b) By reference to the map contour lines, establish the ground elevation above mean sea level (AMSL) at 2, 4, 6, 8, and 10 miles from the antenna structure along each radial. If no elevation figure or contour line exists for any particular point, the nearest contour line elevation shall be employed.

(c) Calculate the arithmetic average of these 40 points of elevation (5 points of each of 8 radials).

(d) The height above average terrain of the antenna is thus the height AMSL of the antenna's center of radiation, minus the height of average terrain as calculated above.

NOTE 1: Where the transmitter is located near a large body of water, certain points of established elevation may fall over water. Where it is expected that service would be provided to land areas beyond the body of water, the points at water level in that direction should be included in the calculation of average elevation. Where it is expected that service would not be provided to land areas beyond the body of water, the points at water level should not be included in the average.

NOTE 2: In instances in which this procedure might provide unreasonable figures due to the unusual nature of the local terrain, applicant may provide additional data at his own discretion, and such data may be considered if deemed significant.

---

¹As may appear in public notices issued by the Commission.

WORD INDEX TO PART 97

A

Abbreviations	97.117
Additional station	97.3, 97.40(b), 97.41
Address	97.13(e), 97.42, 97.47(c)
Advanced class	97.5, 97.7, 97.23(b), 97.27, 97.28
Aircraft, Operation aboard	97.101
Aircraft, radiotelegraph endorsement	97.25(c)
Airport, antenna restrictions near	97.45
Aliens	97.301, 97.303, 97.305, 97.307, 97.309, 97.311, 97.313
Amateur extra class	97.5, 97.7, 97.23(a), 97.25, 97.27, 97.78
Amateur extra first class license	97.25(d)
Amateur radio organization	97.3(c), 97.39
Amplifier oscillator transmitter	97.67(a)
Amplitude modulation	Appendix 3
Antenna	97.3, 97.41, 97.45, 97.67, Appendix 5
Application	97.11, 97.13, 97.28, 97.41, 97.45, 97.47, 97.95, 97.173, 97.305, 97.307
Association, eligibility for license	97.39
Audio frequency tones	97.93
Automatic control	97.3, 97.79, 97.88
Automatic identification device	97.84(g)
Auxiliary link station	97.3, 97.86, 97.103(d), 97.126

B

Bandwidth	97.65
Broadcasting	97.91, 97.113
Business communications	97.114

C

Call sign, amateur	97.13(f), 97.47(c), 97.51, 97.99, 97.103, 97.121, 97.313
Call sign, preferred	97.51
Call sign, specific	97.51
Call sign areas	97.51
Canadian citizens	97.42, Appendix 4
Carrier	97.63, 97.65, 97.73, 97.74, 97.93
Change of mailing address	97.13(e), 97.47(c)
Change of name	97.13(e), 97.47(c)
Ciphers	97.99(b), 97.117
Civil defense (also see RACES)	97.89, 97.161, 97.163, 97.165, 97.169, 97.171, 97.173, 97.175, 97.177, 97.189, 97.191
Class A (see Advanced Class)	97.5
Class B (see General Class)	97.5
Class C (see Conditional Class or General Class)	97.5
Club station	97.3(h), 97.40, 97.41, 97.47, 97.95, 97.112(b)
Code (telegraphy)	97.7, 97.21, 97.25, 97.28, 97.31, 97.65(b), 97.91, 97.112(b), 97.313
Code practice transmissions	97.65(b), 97.91, 97.112(b)
Codes and ciphers	97.99(b), 97.117
Commercial radiotelegraph operator license	97.25(c)
Compensation	97.112, 97.114, 97.193(a)
Conditional class (see General Class)	97.95, 97.7(c)
Control (see Local control, Remote control, or Automatic control)	97.3, 97.41, 97.43, 97.65, 97.79, 97.85, 97.86, 97.88, 97.103(c), 97.126

486

Control link- 97.3, 97.88, 97.103(c)
Control operator- 97.3, 97.79, 97.88, 97.103(b), 97.112(b), 97.114(b), 97.177, 97.181
Control point- 97.3, 97.43, 97.79(b), 97.88, 97.103(c), 97.181
Control station- 97.3, 97.40, 97.41, 97.88, 97.103
Corporations, eligibility for license- 97.39
Credit, examination- 97.19, 97.25

D

Damage to apparatus- 97.127
Direct mechanical control (see Local control)- 97.3(m)
Direct wire control (see Local control)- 97.3(m)
Disability- 97.27, 97.28(d)
Disaster Communications Service- 97.189(a)
Duplicate license- 97.57, 97.59(c), 97.82, 97.83, 97.181(a)

E

Effective radiated power- 97.3(t), 97.67(c)
Electronic key- 97.28(c)
Elements, examination (also see Examination)- 97.21, 97.23, 97.25, 97.28, 97.31
Emergency (also see RACES)- 97.1(a), 97.3(w), 97.61(b), 97.91, 97.107, 97.114(c), 97.185(b)
Emission- 97.7(e), 97.61, 97.65, 97.69, 97.73, 97.85, 97.88, 97.89(b), 97.117, 97.185, Appendix 3
Environmental information- 97.41(c)
Examination, amateur license- 97.11, 97.13(b), 97.19, 97.21, 97.23, 97.25, 97.27, 97.28, 97.31, 97.32, 97.33, Appendix 1

Examinations, mail- 97.27, 97.28
Expiration date, license- 97.59, 97.307
Expired license- 97.13(d), 97.47(b)
Extra class license- 97.5, 97.7, 97.23(a), 97.25, 97.27, 97.28, 97.41

F

FAA- 97.45
False signals- 97.121
Federal Aviation Administration- 97.45
First class commercial radiotelegraph operator
  license- 97.25(c)
First class license, Amateur Extra- 97.25(d)
Fixed operation- 97.3(b), 97.81, 97.83, 97.95, 97.103, 97.313(a)
Foreign government jurisdiction- 97.95(b)
Foreign government representative- 97.9, 97.37
Form 452-C- 97.99
Form 610- 97.11, 97.13, 97.28, 97.41, 97.47, 97.95(a)
Form 610-A- 97.305, 97.307
Form 610-B- 97.41, 97.47, 97.173(a)
Form 660-B- 97.32
Form 714- 97.45
Form 7460-1 (FAA Form)- 97.129
Fraudulent licenses- 97.3, 97.7, 97.61, 97.63, 97.65, 97.69, 97.71, 97.73, 97.74, 97.75, 97.81, 97.85, 97.86, 97.88, 97.95(b), 97.103(c), 97.107, 97.111(b), 97.112(b), 97.131, 97.185, Appendix 2
Frequency modulation- 97.65(c), 97.71, Appendix 3
Frequency shift keying- 97.69(c)
Functional block diagram- 97.103(c)

487

G

General class — 97.5, 97.7, 97.23, 97.27, 97.28
Government radiolocation service — 97.61(b), 97.185(c)
Grace period — 97.13(d), 97.47(b)
Grading of examinations — 97.31

H

Hand key — 97.28(c)

I

Identification, station — 97.79, 97.84, 97.99, 97.313
Indecent [words] — 97.119
Industrial devices, interference from — 97.61(b)
Interference — 97.61(b), 97.101, 97.125, 97.131, 97.185(c)
Interim Amateur Permit — 97.3(d), 97.32, 97.84(f)
International Morse code — (see Code or Telegraphy)
International Telecommunication Convention Regulations (excerpts) — Appendix 2
International Telegraphic Alphabet No. 2 — 97.69(a)
ISM devices, interference from — 97.61(b)

K

Keyboard keyer — 97.28(c)
Kilowatt — 97.67

L

License (also see station license and operator license) — 97.3, 97.5, 97.7, 97.9, 97.11, 97.13, 97.19, 97.23, 97.25, 97.27, 97.28, 97.31, 97.32, 97.33, 97.37, 97.39, 97.40, 97.41, 97.42, 97.43, 97.47, 97.49, 97.57, 97.59, 97.81, 97.82, 97.83, 97.84, 97.88, 97.95, 97.169, 97.171, 97.173, 97.177, 97.179, 97.181
License, commercial radiotelegraph operator — 97.25(c)
Local control — 97.3(m)
Location, station — 97.43, 97.47, 97.95, 97.313
Logs, station — 97.99(c), 97.103, 97.105
LORAN radionavigation system — 97.61(b), 97.185

M

Mail examinations — 97.27, 97.28
Mailing address — 97.41, 97.42
Mailing address, change in — 97.13(e), 97.47(c)
Manual control (See Local Control or Remote Control) — 97.3(m)
Marking and lighting requirements, antenna — 97.45(c)
Measurement of frequencies — 97.74
Medical devices, interference from — 97.61(b)
Military recreation station — 97.3(g), 97.37, 97.41, 97.47, 97.95(a)
Mobile (operation) — 97.3(l), 97.81, 97.83, 97.95, 97.101, 97.313(a)
Model craft, radio control of — 97.89(b), 97.99
Modification, license — 97.11, 97.13, 97.41, 97.47, 97.49, 97.59(c), 97.82, 97.83, 97.95, 97.181, 97.307, 97.309
Modulation — 97.65, 97.93, Appendix 3
Morse code, international — (see Code or Telegraphy)
Music — 97.115

N

Name change — 97.13(e), 97.47(c)
Network identifier — 97.84(a)
Novice class — 97.5, 97.7(e), 97.11(b), 97.23(e), 97.28(b), 97.39

O

Obscene (words) — 97.119
Obstruction marking and lighting — 97.45(c)
One-way communications — 97.85, 97.91, 97.191(c)
One-year grace period — 97.13(d), 97.47(b)
Operation, Amateur radio (see Fixed, Portable or Mobile) — 97.3(l), 97.81, 97.83, 97.95, 97.101, 97.103, 97.313(a)
Operator, Amateur radio — 97.3(c), 97.37, 97.40, 97.79, 97.82
Operator license — 97.3(d), 97.5, 97.7, 97.9, 97.11, 97.13, 97.19, 97.23, 97.25, 97.28, 97.32, 97.33, 97.41, 97.47, 97.59, 97.83, 97.177, 97.179, 97.181
Organization, Amateur radio — 97.3(h), 97.39
Oscillator transmitter — 97.67

P

Permit, Interim — 97.3(d), 97.32, 97.84(f)
Personal Radio Division — 97.25(d)
Phase modulation — 97.65(c), Appendix 3
Phonetic alphabet — 97.84(g)
Photocopy of license — 97.47, 97.83, 97.85, 97.88, 97.103, 97.181, 97.305(b)
Plain language — 97.117
Plate input power — 97.67
Portable operation — 97.3(l), 97.81, 97.95, 97.103, 97.313(a)
Posting requirements — 97.32(d), 97.82, 97.83, 97.88, 97.181(b)
Power — 97.61(b), 97.67, 97.71, 97.73, 97.75, 97.76, 97.77, 97.103(e), 97.111(f), 97.185(c)
Preferred call sign — 97.51(b)
Primary station — 97.3(f), 97.40, 97.41
Profanity — 97.119
Propagation — 97.21(g)
Purity of emissions — 97.73

Q

Quiet hours — 97.131, 97.133, 97.135

R

RACES — 97.161, 97.163, 97.165, 97.169, 97.171, 97.173, 97.175, 97.177, 97.179, 97.181, 97.185, 97.189, 97.191, 97.193
Radiation center (see Antennas) — 97.3(r)
Radio Amateur Civil Emergency Services (see RACES) — Appendix 5
Radio control (of remote objects) — 97.21(g), 97.89(b), 97.99
Radio teleprinter — 97.69, 97.84(a)
Radiotelegraphy — 97.7(e), 97.21, 97.25(c), 97.28, 97.65(b), 97.84, 97.112(b), 97.313(a)
Radiotelephony — 97.21, 97.65(b), 97.84, 97.93, 97.313(a)
Rebroadcast — 97.113
Reciprocal operation (also see Aliens) — 97.95(b), 97.303
Re-examination — 97.33
Relay stations — 97.113
Remote control — 97.3(m), 97.65, 97.85(e), 97.86(a), 97.88, 97.103(c), 97.126
Remuneration — 97.112, 97.114, 97.193(a)

489

Renewal, license — 97.11(e), 97.13, 97.32(f), 97.47, 97.82, 97.83, 97.181, 97.307
Repeater station — 97.3(l), 97.61, 97.67(c), 97.84(c) & (d), 97.85, 97.88(e), 97.103(e), 97.126
Representative of a foreign government — 97.9, 97.37
Restricted operation — 97.131
Retransmission, automatic — 97.3(x),97.85, 97.86, 97.113, 97.126
Revocation, license — 97.135

S

Schools, eligibility of for station license — 97.39
Scientific devices, interference from — 97.61(b)
Semi-automatic key — 97.28(c)
Ship, operation aboard — 97.101
Sideband — 97.63(b), 97.65
Signal keying equipment, teleprinter — 97.69(b)
Space radio station — 97.3(i)
Special identifier (See Interim Amateur Permit) — 97.84(f)
Specific call sign — 97.51(b)
Spurious radiation — 97.73
Station (also see specific type of station) — 97.3, 97.40(c), 97.41, 97.47, 97.51, 97.61, 97.79, 97.83, 97.88, 97.89, 97.95, 97.103, 97.109, 97.110, 97.111, 97.112, 97.113, 97.126
Station license — 97.3, 97.32, 97.37, 97.39, 97.40, 97.41, 97.43, 97.47, 97.49, 97.59(b), 97.81, 97.85, 97.88, 97.95, 97.103, 97.135, 97.169, 97.171, 97.173, 97.181
Station location — 9.7, 43, 97.47, 97.95, 97.103(b), 97.313
Straight key — 97.28(c)
Structures, antenna — 97.3(q), 97.41(c), 97.45, 97.67, Appendix 5
Suspension, license — 97.135, 97.309
System network diagram — 97.3(u), 97.86(c), 97.103(f), 97.126

T

Technician class — 97.5, 97.7, 97.23, 97.27, 97.28
Telegraphy — 97.7(e), 97.21, 97.25(b) & (c), 97.28(b) & (c), 97.31, 97.65(b), 97.87, 97.112(b), 97.313(a)
Telegraphy practice transmissions — 97.112(b)
Telephony — 97.21, 97.65(b), 97.84, 97.93, 97.313(a)
Teleprinter, radio — 97.69, 97.84(a)
Temporary permit — (see Interim Amateur Permit)
Term of license — 97.13, 97.59
Terrestrial location — 97.3(j), 97.7(e), 97.67(d)
Third-parties (third party traffic) — 97.3(v), 97.79(d), 97.103(b), 97.114
Tower, antenna — 97.3, 97.41(c), 97.45, 97.67, Appendix 5
Transmitter — 97.3, 97.40, 97.67(a), 97.71, 97.81, 97.99, 97.137, 97.181
Transmitter Identification Card — 97.99
Treaties and Agreements — 97.49, 97.95(b), 97.137, 97.311(b), Appendix 4
Trustee — 97.39

490

U

Unidentified communications — 97.123
Unlawful purposes — 97.116

V

Vacuum tube — 97.67

V

Violation notices — 97.133, 97.135, 97.137
Volunteer examiner — 97.11, 97.28(b)

W

Watts — 97.3(u), 97.61, 97.67, 97.99

Appendix F

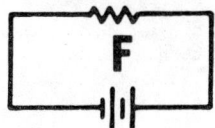

International Prefixes

PREFIX	COUNTRY
AP	PAKISTAN
A2C	BOTSWANA
A35	TONGA
A4X	SULTANATE OF OMAN
A51	BHUTAN
A6X	UNITED ARAB EMIRATES
A7X	QATAR
A9X	BAHRAIN
BV	TAIWAN
BY	PEOPLES REPUBLIC OF CHINA
CE	CHILE
CE9AA-AM	CHILEAN ANTARCTICA TIERRA DE O'HIGGINS, PALMER PENINSULA, GRAHAM LAND
CE9AN-AZ	SOUTH SHETLAND IS.
CE0A	EASTER IS.
CE0X	SAN FELIX IS.
CE0Z	JUAN FERNANDEZ IS.
CM	CUBA
CN	MOROCCO
CO	CUBA
CP	BOLIVIA
CR3	GUINEA BISSAU
CR9	MACAO
CT1	PORTUGAL
CT2	AZORES IS.
CT3	MADEIRA IS.
CX	URUGUAY
C21	REPUBLIC OF NAURU
C31	ANDORRA
C5A	GAMBIA
C6A	BAHAMA IS.
C9M	MOZAMBIQUE
DA-DL	FED REP OF WEST GERMANY
DM	EAST GERMAN DEMOCRATIC REP.
DU	PHILIPPINES
D2A	ANGOLA
D4	REP OF CAPE VERDE
D6	STATE OF COMORO

Prefix	Entity
EA	SPAIN
EA6	BALEARIC IS.
EA8	CANARY IS.
EA9	SPANISH SAHARA, CEUTA Y MELILLA
EI	IRELAND
EL,5L	LIBERIA
EP	IRAN
ET3	ETHIOPIA
F	FRANCE
FB8W	CROZET IS.
FB8X	KERGUELEN IS.
FB8Y	ANTARCTICA
FB8Z	AMSTERDAM & ST. PAUL IS.
FC	CORSICA
FG7	GUADELOUPE
FH8	MAYOTTE
FK8	NEW CALEDONIA
FM7	MARTINIQUE
FO8	FRENCH POLYNESIA, CLIPPERTON IS.
FP8	ST. PIERRE & MIQUELON IS.
FR7	GLORIOSO, JUAN DE NOVA, REUNION, TROMELIN IS.
FS7	ST. MARTIN IS.
FW8	WALLIS & FUTUNA IS.
FY7	FRENCH GUIANA
G	ENGLAND
GD	ISLE OF MAN
GI	NORTHERN IRELAND
GJ	JERSEY
GM	SCOTLAND
GU	BAILIWICK OF GUERNSEY
GW	WALES
HA,HG	HUNGARY
HB	SWITZERLAND
HB0	LIECHTENSTEIN
HC	ECUADOR
HC8	GALAPAGOS IS.
HG	HUNGARY
HH	HAITI
HI	DOMINICAN REPUBLIC
HK	COLOMBIA
HK0	BAJO NUEVO, MALPELO, SAN ANDRES & PROVIDENCIA IS.
HK0	SERRANA BANK & RONCADOR CAY
HM,HL9	KOREA
HP	PANAMA
HR	HONDURAS
HR0	SWAN IS.
HS	THAILAND
HV	VATICAN CITY
HZ,7Z	SAUDI ARABIA
H4	SOLOMON IS.

I, IW	ITALY
IA	TUSCAN ARCHIPELAGO
IC	CAPRI & ISCHIA IS.
IG9	LAMPEDUSA IS.
IH	PANTELLERIA IS.
IM	MADDALENA IS.
IS	SARDINIA
IT	SICILY
JA, JE-JJ, JR	JAPAN
JD1	OGASAWARA, MINAMI-TORI-SHIMA IS.
JR6	OKINAWA (RYUKYU IS.)
JT1	MONGOLIA
JW	SVALBARD IS.
JX	JAN MAYEN IS.
JY	JORDAN
J2	REP OF DJIBOUTI
J3	GRENADA & DEPENDENCIES
K	UNITED STATES OF AMERICA
KA	U.S. PERSONNEL IN JAPAN
KB6	BAKER, CANTON, ENDERBURY, HOWLAND & PHOENIX IS.
KC4	NAVASSA IS.
KC4AA, KC4US	ANTARCTICA
KC6	CAROLINE IS.
KG4	GUANTANAMO BAY
KG6	MARIANA IS.
KG6	GUAM
KG6R	ROTA
KG6S	SAIPAN
KH6	HAWAII & KURE IS.
KJ6	JOHNSTON IS.
KL7	ALASKA
KM6	MIDWAY IS.
KP4	PUERTO RICO
KP6	JARVIS & PALMYRA IS.
KS6	AMERICAN SAMOA
KV4	VIRGIN IS.
KW6	WAKE IS.
KX6	MARSHALL IS.
K25	CANAL ZONE
LA-LJ	NORWAY
LU	ARGENTINA
LU-Z	ANTARCTICA
LX	LUXEMBOURG
LZ	BULGARIA
M1	SAN MARINO
N	UNITED STATES OF AMERICA
OA	PERU
OD5	LEBANON

OE	AUSTRIA
OH	FINLAND
OHO	ALAND IS.
OJO, OHOM	MARKET REEF
OK,OL	CZECHOSLOVAKIA
ON	BELGIUM
OR	ANTARCTICA
OX	GREENLAND
OY	FAEROES IS.
OZ	DENMARK
PA-PI	NETHERLANDS
PJ	NETHERLANDS ANTILLES
PJ2,9	CURACAO
PJ3,9	ARUBA
PJ4,9	BONAIRE
PJ5,8	ST. EUSTATIUS
PJ6,8	SABA IS.
PJ7,8	SINT MAARTEN
PP-PY	BRAZIL
PY0	FERNANDO DE NORONHA IS.
PY0	ST. PETER & ST. PAUL'S ROCKS
PY0	TRINIDADE & MARTIM VAZ IS.
PZ	SURINAM
P29	PAPUA NEW GUINEA
SJ-SM	SWEDEN
SP	POLAND
ST	SUDAN
SU	EGYPT
SV	CRETE,GREECE
SV5	DODECANESE IS.
S2,S3	BANGLADESH
S7	SEYCHELLES IS.
S8	TRANSKEI
S9	SAO TOME & PRINCIPE IS.
TA,TC	TURKEY
TF	ICELAND
TG	GUATEMALA
TI	COSTA RICA
TI9	COCOS IS.
TJ	CAMEROON
TL8	CENTRAL AFRICAN REPUBLIC
TN8	REPUBLIC OF CONGO
TR8	GABON REPUBLIC
TT8	REPUBLIC OF CHAD
TU2	IVORY COAST
TY	PEOPLLES REPUBLIC OF BENIN
TZ	MALI REPUBLIC
UA1,2,3,4,6	EUROPEAN RUSSIAN SOVIET FEDERATED SOCIALIST REPUBLIC
UA9,0	ASIATIC RUSSIAN S.F.S.R.
UA1	FRANZ JOSEF LAND
UA2	KALININGRADSK

UB5	UKRAINIAN S.S.R.
UC2	WHITE RUSSIAN S.S.R.
UD6	AZERBAIDZHAN S.S.R.
UF6	GEORGIAN S.S.R.
UG6	ARMENIAN S.S.R.
UH8	TURKMEN S.S.R.
UI8	UZBEK S.S.R.
UJ8	TADZHIK S.S.R.
UL7	KAZAKH S.S.R.
UM8	KIRGHIZ S.S.R.
UN1	KARELO-FINNISH S.S.R.
UO5	MOLDAVIAN S.S.R.
UP2	LITHUANIAN S.S.R.
UQ2	LATVIAN S.S.R.
UR2	ESTONIAN S.S.R.
VE	CANADA
VK	AUSTRALIA
VK2	LORD HOWE IS.
VK9N	NORFOLK IS.
VK9X	CHRISTMAS IS.
VK9V	COCOS IS.
VK9Z	WILLIS IS.
VK0	ANTARCTICA
VO1	NEWFOUNDLAND
VO2	LABRADOR
VP1	BELIZE
VP2	LEEWARD & WINDWARD IS.
VP2A	ANTIGUA,BARBUDA
VP2D	DOMINICA
VP2E	ANGUILLA
VP2K	ST KITTS,NEVIS
VP2L	ST LUCIA
VP2M	MONSTERRAT
VP2S	ST VINCENT & DEPENDENCIES
VP2V	BRITISH VIRGIN IS.
VP5	TURKS & CAICOS IS.
VP8	FALKLAND,S.GEORGIA,S.ORKNEY, S.SANDWICH,S.SHETLAND IS.,GRAHAM LAND
VP9	BERMUDA IS.
VQ9	CHAGOS
VR1	BRITISH PHOENIX,GILBERT & OCEAN IS.
VR3	NORTHERN LINE IS.
VR6	PITCAIRN IS.
VR7	CENTRAL & SOUTHERN LINE IS.
VR8	TUVALU IS.
VS5	BRUNEI
VS6	HONG KONG
VU2	INDIA
VU7	ANDAMAN & NICOBAR IS.
VU7	LACCADIVE IS.
W	UNITED STATES OF AMERICA
XE,XF	MEXICO

Prefix	Country
XF4	REVILLA GIGEDO IS.
XT2	VOLTAIC REPUBLIC
XU	CAMBODIA/KHMER REPUBLIC
XV5	VIETNAM
XW8	LADS
XZ	BURMA
YA	AFGHANISTAN
YB-YD	INDONESIA,TIMOR IS.
YI	IRAQ
JQ	NEW HEBRIDES
YK	SYRIA
YN	NICARAGUA
YO	ROMANIA
YS	EL SALVADOR
YU	YUGOSLAVIA
YV	VENEZUELA
YV0	AVES IS.
ZA	ALBANIA
ZB2	GIBRALTAR
ZD7	SAINT HELENA IS.
ZD8	ASCENSION IS.
ZD9	TRISTAN DA CUNHA & GOUGH IS.
ZE	RHODESIA
ZF1	CAYMAN IS.
ZK1	COOK & MANIHIKI IS.
ZK2	NIUE IS.
ZL	NEW ZEALAND & AUCKLAND,CAMPBELL,CHATHAM,KERMADEC IS.
ZL5	ANTARCTICA
ZM7	TOKELAU IS.
ZP	PARAGUAY
ZR,ZS1,2,4,5,6	REPUBLIC OF SOUTH AFRICA
ZS1ANT	ANTARCTICA
ZS2	PRINCE EDWARD & MARION IS.
ZR,ZS3	SOUTHWEST AFRICA (NAMIBIA)
3A	MONACO
3B6	AGALEGA IS.
3B7	ST. BRANDON IS.
3B8	MAURITIUS IS.
3B9	RODRIGUEZ IS.
3C	EQUATORIAL GUINEA
3D2	FIJI IS.
3D6	SWAZILAND
3V8	TUNISIA
3X	REPUBLIC OF GUINEA
3Y	BOUVET IS.
4K1	ANTARCTICA
4S7	SRI LANKA
4U	UNITED NATIONS,GENEVA
4W	YEMEN
4X4,4Z4	ISRAEL

Prefix	Country
5A	LIBYAN ARAB REPUBLIC
5B4, ZC4	CYPRUS
5HI	ZANZIBAR, TANZANIA
5H3	TANZANIA
5L	LIBERIA
5N2	NIGERIA
5R8	MALAGASY REPUBLIC
5T5	MAURITANIA
5U7	NIGER
5V	TOGO
5W1	WESTERN SAMOA
5X5	UGANDA
5Z4	KENYA
6O	SOMALI REPUBLIC
6W8	SENEGAL REPUBLIC
6Y5	JAMAICA
7J	OKINO-TORI-SHIMA
7O	SOUTH YEMEN & KAMARAN IS
7P8	LESOTHO
7Q7	MALAWI
7X	ALGERIA
7Z	SAUDI ARABIA
8J	ANTARCTICA
8P6	BARBADOS IS.
8Q6	MALDIVE IS.
8R	GUYANA
8Z4	SAUDI ARABIA/IRAQ NEUTRAL ZONE
9G1	GHANA
9H1,5	MALTA
9H4	GOZO (MALTA)
9I, 9J	ZAMBIA
9K2	KUWAIT
9L1	SIERRA LEONE
9M2	WEST MALAYSIA
9M6	SABAH
9M8	SARAWAK
9N1	NEPAL
9Q5	REPUBLIC OF ZAIRE
9U5	BURUNDI
9V1	REPUBLIC OF SINGAPORE
9X	RWANDA
9Y4	TRINIDAD & TOBAGO IS.

COUNTRY	PREFIX
AFGHANISTAN	YA
AGALEGA IS	386
ALAND IS	OH0
ALASKA	KL7
ALBANIA	ZA
ALGERIA	7X
AMSTERDAM & ST. PAUL IS	FB8Z
ANDAMAN & NICOBAR IS	VU7

Country	Prefix
ANDORRA	C31
ANGOLA	D2A
ANTARCTICA	CE9AA-AM,FBBY,KC4,LU-Z, UA1,VK0,VP8,ZL5,ZS1ANT,3Y,4K1,8J
ARGENTINA	LU
ARUBA	PJ3,9
ASCENSION IS	ZD8
AUCKLAND & CAMPBELL IS	ZL
AUSTRALIA	VK
AUSTRIA	OE
AVES IS	YV0
AZDRES IS	CT2
BAHAMA IS	C6A
BAHRAIN IS	A9X
BAJO NUEVO IS	HK0
BAKER IS	KB6
BALEARIC IS	EA6
BANGLADESH	S2,S3
BARBADOS IS	8P6
BELGIUM	ON
BELIZE	VPI
BENIN, PEOPLE'S REP OF	TY
BERMUDA IS	VP9
BHUTAN	A51
BOLIVIA	CP
BONAIRE	PJ4,9
BOTSWANA	A2C
BOUVET IS	3Y
BRAZIL	PP-PY
BRITISH PHOENIX IS	VRI
BRUNEI	VS5
BULGARIA	LZ
BURMA	XZ
BURUNDI	9U5
CAMBODIA/KHMER REP	XU
CAMEROON	TJ
CANADA	VE
CANAL ZONE	KZ5
CANARY IS	EA8
CANTON IS	KB6
CAPE VERDE, REP OF	D4
CAPRI IS	IC
CAROLINE IS	KC6
CAYMAN IS	ZF1
CENTRAL AFRICAN REPUBLIC	TL8
CEUTA Y MELILLA, SPANISH	EA9
CHAD REPUBLIC	TT8
CHAGOS IS	VQ9
CHATHAM IS	ZL
CHILE	CE
CHINA, PEOPLES REP OF	BY
CHRISTMAS IS	VK9X
CLIPPERTON IS	FO8
COCOS IS	TI9

COCOS (KEELING) IS	VK9Y
COLUMBIA	HK
COMORO,STATE OF	D6
CONGO, REPUBLIC OF	TN8
COOK IS	ZK1
CORSICA	FC
COSTA RICA	TI
CRETE, GREECE	SV
CROZET IS	FB8W
CUBA	CM,CO
CURACAO	PJ2,9
CYPRUS	5B4,ZC4
CZECHOSLOVAKIA	OK,OL
DENMARK	OZ
DJIBOUTI,REP OF	J2
DODECANESE IS	SV5
DOMINICAN REPUBLIC	HI
EASTER IS	CE0A
ECUADOR	HC
EGYPT	SU
EL SALVADOR	YS
ENDERBURY IS	KB6
ENGLAND	G
ETHIOPIA	ET3
FAEROES IS	DY
FALKLAND IS	VP8
FERNANDO DE NORONHA	PY0
FIJI IS	3D2
FINLAND	OH
FRANCE	F
FRENCH POLYNESIA	FD8
GABON REPUBLIC	TR8
GALAPAGOS IS	HC8
GAMBIA	C5A
GERMANY, FED. REP. (WEST)	DA-DL
GERMAN DEM. REP. EAST	DM
GHANA	9G1
GIBRALTAR	ZB2
GILBERT & OCEAN IS	VR1
GLORIOSO IS	FR7
GOUGH IS	ZD9
GOZO (MALTA)	9H4
GRAHAM LAND	VP8,LU-Z
GREECE	SV
GREENLAND	OX
GRENADA	J3
GUADELOUPE	FG7
GUAM IS	KG6
GUANTANAMO BAY	KG4
GUATEMALA	TG
GUERNSEY,BAILIWICK OF	GU
GUIANA, FRENCH	FY7

GUINEA BISSAU	**CR3**
GUINEA, EQUATORIAL	**3C**
GUINEA, REPUBLIC OF	**3X**
GUYANA	**8R**
HAITI	**HH**
HAWAII	**KH6**
HEARD IS.	**VK0**
HONDURAS	**HR**
HONG KONG	**VS6**
HOWLAND IS.	**KB6**
HUNGARY	**HA, HG**
ICELAND	**TF**
INDIA	**VU2**
INDONESIA	**YB-YD**
IRAN	**EP**
IRAQ	**Y1**
IRELAND	**EI**
ISCHIA	**IC**
ISLE OF MAN	**GD**
ISRAEL	**4X4,4Z4**
ITALY	**I,IW**
IVORY COAST	**TU2**
JAMAICA	**6Y5**
JAN MAYEN IS	**JX**
JAPAN	**JA-JR**
JAPAN, U.S.PERSONNEL IN	**KA**
JARVIS IS.	**KP6**
JERSEY	**GJ**
JOHNSTON IS	**KJ6**
JORDON	**JY**
JUAN DE NOVA IS.	**FR7**
JUAN FERNANDEZ IS	**CE02**
KAMARAN IS	**7D**
KENYA	**5Z4**
KERGUELEN IS	**FB8X**
DERMADEC IS	**ZL**
KOREA	**HM,HL9**
KURE IS	**KH6**
KUWAIT	**9K2**
LABRADOR	**VO2**
LACCADIVE IS.	**VU7**
LAMPEDUSA IS.	**IG9**
LAOS	**XW8**
LEBANON	**OD5**
LEEWARD IS.	
ANGUILLA	**VP2E**
ANTIGUA, BARBUDA	**VP2A**
BRITISH VIRGIN IS	**VP2V**
MONTSERRAT	**VP2M**
ST. KITTS, NEVIS	**VP2K**
LESOTHO	**7P8**
LIBERIA	**EL,5L**

LIBYAN ARAB REPUBLIC	5A
LIECHTENSTEIN	HB0
LINE IS, NORTHERN	VR3
LINE IS, CENTRAL & SOUTHERN	VR7
LORD HOWE IS	VK2
LUXEMBOURG	LX
MACAO	CR9
MACQUARIE IS	VK0
MADDALENA IS	IM
MADEIRA IS	CT3
MALAGASY REPUBLIC	5R8
MALAWI	7Q7
MALAYSIA, EAST	9M6,8
MALAYSIA, WEST	9M2
MALDIVE IS	8Q6
MALI REPUBLIC	TZ
MALPELO IS	HK0
MALTA	9H1,5
MANIHIKI IS	2K1
MARIANA IS	KG6
MARKET REEF	0J0,0HOM
MARSHALL IS	KX6
MARTINIQUE	FM7
MAURITANIA	5T5
MAURITIUS IS	3B6
MAYOTTE	FH8
MEXICO	XE,AF
MIDWAY IS	KM6
MINAMI-TORI-SHIMA IS	JD1
MONACO	3A
MONGOLIA	JT1
MOROCCO	CN
MOZAMBIQUE	C9M
NAURU, REPUBLIC OF	C21
NAVASSA IS	KC4
NEPAL	9N1
NETHERLANDS	PA-PI
NETHERLANDS ANTILLES	PJ
NEW CALEDONIA	FK8
NEWFOUNDLAND	VO1
NEW HEBRIDES	YJ
NEW ZEALAND	ZL
NICARAGUA	YN
NIGER	5U7
NIGERIA	5N2
NIUE IS	ZK2
NORFOLK IS	VK9N
NORTHERN IRELAND	GI
NORWAY	LA-LJ
OGASAWARA IS	JDI
OKINAWA (RYUKYU IS.)	JR6
OKINO-TORI-SHIMA	7J1
OMAN, SULTANATE OF	A4K

PAKISTAN	AP
PALMYRA IS	KP6
PANAMA	HP
PANTELLERIA IS	IH
PAPUA NEW GUINEA	P29
PARAGUAY	ZP
PERU	OA
PHILLIPINES	DU
PHEONIX IS	K36
PITCAIRN IS	VR6
POLAND	SP
PORTUGAL	CT1
PRINCE EDWARD & MARION IS.	Z52
PRINCIPE & SAO TOME IS	S9
PUERTO RICO	KP4
QATAR	A7K
REUNION IS	FR7
REVILLA GIGEDO IS	XF4
RHODESIA	ZE
RODRIGUEZ IS	3B9
ROMANIA	YO
ROTA	KG6R
RWANDA	9X
SABAH	9M6
SABA IS	PJ6,8
ST. BRANDON IS	3B7
ST. EUSTATIUS	PJ5,8
SAINT HELENA IS	ZD7
SAINT MARTIN IS	FS7
ST. PETER & ST. PAULS ROCKS	PYO
ST. PIERRE & MIQUELON IS	FP8
SAIPAN	KG6S
SAMOA, AMERICAN	KS6
SAMOA, WESTERN	5WI
SAN ANDRES & PROVIDENCIA	HKO
SAN FELIX IS	CE0X
SAN MARINO	M1
SARAWAK	9M8
SARDINIA	IS
SAUDI ARABIA	HZ, 7Z
SAUDI ARABIA/IRAQ NEUTRAL ZONE	BZ4
SCOTLAND	GM
SENEGAL REPUBLIC	6W8
SERRANA BK & RONCADOR CAY	8KO
SEYCHELLES IS	S79
SICILY	IT
SIERRA LEONE	9LI
SINGAPORE, REP OF	9V1
SINT MAARTEN	PJ7,8
SOLOMON IS	H4
SOMALI REPUBLIC	60
SOUTH AFRICA,REP OF	ZR,AZQ,2,4-6
SOUTH GEORGIA IS	VP8

SOUTH ORKNEY IS	QZ, 2, VP8
SOUTH SANDWICH IS	QZ, 2, VP8
S.SHETLAND IS	CE9AN-AZ,QZ-Z,VP8
SOUTHWEST AFRICA(NAMIBIA)	ZR3,ZS3
SOUTH YEMEN	70

SOVIET UNION:

EUROPEAN RUSSIA SOVIET FEDERATED SOCIALIST REPUBLIC	UA1,2,3,4,6
ASIATIC RUSSIAN SFSR	UA9,0
ARMENIAN S.S.R.	UG6
AZERBAIDZHAN S.S.R.	UD6
ESTONIAN S.S.R.	UR2
FRANZ JOSEF LAND	UA1
GEORGIAN S.S.R.	UF6
KALININGRADSK	UA2
KARELO-FINNISH S.S.R	UN1
KAZAKH S.S.R.	UL7
KIRGHIZ S.S.R	UMB
LATVIAN S.S.R.	UQ2
LITHUANIAN S.S.R	UP2
MOLDAVIAN S.S.R.	UO5
TADZHIK S.S.R.	UJ8
TURKMEN S.S.R.	UH8
UKRANIAN S.S.R.	UB5
UZBEK S.S.R.	UI8
WHITE RUSSIAN S.S.R	UC2

SPAIN	EA
SPANISH SAHARA	EA9
SRI LANKA	4S7
SUDAN	ST
SURINAM	P2
SVALBARD IS	JW
SWAN IS	HRO
SWAZILAND	3D6
SWEDEN	SJ-SM
SWITZERLAND	HB
SYRIA	YK
TAIWAN	BV
TANZANIA	5H3
THAILAND	HS
TIBET	AC4
TIMOR PORTUGUESE	YB-YD
TOGO	5V
TOKELAU IS	2M7
TUNGA	A35
TRANSKEI	S8
TRANDADE & MARTIM VAZ IS	PY0
TRINIDAD & TOBAGO IS	9Y4
TRISTAN DE CUNHA	ZD9
TROMELIN IS	FR7
TUNISIA	3V8
TURKEY	TA,TC
TURKS & CAICOS IS	VP5

TUSCAN ARCHIPELAGO	IA
TUVALU IS	VR8
UGANDA	5X5
UNITED ARAB EMIRATES	A6X
UNITED NATIONS, GENEVA	4U
UNITED STATES OF AMERICA	N,W,K
URUGUAY	CX
VITICAN CITY	HV
VENEZULA	YV
VIETNAM	XV5
VIRGIN IS	KV4
VOLTAIC REPUBLIC	XT2
WAKE IS	KW6
WALES	GW
WALLIS & FUTUNA IS	FW8
WILLIS IS	VK9Z
WINDWARD IS:	
DOMINICA	VP2D
ST. LUCIA	VP2L
ST. VINCIENT & DEPENDENCIES	VP2S
YEMEN	4W
YUGOSLAVIA	YU
ZAIRE, REPUBLIC OF	9Q5
SAMBIA	9I,9J
ZANZIBAR (TANZANIA)	5H1

Appendix

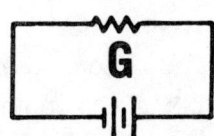

International "Q" Signals

QRA	What is the name of your station?
QRB	How far are you from my station?
QRD	Where are you going and where are you from?
QRG	Will you tell me my exact frequency?
QRH	Does my frequency vary?
QRI	How is the tone of my transmission?
QRK	What is the readability of my signals?
QRL	Are you busy?
QRM	Are you being interferred with?
QRN	Are you troubled by static?
QRO	Shall I increase power?
QRP	Shall I decrease power?
QRQ	Shall I send faster?
QRR	Are you ready for automatic operation?
QRS	Shall I send more slowly?
QRT	Shall I stop sending?
QRU	Have you anything for me?
QRV	Are you ready?
QRW	Shall I inform _____ that you are calling him on __ kHz.
QRX	When will you call me again?
QRY	What is my turn?
QRZ	Who is calling me?
QSA	What is the strength of my signals?
QSB	Are my signals fading?
QSD	Is my keying defective?

QSG	Shall I send _____ telegrams at a time?
QSJ	What is the charge to be collected per word to _____ including your internal telegraph charge?
QSK	Can you hear me between your signals?
QSL	Can you acknowledge receipt?
QSM	Shall I repeat the last telegram which I sent you, or some previous telegram?
QSN	Did you hear me on _____ kHz?
QSO	Can you communicate with _____ direct or by relay?
QSP	Will you relay to _____ free of charge?
QSQ	Have you a doctor on board [or is ... (name of person) on board]?
QSU	Shall I send or replay on this frequency [or on _____ kHz] (with emission of class _____)?
QSV	Shall I send a series of V's on this frequency?
QSW	Will you send on this frequency?
QSX	Will you listen to _____ on _____ kHz?
QSY	Shall I change to transmission on another frequency?
QSZ	Shall I send each word or group more than once?
QTA	Shall I cancel telegram number _____ as if it had not been sent?
QTB	Do you agree with my counting of words?
QTC	How many telegrams have you to send?
QTE	What is my true bearing from you?
QTG	Will you send two dashes of ten seconds each followed by your call sign [on _____ kHz]?
QTH	What is your location?
QTI	What is your true track?
QTJ	What is your speed?
QTL	What is your true heading?
QTN	At what time did you leave _____ (place)?
QTO	Have you left dock (or port)? or Are you airborne?

QTP	Are you going to enter dock (or port)?
	or
	Are you going to land?
QTQ	Can you communicate with my station by means of the International Code of Signals?
QTR	What is the correct time?
QTS	Will you send your call sign for _____ minute(s) on _____ kHz so that your frequency may be measured?
QTU	What are the hours during which your station is open?
QTV	Shall I stand guard for you on the frequency of _____ kHz?
QTX	Will you keep your station open for further communication with me until further notice?
QUA	Have you news of _____ (call sign)?
QUB	Can you give me, in the following order, information concerning: visibility, height of clouds, direction and velocity (place of observation)?
QUC	What is the number of the last message you received from me?
QUD	Have you received the urgency signal, sent by _____ (call sign of mobile station)?
QUF	Have you received the distress signal sent by _____ (call sign of mobile station)?
QUG	Will you be forced to land?
QUH	Will you give me the present barometric pressure at sea level?

Note: If a "Q" signal is sent without a question mark, it becomes a statement.

Appendix

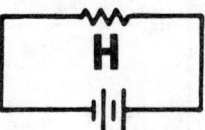

Citizens Band and Amateur Radio

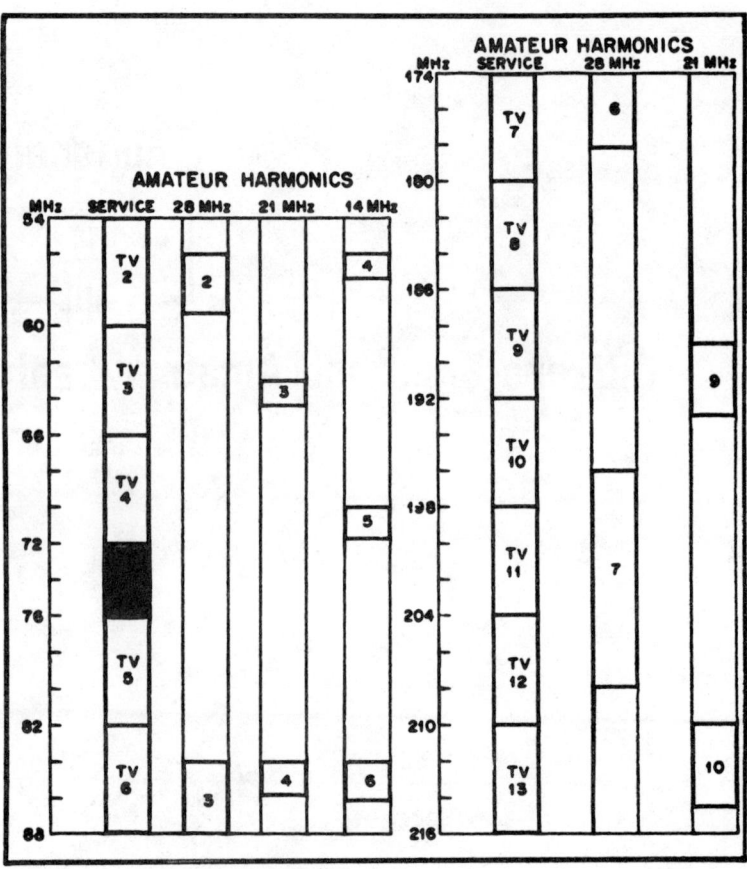

Fig. H-1. The relationship of amateur-band harmonics to the vhf TV channels (courtesy of American Radio Relay League Inc.).

Table H-1. TV Channels and Amateur and Citizens Band Harmonics.

Channel	Freq. Range	Picture Carrier Freq.	Harmonics				
			CB	40 meters	20 meters	15 meters	10 meters
TV I-F	41-47	42	—	—	42-43	42-43	—
2	54-60	55.25	53.9-54.8 (2nd)	56-58.4 (8th)	56-57.3 (4th)	—	56-59.4 (2nd)
3	60-66	61.25	—	63-65.7 (9th)	—	63-64.35 (3rd)	—
4	66-72	67.25	—	70-73 (10th)	70-72 (5th)	—	—
5	76-82	77.25	80.9-82.2 (3rd)	—	—	—	—
6	82-88	83.25	82-82.2 (3rd)	—	84-86.4 (6th)	84-85.8 (4th)	84-89.1 (3rd)

Table H-2. Harmonic Relationship—Amateur VHF Bands and UHF TV Channels.

Amateur Band	Har-monic	Fundamental Freq. Range	Channel Affected	Amateur Band	Har-monic	Fundamental Freq. Range	Channel Affected
144 MHz	4th	144.0-144.5	31	220 MHz	3rd	220.00-220.67	45
		144.5-146.0	32			220.67-222.67	46
		146.0-147.5	33			222.67-224.67	47
		147.5-148.0	34			224.67-225.00	48
	5th	144.0-144.4	55		4th	220-221	82
		144.4-145.6	56			221.0-222.5	83
		145.6-146.8	57	420 MHz	2nd	420-421	75
		146.8-148.0	58			421-424	76
	6th	144.00-144.33	79			424-427	77
		144.33-145.33	80			427-430	78
		145.33-147.33	81			430-433	79
		147.33-148.00	82			433-436	80

Appendix

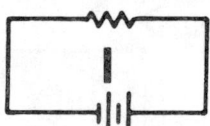

Symbols, Codes and Alphabets

Table I-1. US American Standard Code for Information Interchange

The U.S. American Standard Code for Information Interchange (USACII or ASCII) is an 8-unit code as shown here, used largely with computers.

NUL	Null, or all zeros	DC1	Device control 1
SOH	Start of heading	DC2	Device control 2
STX	Start of test	DC3	Device Control 3
ETX	End of text	DC4	Device Control 4
EOT	End of transmission	NAK	Negative acknowledge
ENQ	Enquiry	SYN	Sychronous idle
ACK	Acknowledge	ETB	End of transmission block
BEL	Bell, or alarm	CAN	Cancel
BS	Backspace	EM	End of medium
HT	Horizontal tabulation	SUB	Substitute
LF	Line feed	ESC	Escape
VT	Vertical tabulation	FS	File separator
FF	Form feed	GS	Group separator
CR	Carriage return	RS	Record separator
SO	Shift out	US	Unit separator
SI	Shift in	SP	Space
DLE	Data link escape	DEL	Delete

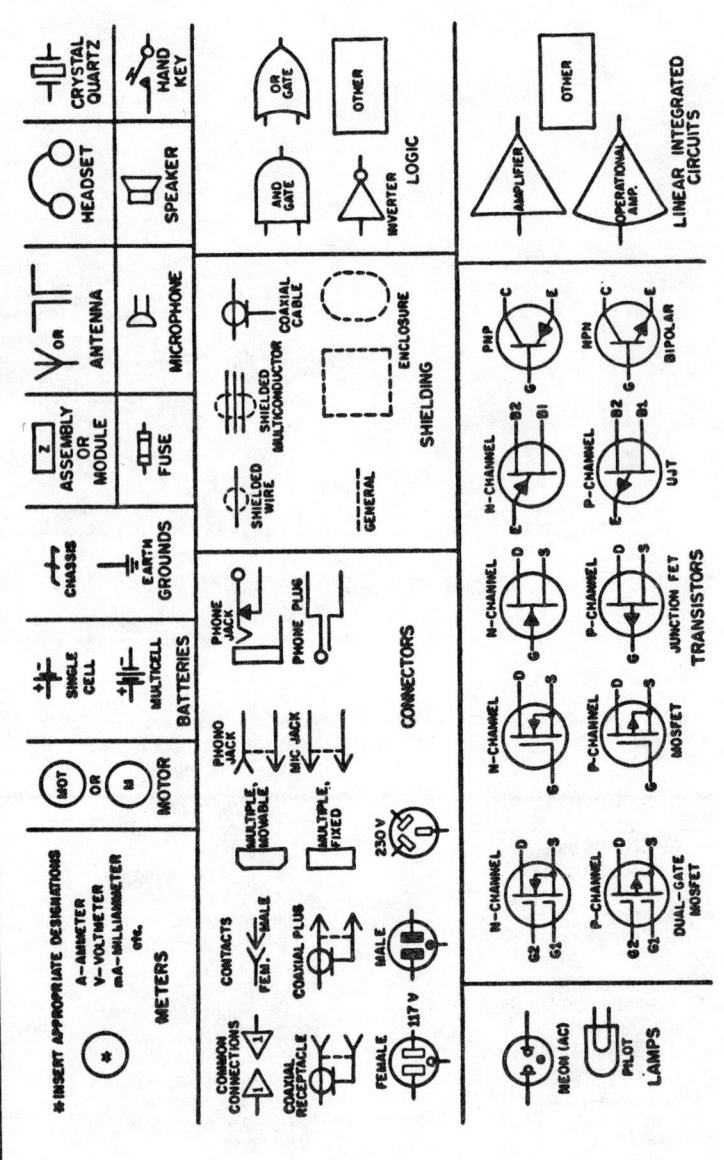

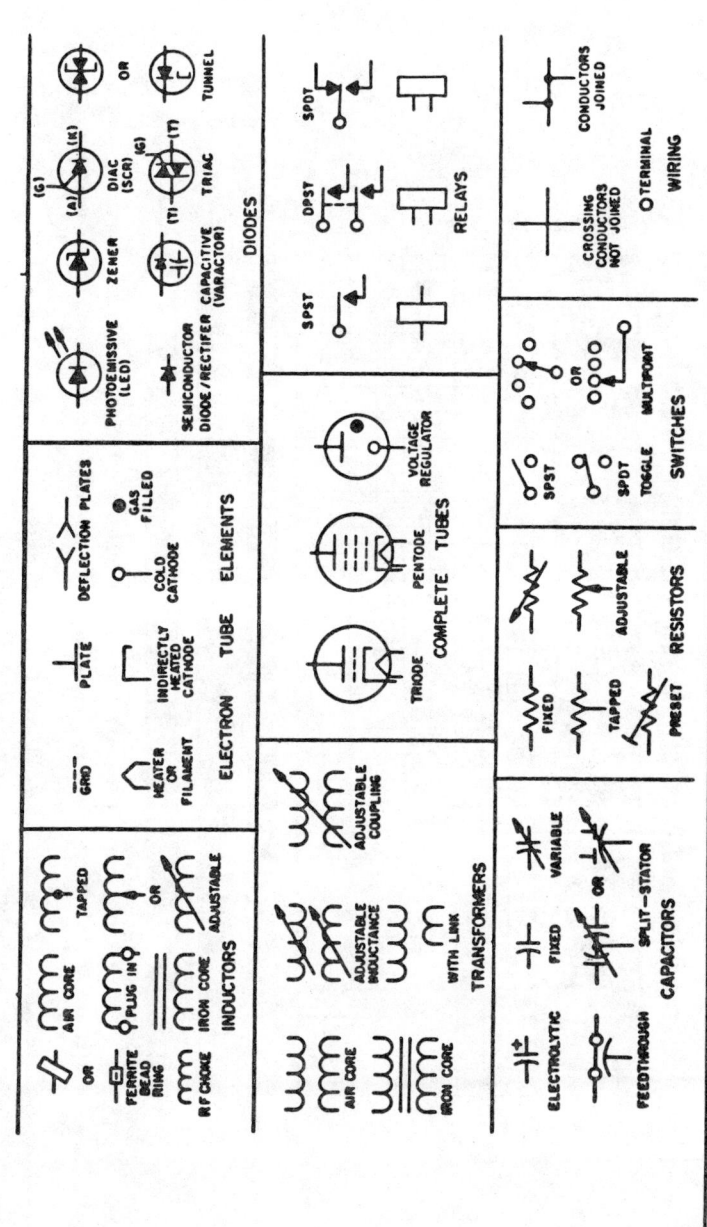

Fig. I-1. Commonly used electronic schematic diagrams.

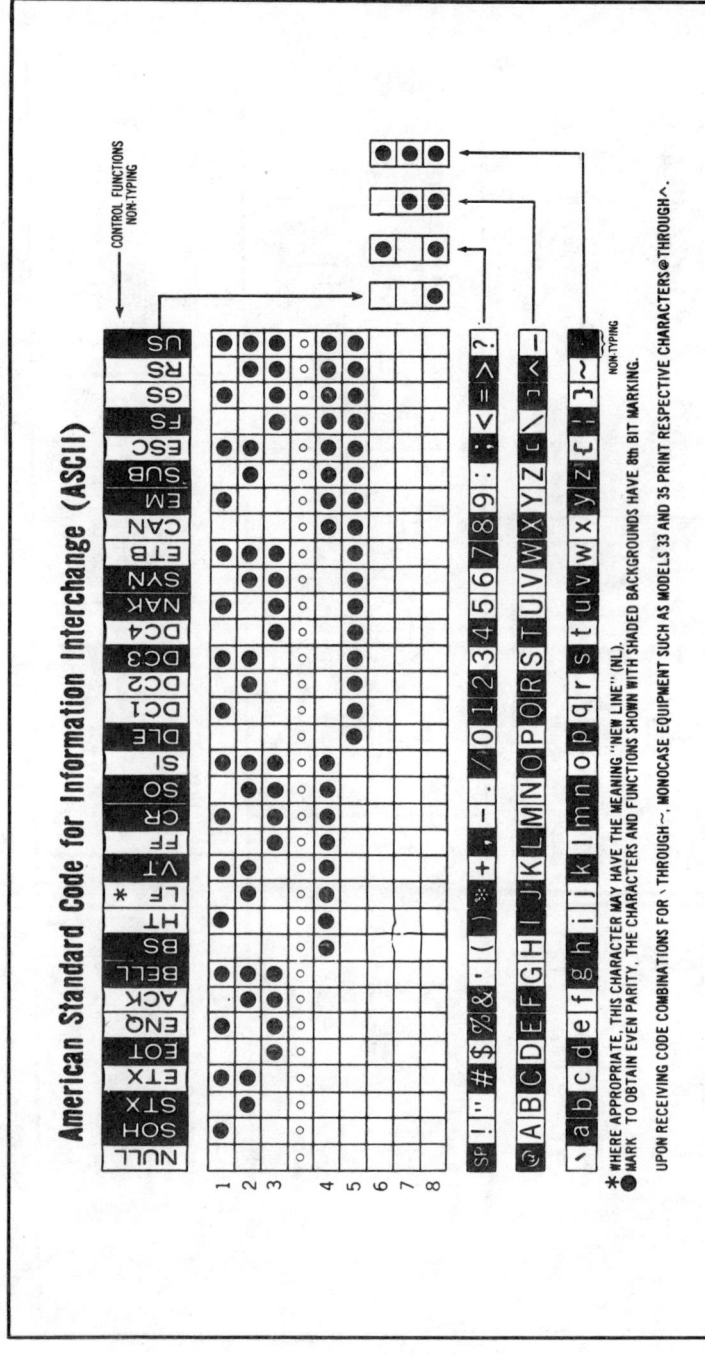

Fig. I-2. American Standard Code for Information Interchange (ASCII) chart.

Table I-2. Phonetic Alphabet.

Letter	Code Word	Pronunciation
A	Alfa	AL FAH
B	Bravo	BRAH VOH
C	Charlie	CHAR LEE
D	Delta	DELL TAH
E	Echo	ECK OH
F	Foxtrot	FOKS TROT
G	Golf	GOLF
H	Hotel	HOH TELL
I	India	IN DEE AH
J	Juliett	JEW LEE ETT
K	Kilo	KEY LOH
L	Lima	LEE MAH
M	Mike	MIKE
N	November	NO VEM BER
O	Oscar	OSS CAH
P	Papa	PAH PAH
Q	Quebec	KEH BECK
R	Romeo	ROW ME OH
S	Sierra	SEE AIR RAH
T	Tango	TANG GO
U	Uniform	YOU NEE FORM
V	Victor	VIK TAH
W	Whiskey	WISS KEY
X	X-ray	ECKS RAY
Y	Yankee	YANG KEY
Z	Zulu	ZOO LOO

Appendix J

Log Sheets

DATE	QSO START TIME	STATION WORKED	REPORT		FREQ	MODE	PWR	TIME OF ENDING QSO	OTHER DATA	QSL	
			SENT	RCVD						S	R

DATE	QSO START TIME	STATION WORKED	REPORT		FREQ	MODE	PWR	TIME OF ENDING QSO	OTHER DATA	QSL	
			SENT	RCVD						S	R

DATE	QSO START TIME	STATION WORKED	REPORT		FREQ	MODE	PWR	TIME OF ENDING QSO	OTHER DATA	QSL	
			SENT	RCVD						S	R

DATE	QSO START TIME	STATION WORKED	REPORT		FREQ	MODE	PWR	TIME OF ENDING QSO	OTHER DATA	QSL	
			SENT	RCVD						S	R

DATE	QSO START TIME	STATION WORKED	REPORT		FREQ	MODE	PWR	TIME OF ENDING QSO	OTHER DATA	QSL	
			SENT	RCVD						S	R

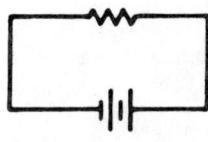

Index

A
AC amplifier	148
Active devices	187
Admittance	50
Alternating current	21, 27
Alternator	27
AM	234
Ampere	9
Amplification	103
Amplification quality	126
Amplifier, class A	123
class B	123
class C	123
final	224
Anode	110
Antenna	83
8JK	286
flattop	286
folded dipole	288
inverted vee	283
log periodic	287
long-wire	278
monopole	290
parallel dipole	283
sterba curtain	286
trap	283
Windom	284
ZL special	286
Antiresonance	54
ARRL	385
ASCII	515, 518
Audio interference	419
Average power	69
Average value	30

B
Balanced circuit	102
Balanced modulator	267
Bandwidth	237
Base	146
BCI	409
Beam power tubes	116
Beat frequency oscillator	262
Beat note	260
BFO	262
Bias voltage	114
Bipolar transistors	152
Bleeder resistor	170
Broadcast interference	409
Buckshot	204
Buffer	224

C
Call signs	493
Capacitance	17
Carrier wave	211
Cathode	110
CB	306, 511
CB interference	422
Charge, electrical	16
Circuit, parallel	32
series	32
series-parallel	34
simple	32
Circulating current	56
Collector	136
Combining reactances	49
Crystal resonators	217
Current	9
alternating	21, 27
circulating	56
lag	39
limiting	184
measuring	30
secondary	29
self-induced	29

gain	138
Cutoff frequency	100
Cutoff voltage	114
CW	257

D

dB	70
Decibel	70
Degeneration	212
Depletion zone	136
Differential amplifier	102
Diode	110
zener	178
Directivity	278
Distortion	122
D layer	303
Double conversion	263
Double sideband	268
Drain	160
Drive	202
Ducting	301
DX	256

E

E layer	303
Edison effet	108
Effective impedance	295
Effective power	240
Effective value	31
Efficiency	278
80 meters	305
Electric fields	18
Electrical device interference	416
Electricity	11
Electrolytic capacitors	172
Electromagnetic field	12
energy	13
propagation	14
Electron theory	108
11-meter band	306
Emitter	146
Envelope	195
Examination points	395
Examination syllabus	401

F

Far-field signal intensity	278
FCC form 610	385
Feedback	127, 212
negative	212
positive	213
FET	152
Field-effect transistors	152
15 meters	305
Filament	110
Filter, capacitor-input	167
choke-input	167
design	95
high-pass	413
method	267
Filters	93
active	93
passive	93
Final amplifier	224
First detector	261
Flat-topping	204
Flip-flop	215
Flux, magnetic	29
FM	234
interference	418
F1 layer	303
40 meters	305
Forward-biased PN junction	136
Frequency deviation	254
Frequency response	148
F2 layer	303

G

Gain	124
Gate	160
Grid leak resistor	202

H

Harmonic suppression	77
Harmonics	74, 512, 513
Heater	110
Henry	29
Hertz	28
Heinrich Rudolph	256
Heterodyne	260
Hi-fi interference	409
High-pass filter	413

I

I-f	260
Ignition interference	416
Ignition noise	208
Image	262
Impedance	45, 278
coupling	190
matching	85
output	175
transformation	61
Impurity	132
Inductance	28
Inertia	29
Instantaneous power	69
Intermediate frequency	260
Intermodulation	205
International Morse code	381, 407
International prefixes	493

J

JFET	152

L

LC resonator	220
Lee DeForest	113
Licensing	385
Lightning protection	297
Linearity	122
Local oscillator	263
Logs	521

M

Magnetic flux	29
Magnetism	11
Matching pads	93
Maxwell, James	11
Mixing	245
Modulation	233
amplitude	234
frequency	234
index	249
percentage of	248
phase	234
MOSFET	152
Multivibrator	215

N

Negative peak	201
Negative resistance	46
Neutralization	127
Neutralizing capacitors	227
N-type semiconductors	134

O

Offset frequency	261
Ohm	15
Ohm's law	19
160 meters	304
Open-loop gain	213
Oscillator, Armstrong	220
Clapp	222
Hartley	221
High-C Colpitts	222
Miller crystal	221
Output impedance	175
Overmodulation	249
Overneutralize	226

P

Part 97	441
Peak clipping	205
Peak inverse voltage	164
Peak power	68
Peak-to-peak	31
Peak voltage	31
Pentode	116
Phase reversal	148
Phasing method	267
Phonetic alphabet	519
Pi-section filter	100
Pitch control	262
Plate	110
PM	234
PN junctions	135
Polarization	280
Power	63
Power factor	60
Power, measuring	66
Power ratio	72
Power supply	161
regulated	182
Prefixes	493
Product detector	262
Propagation	299
tropospheric	301
P-type semiconductor	134
Pulsating dc	166
Pure dc	161
Push-pull amplifier	83
Push-pull multiplier	82
Push-push doubler	81

Q

Q factor	35, 57
Q-multiplier	213
Q signals	507
Quality factor	35, 57

R

Reactance	35
capacitive	37
inductive	38
Receiver	255
Rectifier, full-wave	162
half-wave	162
Regeneration	213
Regenerative receiver	256
Regulation	173
dynamic	174
static	174
Resistance	15
coupling	190
negative	46
Resonance	52
Resonant frequency	35, 100
Resonator	215
Ripple amplitude	176
Ripple frequency	169
Root mean square	31
Rms	31
Rules and regulations	319, 441

S

Schematic symbols	517
Secondary current	29
Self-induced current	29
Semiconductor theory	132
75 meters	305
Sidebands	236
Silicon controlled rectifiers	156
Single sideband	265
6 meters	306
Skin effect	52
Skip distance	301
Source	160
Space charge	110
Spectrum distribution	241
Splatter	204
Spradic E	303
Standing wave	276
Standing-wave ratio	294
Storage, electrical	16
Superheterodyne receiver	256
Susceptances	54
Swinging choke	171
SWR	294

T

10 meters	306
Tetrode	115
Transformer-coupling	190
Transformer matching	91
Transistor	132
Transmitter interference	420
Trapezoid pattern	251
Triode	113
Tropospheric propagation	301
Tuned-plate—tuned-grid	221
TVI	309, 409
TV interference	309
20 meters	305
2 meters	306
220-MHz band	306

V

Vacuum tube	107
Valve	109
Vanishing carrier	247
Volt	9
Voltage	9
Voltage gain	138
Voltage gradient	111
Voltage lag	36
Voltage magnification	58
Voltage-sampling network	182
Voltage-variable resistance	121
Vswr	294

W

Watt	65
James	65
Wave, direct	300
ground	300
sky	300
space	300

Z

Zener diode	178